ROUTLEDGE LIBRARY EDITIONS:
HEALTH, DISEASE & SOCIETY

Volume 17

HEALTH, DISEASE AND SOCIETY

HEALTH, DISEASE AND SOCIETY

A Critical Medical Geography

KELVYN JONES
and
GRAHAM MOON

 Routledge
Taylor & Francis Group

LONDON AND NEW YORK

First published in 1987 by Routledge & Kegan Paul Ltd.

This edition first published in 2022
by Routledge
4 Park Square, Milton Park, Abingdon, Oxon OX14 4RN

and by Routledge
605 Third Avenue, New York, NY 10158

Routledge is an imprint of the Taylor & Francis Group, an informa business

British Library Cataloguing in Publication Data
A catalogue record for this book is available from the British Library

ISBN: 978-0-367-52469-2 (Set)
ISBN: 978-1-032-25400-5 (Volume 17) (hbk)
ISBN: 978-1-032-25508-8 (Volume 17) (pbk)
ISBN: 978-1-003-28301-0 (Volume 17) (ebk)

DOI: 10.4324/9781003283010

Publisher's Note
The publisher has gone to great lengths to ensure the quality of this reprint but points out that some imperfections in the original copies may be apparent.

Disclaimer
The publisher has made every effort to trace copyright holders and would welcome correspondence from those they have been unable to trace.

Health, Disease and Society:

A CRITICAL MEDICAL GEOGRAPHY

Kelvyn Jones *and* Graham Moon

Portsmouth Polytechnic

ROUTLEDGE & KEGAN PAUL
London and New York

First published in 1987 by
Routledge & Kegan Paul Ltd
11 New Fetter Lane, London EC4P 4EE

Published in the USA by
Routledge & Kegan Paul Inc.
in association with Methuen Inc.
29 West 35th Street, New York, NY 10001

Set in Times
by Columns of Reading
and printed in Great Britain
by TJ Press (Padstow) Ltd
Padstow, Cornwall

Library of Congress Cataloging in Publication Data
Jones, Kelvyn.
  Health, medicine & society.
  Includes bibliographies and index.
  1. Medical geography. 2. Social medicine.
3. Epidemiology. I. Moon, Graham. II. Title.
III. Title: Health, medicine, and society.
[DNLM: 1. Delivery of Health Care. 2. Epidemiology.
3. Geography. WB 700 J77h]
RA792.J66 1987   614.4'2   87–4551

British Library CIP Data also available
ISBN 0-7102-0063-3 (c)
     0-7102-1219-4 (p)

Contents

Figures

Tables

Preface

There are two clear research traditions within medical geography. Historically the subject has been concerned with the cause and spread of disease; more recently attention has expanded to include the provision and consumption of health care. Available textbooks on medical geography have tended to perpetuate this division. Indeed, it has been reflected in the structure of the recent successful international conferences run by medical geographers from the Institute of British Geographers and the Association of American Geographers; two separate groups have met simultaneously in different rooms! One of our major intentions has been to try to indicate the connections between these two separate research traditions, and to provide a true introduction covering both facets of the subject.

Our aim has thus been not only to review the traditional subject matter of medical geography, but also to indicate the areas in which we feel the discipline will be developing during the coming years. To this end, we stress the need to build upon the cartographic and descriptive heritage of medical geography and develop a real understanding of reasons underlying the geographies of health and health care. In particular, we underline the need for a social approach which considers both why we become ill and what treatments we receive for our illnesses. Inevitably we have been selective in our treatment of the issues involved, but we hope that the reader will gain some feeling for the excitement and interest which is currently being generated. Our treatment cannot be encyclopedic, but we have tried to show the breadth of the subject, while concentrating on the diversity of approaches that are the hallmark of contemporary inquiry. We provide few solutions but pose many questions.

We intend this textbook to serve as a student's guide to the content and nature of medical geography. It will be suitable for option classes at first degree level, and we also hope that it will prove useful to graduate students who wish to review the subject.

Students in the disciplines of medical sociology and health studies should find much to interest them in introducing a geographical perspective to their studies.

This book is an outgrowth of a course in medical geography at Portsmouth, and parts draw on our joint research and other work in which we have been involved. The book is essentially a joint effort with each of us substantially amending the other's drafts. It has taken much longer to write than we would have wished; the patience of Liz Fidlon of Routledge & Kegan Paul and Neil Wrigley the editor of this series during their long wait for our manuscript has been immense. Other thanks are due to the Department of Geography and the School of Social and Historical Studies at Portsmouth Polytechnic for providing friendly working environments. We would like to acknowledge the many people who have helped our work, and in particular Alan Burnett, Christine Hay, James Bird (for 'swimmer's itch'), Jack Render, Karen Bramley and Stephen Pinch. Finally, to thank adequately Tina and Liz leaves us unusually lost for words. . . .

CHAPTER 1

The social context of disease, health and medicine

Introduction: Medical geography

We have been asked many times 'What are geographers doing looking at disease and health?', the implied criticism being that we should leave these matters to relevant experts such as physicians, surgeons and biochemists. Implicit in such a view is the belief that the causes of disease will be found in biomedical research and that the only effective treatment is intervention by chemical, electrical or physical means to restore the body to its normal biological functioning, that is technological medicine. Our reply is that this dominant 'biomedical' viewpoint is both flawed and limited. There is an urgent need to 'go outside the body' to develop an alternative social and environmental perspective on health in which geography can play an important part along with other social sciences.

To provide an appetiser for the rest of this book, and to illustrate the diversity of interest of medical geographers, it is useful to begin with the changing nature of the subject by examining both the content of the discipline and the differing approaches that have been used. In terms of content, two distinct areas of study, or 'traditions', can be recognised. First, there is the disease ecology tradition with the geographer attempting to elucidate the social and environmental causes of ill-health; such work is closely allied to epidemiology, which itself is trying to discover the patterns and determinants of disease. The second, and more recent, area of study is the geography of medical care which is concerned with the consumption of care in respect of such matters as distribution and accessibility. This tradition draws on parallel work in economics and sociology.

It is possible to recognise several differing approaches to the study of medical geography but, as it is not uncommon to find individual researchers who use several approaches simultaneously, published work often cannot easily be forced into any rigid classification. For our purposes, and following Pyle (1979) and Phillips (1981), five approaches can be recognised. The cartographic approach is concerned with the mapping of spatial data; modelling attempts to quantify the relationships between variables; while the behavioural approach seeks to understand individual decision-making. The welfare approach attempts to answer the questions of who gets what and where, and how improvement in the quality of life can be achieved by the gradual reform of existing society. In contrast, a structuralist perspective stresses the need to consider phenomena in relation to the totality of society. Emphasis is given to societal constraints, and in a marxist version of this approach, it is presumed that genuine improvements can only be achieved by revolutionary change to a new form of society.

Using these two frameworks of 'content' and 'approach', the nature of medical geography is readily described by giving exemplars. The earliest work in the disease ecology tradition employs the cartographic approach to map rates of disease incidence. For example, Gilbert (1958) reviews Victorian pioneers in medical mapping, while the approach reaches its peak in the national atlases of mortality such as that of Howe (1963). Of course, once the diseases were mapped, other possible causal variables were also portrayed, and it was then but a short step to the quantitative calibration of models that relate disease rates to social and environmental variables. The quantitative approach reaches its most sophisticated in the modelling of the spread of communicable diseases with, for example, Cliff *et al.* (1981) attempting to predict measles epidemics in Iceland. Research using the behavioural approach has focused on the individual attributes that may determine health outcomes. For example, Girt (1972) used the 'lifestyle' variable of cigarette smoking to develop an ecological model of chronic bronchitis in Leeds. While the welfare geographers have not been greatly concerned with the development of explanations for disease patterns, they have used infant and adult mortality rates, along with many other variables, to derive indices of the quality of life. Using this approach, Knox (1975) examined geographical variations within England and Wales; Smith (1977, Chapter 10) provides equivalent results for the USA. Surprisingly, there has been little work by geographers that uses the structuralist approach to illuminate how health outcomes are related to the nature and organisation of society, and even commentaries which are sympathetic to societal explanations, such as Eyles and Woods (1983, Chapter 3) have to use examples from the work of non-geographers.

The earliest work in the medical-care tradition also used the cartographic approach with, for example, Coates and Rawstron (1971) mapping the geographical provision in Britain of such services as dentists and general practitioners; they found marked inequalities. The modelling approach has been extensively used to examine the geographical location of medical provision, with Shannon and Dever (1974) using central-place theory to explore the efficiency, in terms of distance minimisation, of the locations of hospitals. The behavioural approach has been used to study the decision of individuals of whether or not to attend for treatment; for example Girt (1973) examined the influence of distance in determining consultation at a general practice. A considerable amount of work using the welfare approach has been concerned with medical care and Knox (1982, Chapter 7) revealed the inequitable provision of primary care in Scottish cities and considered the measures needed to achieve a gradual improvement in effective assessibility. It is only in relatively recent years that the structuralist approach has been used by medical geographers but there are now a number of examples such as Mohan and Woods's (1985) discussion of how the geography and type of medical provision are being restructured in Britain by the decisions of the Conservative government with regard to the needs of capital.

Before exploring further the nature of medical geography, we need to consider the social perspective that is the distinguishing characteristic of this book. It is only by providing this essential underpinning that current research can be described and evaluated. The remainder of this chapter is organised into four sections. First, a social constructionist viewpoint is developed which contrasts the scientific, biomedical approach to knowledge with a social perspective. Second, a social history of medicine is undertaken to reveal the way in which the biomedical approach became dominant and what alternatives were discarded on the way. Third, the major critiques that have developed of the biomedical approach in the last thirty years are outlined. The final section provides a framework for the remaining chapters.

Social construction

In an influential text Berger and Luckmann (1967) have argued that reality is constructed through human action and does not exist independently of it. We learn how to see, structure and organise the world from our parents, teachers, the media and our general social environment. Our attitudes and judgements may appear to be personal and individual but are, in fact, derived from societal viewpoints. In terms of medicine, so successful is this learning process that many people regard the biomedical view as reality and not as one view of reality. Here we try to expose the biomedical viewpoint as a belief system through a consideration of notions of disease.

Biomedical knowledge

For medical scientists, disease is biological fact and not something that human beings have thought up. Diseases are seen as objects which exist independently and prior to their discovery and description by physicians. Disease is seen as abnormal biological functioning and the physician, as expert, can make allowances for normal variation and spot these abnormal deviations. While laypeople are able to describe their illness in subjective terms, the doctor is expected to provide a scientific, objective diagnosis. The key act in western medicine is this diagnosis for it allows the physician to go back to the genesis of the disease and forward to the treatment. Diagnosis is based on the combination and severity of symptoms and the use of diagnostic tests. Such an approach is based on the presumption of *generic* diseases, that is each disease has specific and distinguishing features which are universal to human species. Any failure to diagnose is attributed to the current lack of medical knowledge or appropriate technical tests. At the same time, western scientific taxonomy is regarded as the best available and universally applicable codification of medical knowledge, having been refined by centuries of empirical scientific investigation. Medical knowledge is regarded as having unfolded in a linear fashion towards a more comprehensive and accurate understanding of reality. Indeed, histories of medicine are literally peppered with such phrases as victories, breakthroughs and remarkable advances (see Cox, 1983, Chapter 2).

Social knowledge

In contrast, from a constructionist viewpoint, diseases are seen as human constructs which would not exist without someone describing and recognising them. Disease may be defined as abnormal biological deviation but we must ask who decides what is normal, who decides what is normal variation, and who decides what is abnormal deviation? What is normal and abnormal is a social and moral judgement and this will vary according to society's own norms, expectations and culturally shared rules of interpretation. Normality is what prevalent social values hold to be acceptable or desirable, and in contrast to the biomedical notion of universal generic diseases, the social view accepts that what constitutes disease can vary temporally, culturally, and indeed, geographically.

To illustrate the importance of belief in present-day medicine, we will examine through the eyes of a social anthropologist a disease which appears to be defined with considerable biological exactitude. Posner (1984) has examined the treatment of diabetics in a number of London clinics, and concluded that today's treatment has much in common with magic ritual. Many adults are diagnosed as diabetic on the basis of routine tests showing high blood sugar levels in comparison to the general population. It is accepted that in such cases even though the symptoms of diabetes are not currently present, they will develop in time together with other complications including loss of sight and death from gangrene or thrombosis. Treatment should, therefore, start at once to reduce sugar levels to normality and continue for the rest of life. The choice of treatment is some combination of tablets, insulin and a change of diet.

Posner questions both the disease definition and the appro-priateness of the treatment. High sugar levels are not the disease but they are equated in practice and this is despite the finding that successive tests with the same subject can vary as much as between different subjects. Indeed, what can be regarded as abnormally high in terms of the general population may be perfectly normal for an individual. While diabetes is a metabolic disorder which involves a complex of biochemical parameters, one easily measured parameter, blood sugar, takes on a *symbolic* meaning; as one doctor commented to Posner, 'diabetes is what I say it is'. The scientific evidence linking high sugar levels with the disease and its complications is not strong, but there is evidence that drug treatment can produce dangerously low levels and even comas. The decision to treat and control high sugar levels in this context must be seen as an *act of faith* rather than as a purely rational scientific decision. Diet plays an important part in the theory of diabetes but in treatment it is downplayed. The doctors recognise that for many patients the sugar level could be better controlled by diet and yet they persist in their *ritual* use of interventionist medicine, believing that patients expect real medicine, tablets and injections and not menus.

The worst possible case that could occur is when a healthy person is found to have a high sugar level, is labelled diabetic and is given treatment that produces severe complications. Why can we even contemplate such an outcome? Posner argues that the medical treatment of diabetes is a belief system which is based on certain assumptions and is sustained by social and cultural factors. Faced with great uncertainty and doubt, the physician places great faith in objective tests; if these reveal abnormality then the doctor feels morally and ethically bound to intervene to try and restore normality. Moreover, the physician believes that this is what the patient expects to be done. Posner (p. 54) concludes that, 'the application of strictly scientific criteria to diagnosis and treatment would change the whole nature of the disease and the specialism that has grown up to treat it'.

Social control

Society has conferred the power on physicians to define disease and it allows physicians to make such decisions as 'she is mentally ill and cannot be responsible for her actions', 'he has a genuine bad back and cannot be expected to work' or 'in attacking the state his thoughts are so abnormal that he needs to be locked up for everybody's good'. Disease is therefore not only a biological state but also a social status which physicians have the power to confer or withhold. In taking such decisions and by following the norms of society, physicians are acting as agents of social control (Zola, 1972). The medical profession can be seen as promoting the existing order of society. For example, the bored housewife who is fed up with the tedium of her life may be diagnosed by her doctor as depressed and requiring tranquillisers. Her real problem, it may be argued, lies not within her mind but with the social and economic order which removes independence and ties her to the home. The doctor does not treat this cause but merely eases the pain of her position and thereby prevents the dissatisfaction she feels being expressed in political or social terms. As Kennedy (1983, p. 12) puts it, 'the social and economic status quo is maintained through the agency of the doctor . . . an agent of a particular form of social system that ensures that its values persist'. This control is achieved not by deliberate conspiracy but by the very ideas and practices of medicine.

To take another example, homosexuality was until recently unambiguously classified by western medicine as a disease. Today, there are medical professionals who still believe it is, with, for example, Edmund Bergler (1956), a prominent New York psychiatrist, insisting that there are no 'healthy homosexuals'. In a similar vein, the East German endocrinologist Gunther Dorner (1976) believes that homosexuals are in need of a cure. For him homosexuality is a gonadic disturbance with a low level of oestrogen in a male baby resulting in the adult being sexually oriented towards males. In marked contrast are those who see homosexuality as nothing more than normal human variation on a par with having brown eyes instead of blue. As Gold (1973, p. 1211) emphatically argues

I have come to an unshakeable conclusion: the illness of homosexuality is a pack of lies, concocted out of myths of a patriarchal society for a political purpose. Psychiatry dedicated to making sick people well, has been the corner-stone of a system of oppression that makes gay people sick.

Conrad and Schneider (1980) discuss how homosexuality has been seen in different cultures and at different times as a sin, a crime, a disease and a matter of personal choice of lifestyle. In the USA homosexuality was officially a disease until 1974 when the American Psychiatric Association decided on a vote that it was not; a very clear example that normality can be re-defined and that conflicts in society over social acceptability can result in changes in disease status.

This example must be seen as a fundamental challenge to the view that perceives medical science as unfolding in a progressive manner leaving beneficial advances in its wake. Changes in what constitutes a disease represent much more than an increase in knowledge and diagnostic accuracy for they reflect changes in the very nature of society itself. Indeed, from the social construc-tionist viewpoint, what today is accepted as fact may never have been fact before and may in the future be fiction again. Thus, there may not always be progress to a better reality but merely the replacement of one reality or belief system by another.

In concluding this discussion on social construction, we must emphasise a number of points. Every disease can be examined from this viewpoint and even such conditions as genetic diseases and leprosy have been socially constructed by Yoxen (1982) and Waxler (1981) respectively. We are not arguing that biological measurements have no reality, but that the way they are interpreted must be seen in a social context; diseases are what physicians choose to call diseases. It is not our contention that some diseases are somehow bogus but that they are all the more real for being socially constructed. Accordingly medicine is not unscientific, but medicine, like all science, is essentially a social enterprise (Barnes, 1982). Diseases must not be seen as exclu-sively social or biological phenomena for they are simultaneously both. If an exclusively biological view is taken of illness then one naturally adopts a medical solution by intervening to restore the body to normal functioning. But if one adopts the WHO (1948) definition of health as 'a state of complete physical, mental and social well-being and not merely the absence of disease or infirmity', or appreciates the Brent Community Health Council's rather utopian definition (1981),

> good health is possible if you are able to choose to do a job you enjoy in a pleasant and safe environment, to live in a warm house with enough space so that people do not get on top of one another, and with a safe place for children to play, to be able to have your children looked after during the day, eat the food you like best, to have a garden, ... it is having time with people you love and time on your own.

then we see the need for a social approach to be developed.

A social history of Western medicine

This section provides a brief account of medicine not as a history of the past but as a history, and essential background, of the present. Unlike other medical histories we will not concern ourselves with the dates of medical discoveries nor the people who made them, but rather we try to provide an alternative to such an 'onward and upward' account that sees medical history merely as scientific progress. We consequently focus on changing conceptions of health and medicine in relation to the society in which they were developed. We do not consider such systems as Chinese or Arabian medicine but deal only with western medicine, for it was in Europe and the USA that the biomedical viewpoint developed and ultimately gained hegemony. For convenience we have labelled the different eras of medicine as Greek, Roman, Dark Age, Bedside, Hospital, Laboratory and Industrial medicine.

Greek, Roman and Dark Age medicine

It was in Greek society that medicine really began to emerge with its own theories and practitioners. Health was achieved when the four humours – blood, phlegm, yellow bile and black bile – were in balance. Medical treatment consisted of two forms. The first was associated with the male god Asclepius who according to Greek legend was the first physician, a master of the knife and the curative use of plants. In what we would now call medical intervention, a cure was attempted by trying to produce major systemic change in the body. For example, if the patient was feverish this was attributed to a blood imbalance and the appropriate treatment was bleeding. The second approach was associated with a goddess, Hygeia, who was the guardian of health and symbolised that people could be healthy if they lived according to reason. Health was produced by way of life and is clearly seen in such Hippocratic writings as *On Airs, Waters and Places* (quoted in Howe, 1963, p. 8):

> Whoever wishes to investigate medicine properly should proceed thus: in the first place to consider the seasons of the year, and what effects each of them produces. Then the winds the hot and cold, especially such as are common to all countries, and then such as are peculiar to each locality. . . . One should consider most attentively the waters which the inhabitants use . . . the mode in which the inhabitants live, and what are their pursuits, whether they are fond of drinking and eating to excess, and given to indolence, or are fond of exercise and labour.

This passage has been often quoted by medical geographers and it is not surprising when one can see scope for climatology, micro-climatology, hydrology, social and economic geography, and the stress on locality. The Greek view was that both internal medical intervention and prevention in the social and physical environment were necessary for health.

Roman medicine accepted the Greek pattern but under the great physician Galen there was an increase in the empirical content of medicine with the rising power of Ascelepian medicine and the waning of Hygeia; in fact (in an early example of patriarchy in medicine) her Roman equivalent had become a handmaiden to the male god. With the collapse of the Roman Empire in the fifth century this knowledge of medicine was effectively lost in the west; indeed for the early Christians, medical intervention was heretical. Ill-health was explained by supernatural causes and disease was seen as God's punishment for past sin and a signal of the need to repent. The main sources of healing were prayers and repentance aided by priests, and herbal remedies which were usually provided by women.

Bedside medicine

In the period following the Dark Ages, Jewson (1976) has recognised three major viewpoints that succeeded each other: bedside, hospital and laboratory medicine. The first of these viewpoints lasted from the Middle Ages until the late eighteenth century and was based on the rediscovery of the classical humoural theory. Bedside medicine was based on patronage and each physician had a small coterie of patients drawn from the ruling aristocracy. It was the patient and not the doctor who was all powerful, with the patron retaining ultimate control over treatment. Patients selected the practitioners on a personal assessment of their skills, and a successful physician was required to be a 'gentleman'; wit, elegance and integrity were of more importance than technical skill. Diagnosis was done at the bedside and was based on the total condition of the patient. The physician had to make his decisions on the basis of what the patient told him and on external manifestations, for there was very little manual examination. Moreover, there was little or no technical development with rich clients unwilling to act as 'guinea pigs'. Each individual had a unique pattern of body events which the physician had to discern in each case.

The mechanisation of the world picture

Although bedside medicine continued to be practised until the end of the eighteenth century, theoretical developments and social changes were occurring during the period 1500–1700 that were to allow new forms of medicine to be developed. Before proceeding to these different forms, we need to consider what Dijksterhuis (1961) has dubbed the 'mechanisation of the world picture', the process by which mechanistic philosophy gained dominance over vitalism. In vitalism, nature was the result of the harmony of God's plan and all events were manifestations of the vital spirit that was inherent in all matter. Explanations were based on the nature of things; an apple falls because that is its nature, and that is God's plan. Nature was perfect as it was and should not be changed. Direct biological experimentation and investigation was discouraged and the publication in 1543 of the first modern anatomy textbook by Vesalius was denounced by the church.

In contrast, the mechanistic or Cartesian philosophy (in honour of René Descartes, 1590–1650) separated the spiritual world from the material world and held that the material world could be conceived as a vast machine. Nature was like a piece of clockwork which was governed by exact underlying mathematical laws. The main method of this philosophy was the *reductionist* one of analysing the world by dividing it up into pieces and by arranging the parts into logical order. The approach was empirical and only measurable things were deemed scientific. Consequently, aspects of life such as values, feelings, motives and the soul were all placed outside of science. The aim of research was to discover the underlying general causal mechanisms, while the aim of science was no longer to live in harmony with nature, but to control, predict and dominate. For Francis Bacon (1561–1626) nature was a woman whose secrets could only be revealed by torturing her with mechanical devices.

Descartes, in his *Treatise on Man*, applied these ideas to medicine. He conceived of the body as a machine with component cogs and pulleys driven by a heart pump. People were machines and he compared a sick man with an ill-made clock. The body machine could be analysed into separate parts and in 1761 Morgagni identified organs as the seats of disease with such categories as angina pectoris, sinusitis, tonsilitis, and urethritis. Descartes separated the mind from the body and it was the role of medical science to research the corporeal body, while it was the role of the physician as a body-mechanic to restore the particular part of the body that was malfunctioning to normality.

Such scientific development did not take place in a social vacuum. The feudal view of the world was essentially static and unchanging while the mechanistic philosophy reflected a progressive and changing perspective. The former view did not suit the growing mercantile and capitalist order and as Rose *et al.* (1984, p. 41) have argued, 'even the most abstract pronouncements of physics . . . could be seen as arising out of the social needs of an emergent class'. To take an example, Descartes as a Catholic could not turn the whole of man into a machine for the mind had a self-consciousness, or soul, which had been touched by the breath of God. Consequently, he divided man into the corporeal body, governed by mechanics, and the soul or mind which was the immortal fragment. The two parts were only linked by the pineal gland in the brain; *dualism*, the separation of the body from the mind, was born. This division was compatible with the developing capitalist social order for it enabled the workers to be treated as physical mechanisms during the week, 'while on Sundays ideological control could be reinforced by the assertion of the immortality and free-will of an unconstrained incorporeal spirit unaffected by the workaday world to which the body had been subjected' (Rose *et al.*, 1984, p. 46). According to this social theory of science, it was not the mechanisation and associated inventions that produced social change but it was the new social order that used already known technological development (Braverman, 1974).

Hospital medicine

The mechanistic medical approach began to be applied in the early years of the nineteenth century particularly in the Parisian medical schools. Again this can be related to social change for, following the French revolution (1789), there developed strong, centralised control of hospital medicine as part of a plan to expand health and welfare facilities. Hospital doctors became the elite group of medical practitioners. The patients were the sick and poor. Hospitals were not places one went to by choice and they housed those who could not afford the physician to attend their homes, and the homeless. This major transformation of the social status of the patient meant that there were literally thousands of cases which formed an inexhaustible supply of acquiescent research material (Waddington, 1973). Moreover, autopsies were allowed and encouraged in the Paris schools, and such investigation of the body allowed the physician as researcher to correlate the external symptoms with the internal pathology.

The development of hospital medicine also saw the beginnings of professionalisation with physicians specialising in the different organs of the body. A new type of career structure was developing in which peer review and promotion was achieved not by treating the sick but by discovering new things and publishing them. The overall result was a major transformation in medicine: the patient was no longer the centre of medical attention which was transferred to the diagnosis of disease in particular organs by the physician independent of the patient. The sick man had been replaced by a collection of organs, one or more of which had a disease; the patient had become less of a person and more of an object. Important as these developments were for disease classification, anatomy and pathology, it is fair to say that there was not a concomitant improvement in treatment and cure.

Sanitary reform

Contemporaneous with the rise of hospital medicine, but unrelated to it, was considerable pressure for environmental reform; a fleeting return to Hygeia via hygiene. In Britain, industrialisation and urbanisation had resulted in a deteriorating physical environment in the early nineteenth century and after 1816 there was a discernible rise in the mortality rate (Doyal, 1979, p. 50). Britain at the time was a predominantly free-market society with minimal state intervention. But, following Chadwick's report, *The Sanitary Conditions of the Labouring Population of*

Great Britain (1842), there was a widespread belief that disease caused poverty (and not vice versa) and that there was a growing burden on society in terms of poor relief. This economic fear was reinforced by fears of working-class unrest and sedition, and by capital's requirement of a physically healthy labour force to exploit the new machines and technology. The result was a number of reforms that culminated in the Public Health Act of 1848.

The sanitary reforms were based on the miasmic theory that postulated that disease was spread by foul odours emanating from decaying bodies, decomposing rubbish and excreta. The 1848 Act allowed local authorities to introduce a range of sanitary reforms including 'pure' water supply, refuse and sewage disposal. While the reforms did meet opposition (especially from the rearguard action by the gentry against such 'centralising' measures and by the commercial concerns that stood to lose by municipalisation), nearly 800 towns had adopted the measures by 1868. This paved the way for the more comprehensive Act of 1875 which resulted in a substantial improvement in the health of the British population.

Laboratory medicine

The third category identified by Jewson is laboratory medicine. This had its beginnings in the middle decades of the nineteenth century in Germany. The approach represents the full flowering of the Cartesian reductionist viewpoint. While Morgagni had claimed that organs should be the basic unit of analysis, Bichat stated in 1800 that it should be tissues, and in 1839 Schwann went smaller still, saying that it was the cell. While bedside medicine had seen the totality of the sick man, and hospital medicine saw malfunctional organs, laboratory medicine saw life as a succession of cells and disease as a disruption in the process. Medical knowledge was not gained from the patient, but in the laboratory. This lead in turn to more specialisation based on the various scientific disciplines from which knowledge was derived, and clinical diagnosis itself was reorganised around the application of tests and technical procedures.

The ultimate dominance of this approach was considerably aided by the apparent success of the 'germ theory' which postulated that disease could be produced by the introduction of a single specific factor, a virulent micro-organism, into the body. Pasteur published his germ theory of disease in 1861 and in 1880 showed that injection of cholera into chickens produced the disease. In 1882 Koch discovered the tubercle bacillus which was pronounced as the cause of tuberculosis. Two important associated ideas were the doctrine of 'specific aetiology' and the concept of the 'magic bullet'. According to the former, each disease had a single, specific, objectively identifiable cause. The cause of cholera was the cholera vibrio and this vibrio caused no other disease. Researchers aimed to find such causes and then destroy them with magic bullets leaving other organisms unharmed. In 1909 Ehrlich found such a particular curative agent, Salvarsan, for the treatment of syphilis. The claims for the germ theory reached their most extensive form with the German bacteriologist von Behring claiming in 1898 that all disease was caused by bacteria, dismissing the idea that poverty was a breeder of disease as unscientific and antiquated.

After 1900 there was an obvious decline in the mortality rates in the countries of Europe and it was scientific medicine and especially the germ theory and the other medical 'breakthroughs', such as anaesthesia, that received the credit. It was accepted that social and environmental causes could be ignored and that disease was produced by the entering of germs into the body. It was cure, treatment and repair that was all-important; action was required in terms of internal medical intervention, while the emotional, social, environmental and psychological context could be removed.

Allopathy gains hegemony: the Flexner report

In the mid-nineteenth century there was a wide variety of treatments and approaches to disease, and the public could choose from a range of professionals who could provide allopathic, chiropractic, and homeopathic medicine. Allopathic medicine is the precursor of modern orthodox medicine and its distinctive feature is the use of a large amounts of drugs as antidotes to counteract symptoms. Chiropractice aims to restore body equilibrium and considerable importance is attached to body manipulation. Homeopathy was developed empirically by Hahnemann (1755–1843) who tried to use drugs in a more gentle, precise way. Drugs are not used as counteracting antidotes but as a means of curing like by like. The general principle of the treatment is to administer minute doses of a substance which produces symptoms in a healthy person similar to those produced by the disease.

In the USA at mid-century, there was a surfeit of allopathic physicians as it was relatively easy to purchase a medical degree;

there was also considerable consumer choice of modes of care. To combat this severe competition, a small group of allopaths formed a professional organisation in 1847 (the American Medical Association). This select group, the majority having trained in postgraduate medicine in Europe, believed implicitly in the importance of scientific medicine. They involved themselves in examining and educating, and their influence is to be found in new licensing laws of the 1890s whereby allopaths gained effective control of this process. All students had to pass a paper in scientific medicine and there was only a special option in homeopathy. In 1893 the Johns Hopkins University followed the European examples of the Pasteur and Koch Institutes by putting laboratories in its medical school and staffing them with European-trained scientific researchers. But despite these victories the allopaths' position was still not secure; the *Journal of the American Medical Association* stated bluntly in 1901 'Growth of the profession must be stemmed if individual members are to find the practice of medicine a lucrative profession'.

It was at this time that the corporate capitalists, such as Carnegie and Rockefeller, enter the picture for their fortunes were used to create a society in their image and with it an industrialised medicine. In 1892 Rockefeller appointed Frederick Gates as his administrator of philanthropy. Gates believed that for industrial capitalism to thrive there was a need to produce a healthier, and therefore, happier society. He required technical solutions that would cure all diseases but he discovered that medicine as then practised was largely ineffective (curing no more than four or five diseases). He became convinced that such cures could only be developed by scientific research. On the advice of Gates, Rockefeller funded an Institute of Medical Research in 1901 and allocated considerable sums to the development of laboratory medicine.

In 1907 the AMA requested the Carnegie Fund to mount an investigation into the standards of all medical schools. An independent (*sic*) investigator was chosen, a non-medical man, Abraham Flexner; he was the brother of the head of the Rockefeller Institute. Flexner visited all the American medical schools as well as European ones, and his report was a complete vindication of the scientific approach; he recommended that medical education should be based on the Johns Hopkins pattern. Scientific medicine was judged better but the alternatives were not subjected to scientific comparison. (For example, the death-rate in the 1854 cholera epidemic was only 16 per cent in the London homeopathic hospital as compared to 50 per cent in the allopathic ones. This was suppressed by the Board of Health lest the figures would hinder the 'progress of science' (Pinchuck and Clark, 1984, p. 50.) Most importantly, Flexner's recommendations were funded by the corporate capitalists, and over the next twenty-five years $600 million was spent restructuring medical education (Berliner, 1977). The result was a highly specialised medicine that required four years of college, four years of medical school (the first two years of this studying the basic sciences and anatomy) followed by hospital training. Medical education was built around the physical and biological sciences and social, psychological and environmental concerns were all but eliminated from the training of physicians.

This transformation was not restricted to the curriculum, for it reinforced the class composition of those who practised medicine. The cost of biomedical education meant that the poor, blacks and women were effectively excluded from training. Five out of seven black medical schools were closed and all but one of the women's schools were shut; in the ten years following the report a third of the pre-existing schools had closed. The effects of this are felt today for the AMA has continued a policy of professional birth control and in 1973 only 8 per cent of medical students were black, only 17 per cent were female and only 12 per cent came from the poorer half of the population. After the First World War, American philanthropy helped reform medical education in Europe along scientific lines. The effect was a 'doctatorship' in which medical power was concentrated in the hands of a relatively small group of white medical-class men and the legitimacy of alternative modes of healing was effectively destroyed.

Worldwide industrial medical complex: modern medicine

According to Berliner (1982), medicine in the early years of this century adopted an industrial mode of production. Medicine became a product that could be delivered over and over again with product differentiation producing a vast array of drugs and treatments. Research and design began to take place in an industrial laboratory environment staffed by medical engineers, and the final product was delivered by a highly developed distribution network of hospitals and surgeries. What was until fairly recently an essentially cottage industry had become the province of multinationals (Kelman, 1971). In McKinlay's (1984) term there is now a worldwide industrial medical complex which is dominated by profit-seeking corporations.

Medicine in the twentieth century has experienced three major processes: the increasing use of technology, a rising degree of specialisation, and an expansion of the problems that medicine is supposed to solve. The whole of medicine has become a technical subject: whether a person is ill or not is a technical decision based on diagnostic tests; treatment is a technical decision in terms of choosing the appropriate technique for cure; and success is judged in relation to technical advance and the patient's compliance with the imposed technical regime. Kennedy (1983, p. 184) has argued that medicine is gripped by a technological imperative which states that if you can do it then do it (irrespective of the benefit it produces in relation to cost). Strongly related to the increasing use of technology has been the trend towards specialisation. In the USA, sixty-three different specialities were recognised by 1970, and over 75 per cent of all physicians now call themselves specialists. In the UK, the process has been slower but even in this country there are over fifty clinical specialities. Moreover, these specialisms are effectively ranked in terms of professional acclaim, reward and teaching time, with the most technical, such as heart surgery, placed at the top. At the bottom come such specialisms as geriatric medicine, psychiatry and community medicine. The training of physicians, as Inglis (1981, pp. 304–14) makes clear, is now based on the technical specialisms, for after the pre-clinical years studying the core subjects of biochemistry, and physiology, the emphasis is placed on acquiring scientific, technical skills. The training is mostly done in teaching hospitals which, as befits technical 'centres of excellence', are full of sophisticated medical technology waiting to be used. Home visits where it may be possible to see the social conditions in which the patient lives are very rare. The overall result is that resources and training are concentrated on the technically interesting conditions while the insane, the handicapped and the geriatric have to endure poorer conditions. Moreover, by stressing technical, specialist knowledge, the physician removes his work from the competence of ordinary people and becomes an expert whose judgement can only be assessed by other specialists. The role of the patient and the public has been reduced and the decision-making bodies on health and medicine are dominated by the physician as medical expert.

Throughout the twentieth century, medical practitioners have been actively engaged in expanding their sphere of influence. It is in the area of 'social problems' that the medical solution is being most actively promoted. Conrad and Schneider (1980) show that what was once defined as badness and sin (and hence treatable by priests) has become crime (to be treated by the officers of the law) and is increasingly becoming sickness to be treated by medical intervention. A clear illustration of this tendency can be seen in a famous letter to the *Journal of the American Medical Association* by Sweet, Mark and Ervin (1967). They argued, in the wake of urban rioting in the USA, that it could not be unemployment and slum housing that was producing the riots because of the relatively small numbers who had rioted. Therefore, it must be something in the violent rioter that distinguished him from his peaceful neighbours, and they claimed that this was brain damage. They estimated that some 11 million Americans are suffering from obvious brain damage while another 5 million are subtly damaged. Treatment would consist of mass screening to identify those affected and then removal of part of the brain's limbic system (Rose *et al.*, 1984, p. 170).

While this example may be seen as outrageous (although it was subsequently funded by the US Law Enforcement Assistance Agency), it must be accepted that the jurisdiction of medicine now extends from controlling life at death (the decision to switch off the machine), to controlling life at birth (in terms of abortion) to whether life should be allowed to begin at all (genetic counselling). Conrad and Schneider (1980) suggest that a number of conditions such as boredom, lethargy and gambling may be taken into the medical orbit in the future. Such proposals effectively strip away the social and political context from these problems as they became recast as individual deviations, soluble, or at least controllable, by medicine. This technical fix represents what Illich (1975) has called 'the medicalisation of life' and as Frances Crick has written (quoted in Carlson, 1975, p. 62) this process has gone furthest in America: 'Americans have a peculiar illusion that life is a disease which has to be cured, a widespread belief that for every problem of life there is a pill, a bottle of medicine or some other cure'. The public have increasingly come to expect breakthroughs and both the medical profession and the drug and equipment manufacturers have had much to gain in dominance, prestige and profit from this trend.

Berliner (1984) regards the period 1920–50 as the most successful and popular one for medicine. It was a time of a decreasing mortality rate, and life-expectancy in the developed countries approached the biblical three-score and ten years. Technical developments, including antibiotics, blood transfusions, improved surgical procedures and vaccines for diseases like polio, resulted in a high level of public confidence in scientific medicine. In the last thirty years, however, there has been a growing unease about medicine's ability to deliver the goods, with some people, albeit a minority, turning to alternative medicine (Salmon, 1984). In 1959 the first cases of drug-resistant staphylococcus were found and by the 1980s three-quarters of American cases of this micro-organism were resistant to penicillin. Not only were standard treatments losing their effectiveness but new treatments were found to be iatrogenic, that is disease-producing, an infamous example being the thalidomide cases in the early 1960s. While the newspapers continue to trumpet medical breakthroughs, the medical profession has revealed its considerable self-doubt. For example, Henry Simmons, Director of the US Food and Drug Administration, stated bluntly in 1973, 'The drugs age began to decline in 1956. There have been hardly any effective new drugs since then.' This is echoed by Sir Frank Macfarlane Burnet (1971, p. 226), an eminent microbiologist for whom 'the contribution of laboratory science has come to an end . . . almost none of modern basic research in the medical sciences has any direct bearing on the prevention of disease or on the improvement of medical care.'

Moreover, while technical developments remain widely acclaimed, there has also been something of a backlash against technology. For example, there has been considerable public disquiet over medicine's ability to keep patients artificially in the twilight zone between life and death, and from the women's movement there has been concern over the detrimental effects of excessive obstetric technology (Day, 1980). In the area of mental care there is again public concern, reflected in such novels as *One Flew Over the Cuckoo's Nest*, that such treatment is primarily control and not therapeutic assistance.

This growing worry about medical effectiveness is taking place against a backcloth of the changing nature of disease. As Figure 1.1 shows, the major killers in developed countries are no longer infectious diseases but chronic diseases such as the cancers, heart disease and strokes, and curative medicine has not achieved much success against such diseases. Lung cancer is the commonest form of cancer in the UK and the USA today. Treatment is usually by biomedical intervention and the patient faces control either by poisoning through chemical means, torture by radiation, or radical surgery. Yet, less than 10 per cent of lung-cancer cases survive more than five years following diagnosis, and this proportion has not appreciably improved over the last twenty-five years, a period of substantial rise in the number of cases. The biological causes are still largely unknown, and this is despite massive research expenditure. There is a growing recognition of the role of external factors in determining the disease, and Harris (1970, p. 59) estimated that over 80 per cent of cancers are environmental in origin. There is, however, comparatively little research on these environmental factors and preventative action remains minimal (Doyal and Epstein, 1983).

This rather meagre success is being bought at an ever increasing cost. According to Wildavsky (1980, p. 10) the marginal value of one, or one billion dollars spent on medical care would be close to zero in improving health. But the amount actually spent in the world on medicine is increasing so fast that there was a 50 per cent increase between 1976 and 1984. Abel-Smith (1967) found that core-capitalist countries transferred an additional 1 or 2 per cent of GNP to health-related expenditure every year. According to McKinlay (1984) there are parallels between medicine and the car industry. Fifty years ago cars were '90 per cent transport and 10 per cent frills' but, as Henry Ford discovered, this was not the way to maximise profits so the car developed until the percentages were reversed. The extent of the

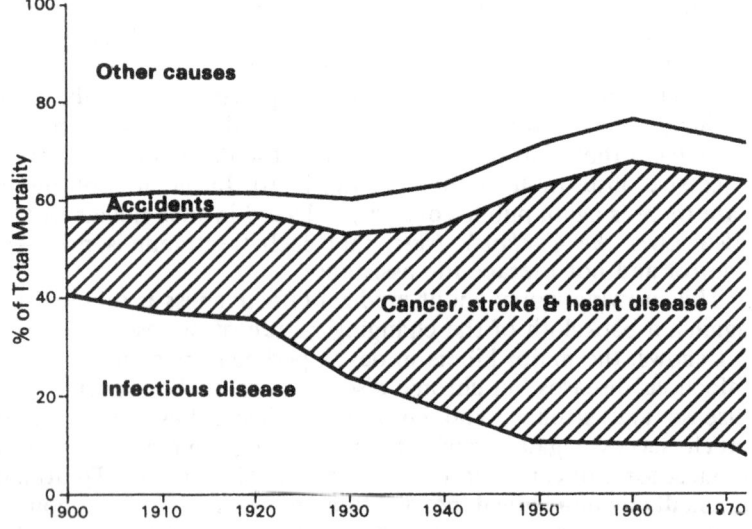

Figure 1.1 The increasing importance of chronic disease: USA 1900–73
Source: McKinlay and McKinlay (1977)

Table 1.1 The 'top ten' multinational drug manufacturers, 1980

Multinational	Country	Sales ($ million)	Profit margin %	Estimated expenditure on publicity & information ($ million)
Hoffman-La Roche	Switzerland	3,100	19	300
Merck & Co	USA	2,200	28	230
Hoechst	Germany	2,000	18	280
Ciba-Geigy	Switzerland	1,800	21	220
Bayer	Germany	1,600	16	200
American Home Products	USA	1,500	24	150
Sandoz	Switzerland	1,400	17	200
Bristol-Myers	USA	1,400	18	210
Warner Lambert	USA	1,300	14	220
Pfizer	USA	1,300	23	170

Source: Melville and Johnson (1983, p. 45).

'big business' nature of modern medicine can be seen from Table 1.1 which gives the top ten multinational drug manufacturers in 1980; also shown are the high levels of profit and the high percentage of total expenditure devoted to marketing and advertising; both figures are considerably higher than for other industries. There can be no doubt that medicine today is truly a worldwide industrial complex which is dominated by multi-national organisations which actively support, promote and extend specialisation, technological medicine and the biomedical viewpoint.

The biomedical model and its critics

Before discussing the critiques of the biomedical model that have been developed over the last thirty years, it is important to consolidate our understanding of what the approach entails. The metaphor that underlies the biomedical position is the body as machine and disease as internal malfunctioning. Machines generally operate as a linear causal sequence so that A causes B which causes C and so on; if a breakdown occurs, it is usually traceable to one component part. Similarly, the body-machine is seen to be made up of interacting but separatable parts; the physician, as body-mechanic, must find the diseased part and the single cause of this disease, and that is why we need specialists in particular organs or processes. Today's equivalent of the germ theory is the concept of unifactorial aetiology in which each disease is produced by a specific biological process. All individuals are conceived as having similar body machines and, in accordance with the concept of generic diseases, all individuals with the disease are to be treated in the same way. Continuing the metaphor, hospitals are garages in which the body-machine can be temporarily immobilised to permit the technician to get to work with all the sophisticated machinery that is housed there. Medicine is applied science and the answers ultimately lie in research done in the laboratory.

Two quotations summarise the biomedical position with unusual clarity. The first comes from the British former Chief Medical Scientist, Sir Douglas Black (1968, p. 5), for whom

> The most characteristic function of a doctor lies in the
> diagnosis and treatment of disease in the individual patient
> . . . the great majority of doctors will remain concerned with
> disease and not with positive health or community medicine or
> social medicine.

The second is by an eminent American researcher, Lewis Thomas; he wrote (1979, p. 168):

> For every disease there is a single key mechanism that
> dominates all others. If one can find it and then think one's
> way around it, one can control the disorder. In short, I believe
> that the major diseases of human beings have become
> approachable biological puzzles, ultimately solvable.

Such views are presented with the full authority of science, but to reiterate our earlier arguments, the biomedical approach is not reality but merely one way of looking at reality, and it is not necessarily the 'best' or the most rational. Indeed, a number of commentators have attacked the biomedical model, and we will first discuss those criticisms made from within medicine by three 'radical doctors' (Stark, 1982): Cochrane, Dubos, and McKeown. We will then proceed to the critiques of Illich (a theologian) and Doyal (a marxist sociologist). The discussion is organised so that the questioning of the biomedical approach becomes more radical as the arguments are developed.

The 'radical doctors'

One of the more narrowly focused of the 'critiques from within' is that of A. L. Cochrane in his 1972 book *Effectiveness and Efficiency*. His basic argument is that given the massive increase in input to the British National Health Service since its inception in 1948 (in terms of money, time and expertise), there has been no concomitant improvement in the nation's health. He argues for an evaluation of medical treatments, by randomised control trials (RCT), to weed out ineffective practices. In such a trial, patients with identical medical problems are randomly allocated to two groups; one group is given the old or no treatment and the other is given the new treatment. Some of the results from RCTs are striking. For example, Mather (1971) found that there was no difference in outcome after a heart attack between resting at home and being treated in a hospital intensive coronary care unit; if anything, the patient recovered quicker at home (the results have been confirmed by Hill *et al.*, 1978). Similarly, so standard was the diabetic drug treatment of 'insulin, potassium and glucose' that it was argued that it would be unethical to perform an RCT and deprive patients of this valuable treatment. The RCT was, however, performed, the combination was found to be ineffective and the practice has disappeared.

Cochrane argues that a considerable amount of modern medicine has not been scientifically evaluated and therefore its application is based on a *belief* in its effectiveness. Cochrane wants to stop ineffective treatments and to transfer the money to more effective forms of clinical medicine, and to improve the care for the long-term geriatric and mentally ill patients. These are sobering thoughts about the effectiveness of medicine particularly when it is realised that of the 3 billion dollars spent on health care in the USA, 60 per cent is spent on care in the last year of life (Kennedy, 1983, p. 36) while Ferguson and MacPhail (1954) found in an RCT that two years after discharge from an acute general hospital, the majority were unimproved and more than a third were dead! It is on such evidence that Cochrane (1972, p. 8) considers the twentieth century as a 'straightforward story of the ineffectiveness of medical therapy'.

Our second critic is the distinguished microbiologist and eminent medical historian, René Dubos. He was one of the earliest of the modern critics and we will discuss his 1959 book, *Mirage of Health*. The book is a magnificent panorama on health and disease from pre-history to the present day. By being very selective we can identify four major themes. First, he argues that it is a myth that the decline in the death-rate in the last hundred years has anything to do with laboratory medicine, for diseases like leprosy and typhus had virtually disappeared from Europe before the germ theory had been developed and applied. He writes (p. 23) 'when the tide is receding it is so easy to have the illusion that one can empty the ocean by removing water with a pail', and argues that it was sanitary measures that had, in reality, produced the decline in the prevalence of infectious diseases. Second, he contends that it is a delusion to proclaim that the present state of western health is the best in the history of the world when the expectancy of life after the age of 45 is hardly any greater than it was several decades ago, when one in four Americans spend at least some months in a mental asylum, and when an increasing number of people depend on drugs to cope with the everyday problems of life. Third, he believes that the concept of unifactorial aetiology is mistaken and that infectious and chronic diseases are (p. 102) 'the indirect outcome of a constellation of circumstances rather than the direct result of a single determinant factor'. Accordingly, he chides the frantic medical efforts to find the cause of cancer, heart disease and mental illness. Fourth, and finally, adaption is a key concept for him, and while he believes people can adapt to virtually anything, this requires time which the increasing tempo of modern life no longer permits. In particular, he implicates the modern environment in determining chronic diseases, citing evidence for occupational carcinogens and diet as causes of disease. He also suggests a connection between increasing automation, boredom, social stress and mental illness. He argues for a refocusing of research away from finding a cure to defining more precisely the environmental causes of disease. However, while he clearly implicates the organisation of society in the production of illness, he stops short of radical reform, stating that 'it is not the function of medicine to become identified with political action' (p. 214).

Our third radical doctor is Thomas McKeown who, in his 1979 book, *The Role of Medicine*, argues that medicine is mistaken in its concentration on internal intervention and should, instead, afford greater attention to external influences and personal behaviour which he sees as the major determinants of ill-health. In the first part of the book, McKeown examines the decline in mortality from specific diseases in England and Wales since the mid-nineteenth century. He empirically verifies the ideas of Dubos and concludes that medical intervention contributed little to the reduction of infectious disease before 1935, and that over the whole period medicine was less important than other influences, especially improvements in nutrition and hygiene. As Figure 1.2 shows, effective medical procedures were applied too late to have much influence on two important causes of death:

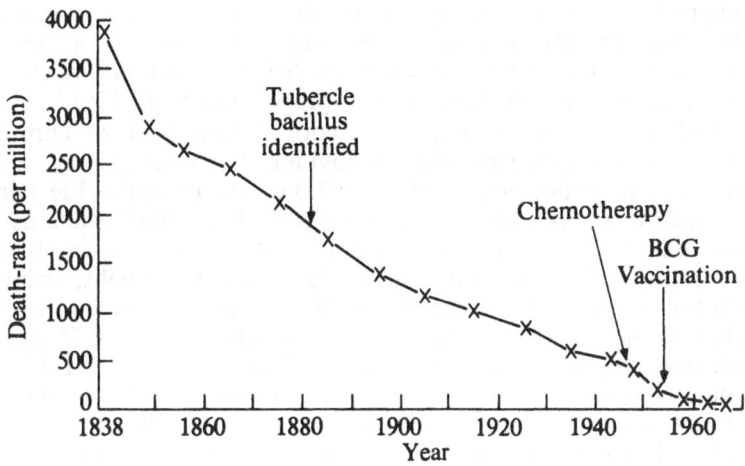

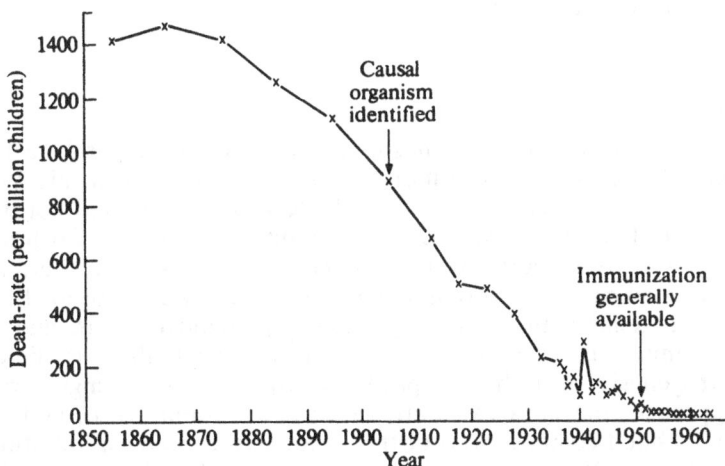

Figure 1.2 Declining mortality from tuberculosis and whooping cough: the effect of medical intervention (a) Respiratory tuberculosis: mean annual death-rates (b) Whooping cough: death-rates for children aged under 15
Source: McKeown (1979)

tuberculosis and whooping cough. Moreover, he is sceptical of the achievements of modern medicine, believing that 'an exact evaluation of twentieth century medicine would do much to restore nineteenth century faith in prayer' (1979, p. 176).

McKeown contends that the major determinant of chronic disease is individual personal behaviour. For example, he cites the work of Belloc and Breslow (1972) on the effects of lifestyles on health. These authors conducted a study of 7000 adults over five years in relation to seven rules of healthy living: no smoking, sleep seven hours a night, eat breakfast, control weight, imbibe alcohol in moderation, exercise daily, no eating between meals. They found on a series of tests that people aged over 75 years following all seven rules were as healthy as those aged 35–44 following only three, and that life-expectancy at 45 was eleven years longer for those following six or more rules than those who adhered to less than four. According to McKeown, if there is to be a future improvement in the public's health then it must come from individuals taking responsibility for their own health. He concludes that we need epidemiology to elucidate the link between personal behaviour and ill-health, and health education to disseminate these findings so as to persuade individuals to change their behaviour.

Illich

Ivan Illich is a moral philosopher and theologian, and he has provided an extensive critique of modern medicine in his 1975 book, *Medical Nemesis*. While McKeown has argued for the historical ineffectiveness of medical intervention and Cochrane for its current ineffectiveness, Illich goes one step further: for him, medicine is downright dangerous. The key concept for Illich is iatrogenesis, that is damage done by medicine itself, and he recognises that this occurs at three levels. First, there is clinical iatrogenesis or ill-health produced by medicine's engineering approach, the physician's arrogance and incompetence, and the unintended complications of treatment. Using medical literature, Illich provides considerable documentation of such effects which has been updated and extended by Melville and Johnson (1983). The second type is social iatrogenesis and this refers to the medical-industrial complex's role in maintaining a sick society, promoting the medicalisation of life and emphasising the technical fix. Structural iatrogenesis is the third type and this refers to the loss of capability for self-care; people have become sickly because of their dependence on others to do something for them.

Illich theorises that the social organisation of society is determined by the 'culture of industrialisation' whereby technology defines social organisation. Pre-industrial society is seen as intergrated and self-sufficient, while industrialisation necessitated strong and rigid hierarchies to control large-scale production. Industrial society therefore has an elaborate organisation to manage industrial structures, with the receivers and consumers of goods and services at the bottom, and experts, technocrats and bureaucrats at the top. The latter are supposed to control for the good of all, but in fact manipulate the system for their own benefit. To maintain this hierarchy, the corporate, centralised culture of industrialisation must be promoted and it is this that erodes individual autonomy and creates the medicalisation of life.

As a solution to these iatrogenic problems Illich rejects consumer control over the medical profession because the experts would keep medical knowledge to themselves, and consequently the patient would not be able to judge the quality of care in any rational way. He also rejects an 'environmental' approach because of the danger of expanding medicine's domain so that there is a 'sanitised hell' from birth to death, thereby further reducing individual autonomy. He also rejects a more egalitarian distribution of medical care believing that this will only spread the damage further. His solution is to change the nature of society by de-industrialisation and de-professionalisation. The licensing and regulation of medical practitioners are removed, giving the individual freedom to choose healing when, where, how and from whom he or she likes. This needs to be accompanied by the active promotion of self-discipline, self-interest and self-care. For Illich, health and freedom are intertwined: a person is only healthy when he or she has sufficient autonomy to cope with the reality of suffering and death.

Doyal

Our final critic is the British marxist sociologist, Lesley Doyal; we will examine some of the arguments presented in her 1979 book, *The Political Economy of Health*. Here we will introduce two of her themes: one is the contention that society is not industrial and composed of individuals (as presumed by McKeown and argued by Illich) but is capitalist and consists of social groups and classes; the second is that the capitalist nature of society determines both ill-health and the current nature of medicine.

Before dealing with these themes we need to outline some basic propositions of marxism.

According to Marx, for thousands of years society has been composed of antagonistic classes that are in mutual conflict. The form of this conflict has changed over time according to the dominant mode of production which consists of the forces and the relations of production. The former consists of the materials worked on, the tools and techniques that constitute the means of production, and the people who make the material goods. The latter consists of the social organisation of the labour power, that is the relationship between who produces the goods and who controls the means of production and the use of the product. For Marx, epochs differ not so much in what they produce but in how production is organised. In a capitalist system, the wealthy, minority bourgeoisie class has control of the means of production (e.g. the machines) and owns the product of labour, while the workers own nothing in terms of productive goods except their ability to work. Marx also predicted a future communist society in which there would be common ownership of the means of production and the resultant product. This society is so arranged to satisfy the dictum 'from each according to ability, to each according to need'; class conflict no longer exists.

The dominant mode of production is the economic base of society and above it lies the superstructure of social, political and cultural phenomena such as the state, science and medicine. For Marx, the character of the superstructure depends on the nature of the base and particularly on the economic interests of the dominant class. The marxist view of society is known as historical materialism because causal priority is given to the production of material goods in determining social forms, and because society is in the process of historical development as one mode of production is replaced by another. Crucially, there can be no genuine change in the societal superstructure unless the economic base is also changing, for the nature of society is constrained by the mode of production.

Capitalism continues to survive and reproduce itself not by force but largely by legitimation and ideological control. Ideology refers to a set of ideas about a society generated by that society and according to marxists, ideology is dependent on economic requirements and represents class interests. Marx called the particular ideology of capitalism 'commodity fetishism' whereby capitalism presents itself in a distorted or fetishised way in a commodity form that hides or mystifies what it is about. Above all it presents itself in a reified way, whereby social relations between people are presented as natural relationships between things so that the current organisation of society appears to be the 'natural order of things'. For example, the relationship between capitalist and worker takes a commodity form and it appears fair and 'legitimate' that workers receive a wage determined by a free market. This hides the exploitive form of the relationship whereby the surplus product is siphoned off to the capitalists' benefit. It may appear that this is fair reward for risking their capital, but as wealth could only have come from previous exploitation of workers, it is not theirs to risk!

Such distortions pervade capitalist society and consequently superficial appearances do not always conform to underlying reality and, in particular, many aspects of society exist in what is called 'contradictory unity'. The state, for example, is not a victory for the working class nor is it merely an appendage of capitalism which provides the necessary infrastructure for capitalism that permits the social control of potentially rebellious groups. It is both simultaneously, and it achieves its negative feature, the preservation of existing exploitive social relations, through its positive feature of providing welfare for the needy. According to Doyal, medicine is a similar contradictory relationship that provides help for the ill but, by providing it in a fetishised form that concentrates on individual body pathology, it hides and exonerates the role of society in determining health. To achieve real understanding of the capitalist society, the marxist argues that we must drop our current theoretical apartheid (economics, sociology, geography, etc.) and develop a political economy which seeks to examine the totality of the system.

Following the tenets of marxist political economy, Doyal argues that health and illness in Britain are inextricably bound-up with capitalist development. In the early period of capitalism, there was worsening health, particularly in urban areas, but the subsequent development of capitalism also made possible dramatic improvements in the life-expectancy of the entire British population. Nevertheless, many people are still dying prematurely and there remain major inequalities in health. These inequalities are attributed to the class nature of society and according to these arguments 'capitalism makes you ill' (Thunhurst, 1982).

It is in this context of the societal determination of ill-health that Doyal develops her criticism of the arguments of McKeown and Illich. Both these authors place the responsibility for ill-health with the individual. For McKeown, people become ill because of their individual failings, be it lack of exercise, or unthinking consumption of cigarettes. But for Doyal such advice is tantamount to 'blaming the victim' for people are not free to choose their behaviour, their occupation and the products they consume. Knowing that taking certain action may lead to improved health does not mean that the individual can do anything about it. Moreover, this stress on a lifestyle explanation can be seen as another example of capitalism's mystifying ploys for it hides the debilitating exploitive relationships in society, and it shifts the debate away from the world of work to the world of consumption.

Doyal's criticisms of Illich are wide-ranging and she employs the ideas of Navarro (1975) to argue that Illich's proposals are not radical but profoundly conservative. Illich contends that it is the industrial, technological nature of modern society that has created class divisions, but according to marxist doctrine, technology did not create class divisions, it merely reinforced divisions that were already there (Braverman, 1974). Medical bureaucracy is not the controller but the servant of the dominant class.

Most importantly, the demand for and dependency on medicine are not just created by the medical establishment for its own good, as Illich argues, but are part of the wider commodity fetishism of capitalism. Capitalism has to promote dependency on the consumption of goods and services, for without artificially created demands and dependencies, and a large passive consumer society, there would be a decrease in consumption and the whole system would collapse. Consequently, Illich's proposals of de-regulation and de-professionalisation would achieve little; addictive demand would remain unchanged as would the social and economic conditions that are producing ill-health. Indeed, Illich's proposals, by concentrating on individual self-help, deflect attention away from societal causes, thereby helping to conserve the current problems of health inequalities in society. Doyal also contends that Illich has overstated his case in relation to the ineffectiveness of medicine by confusing care and cure. She argues that while medicine as currently constituted is poor at cure, it can be effective at care, reducing disability and pain. It is against this background that Doyal makes her proposals for change. She wants equality of access to medical care, demystification of medical knowledge, and a fundamental transformation to a new mode of production so that 'the demand for a healthier society is, in itself, the demand for a radically different socio-economic order'.

Chapter organisation

Having explored the social approach to health and disease, we can now outline the organisation of the rest of this book. Essentially, it is composed of two parts that diverge further and further from the biomedical model. The first part considers disease ecology and the second deals with medical care. Chapter 2 discusses the collection of empirical data that can be used to reveal the geographical patterning of disease. It considers both the 'official' routine aggregate data and the collection of individual survey-based data which have been favoured by those who adopt a behavioural approach to ill-health. Chapter 3 is concerned with the analysis of such epidemiological data and how quantitative causal models can be developed to relate disease patterns to social and environmental variables. Chapter 4 is devoted to communicable diseases; as well as looking at the quantitative aspects of modelling disease diffusion, we will also examine the social condition of the host as an important factor in determining who becomes infected. Another type of 'disease', mental illness, is considered in Chapter 5 and we discuss the attempts by geographers and others to explain the occurrence of this illness. Chapter 2 accepts the biomedical position on what constitutes disease, Chapter 3 extends the biomedical model by emphasising the role of factors external to the body in determining ill-health; Chapter 4 argues that communicable diseases are not just biological entities for they are also social problems; and by Chapter 5, the whole biomedical definition of what constitutes disease is examined and criticised.

In the second part of the book, we consider geographical research on health care. Chapter 6 considers the inequality of health care provision and provides evidence that there are marked social and spatial inequalities in access to care. Particular attention is paid to the spatial scale at which these inequalities are to be found. Chapter 7 attempts to discover the root cause of these inequalities and we argue that a satisfactory explanation must be based on the way society is structured. In Chapter 8 we consider a number of policy initiatives that have been introduced in recent years to ameliorate these inequalities. However, in the context of the primacy of structural inequalities, these initiatives are seen as partial and somewhat ineffective. The final chapter re-examines both the production of ill-health and the consumption of medical care in relation to societal organisation. Much of what has gone before is reviewed in a critical light and is found to be wanting and limited. In preceding chapters, the various approaches used by medical geographers are presented on their own merits and in their own terms so as to provide a 'technical' appraisal before returning in the final chapter to present a critical perspective on the totality of medical geography.

Guided reading

This chapter covers a vast field and it consequently requires extensive reading to develop the issues that have been outlined here. Three relatively short books on the sociology of health and medicine (Armstrong, 1980; Hart, 1985; Patrick and Scrambler, 1982) can be recommended as general introductions to the social approach. The area is also well served by collections and Black *et al.* (1984), Conrad and Kern (1981) and Ehrenreich (1978) contain many useful articles in addition to the ones cited in this chapter. Eyles and Woods (1983) provide an important geographical commentary on this literature.

The most comprehensive guide to the social construction of disease is Conrad and Schneider (1980) which covers mental illness, alcoholism, opiate addiction, child abuse, hyperactivity, homosexuality and crime; it is a fascinating read. For other specific diseases, see Wright and Treacher (1982) who deal with asthma, mental illness, genetic diseases, and miner's nystagmus; Mishler (1981a) who discusses diabetes, alcoholism, mental retardation and blindness; and Open University (1985) which deals with hysterectomy and hysteria. A discussion of social control and professional dominance is given by Friedson (1970).

It is difficult to recommend a single volume that examines the history of medicine in a social context, but certainly the most entertaining is a 'cartoon' book by Pinchuck and Clark (1984). There is considerable detail on the development of the Cartesian viewpoint in Capra (1982, Chapter 4) and in Osherson and Amara (1981), while Rose *et al.* (1984, Chapter 3) consider the growth of science in relation to the rise of bourgeois society. Tesh (1982) discusses sanitary reform. The Flexner report is set in its wider context by Berliner (1975, 1985); Brown (1980) explores the role of corporate capitalists, while Starr (1982) provides an extensive account of the development of American medicine which should be read in conjunction with Navarro's (1984) review.

A clear and sustained attack on the biomedical model is presented by Mishler (1981b). Two lawyers (Carlson, 1975 and Kennedy, 1983) have 'prosecuted' the biomedical model, while 'popular' accounts are provided by Inglis (1981, particularly

Chapters 10 and 11) and Capra (1982, Chapter 10). An early critique is that of Powles (1973).

For the specific critics discussed in the text, their books provide the best starting point. A short and readable account of the marxist position is given by Waitzkin (1981), and a critique by Hart (1982) which is, in fact, a review of Doyal's (1979) book.

The following journals cover recent developments in the social approach to health: *Social Science and Medicine*, *Sociology of Health and Illness*, *International Journal of Health Services* and *Radical Community Medicine*. The last two on this list are specifically concerned with developing a marxist and radical approach to health and medicine.

References

Abel-Smith, B. (1967), 'An international study of health expenditure and its relevance for health planning', *WHO Public Health Reports*, vol. 32.

Armstrong, D. (1980), *An Outline of Sociology as Applied to Medicine*, Bristol, John Wright.

Barnes, B. (1982), *T. S. Kuhn and Social Science*, London, Macmillan.

Belloc, N. B. and Breslow, L. (1972), 'Relationship of physical health status and health practices', *Preventive Medicine*, vol. 1, pp. 409–23.

Berger, P. and Luckmann, T. (1967), *The Social Construction of Reality*, New York, Doubleday.

Bergler, E. (1956), *Homosexuality: Disease or Way of Life?*, New York, Hill & Wang.

Berliner, H. (1975), 'A larger perspective on the Flexner report', *International Journal of Health Services*, vol. 5, pp. 573–92.

Berliner, H. (1977), 'Emerging ideologies in medicine', *Review of Radical Political Economy*, vol. 8, pp. 116–24.

Berliner, H. (1982), 'Medical models of production' in P. Wright and A. Treacher (eds), *The Problem of Medical Knowledge*, pp. 162–73, Edinburgh, Edinburgh University Press.

Berliner, H. (1984), 'Scientific medicine since Flexner' in J. W. Salmon (ed.), *Alternative Medicines: Popular and Policy Perspectives*, pp. 30–56, New York, Tavistock.

Berliner, H. (1985), *A System of Scientific Medicine*, New York, Tavistock.

Black, D. A. K. (1968), *The Logic of Medicine*, Edinburgh, Oliver & Boyd.

Black, N., Boswell, D., Gray, A., Murphy, S. and Popay, J. (1984), *Health and Disease: A Reader*, Milton Keynes, Open University Press.

Braverman, H. (1974), *The Degradation of Work in the Twentieth Century*, New York, Monthly Review Press.

Brent Community Health Council (1981), *Health in Brent*, London, Brent CHC.

Brown, E. (1980), *Rockefeller Medicine Men: Medicine and Capitalism in America*, Berkeley, University of California Press.

Burnet, Sir F. MacFarlane (1971), *Genes, Dreams and Realities*, Aylesbury, Bucks., Medical and Technical.

Capra, F. (1982), *The Turning Point: Science, Society and the Rising Culture*, London, Fontana.

Carlson, R. J. (1975), *The Frontiers of Science and Medicine*, London, Wiley.

Cartwright, F. F. A. (1977), *A Social History of Medicine*, New York, Longman.

Cliff, A. D., Haggett, P., Ord, J. K. and Versey, G. R. (1981), *Spatial Diffusion: An Historical Geography of Epidemics in an Island Community*, Cambridge, Cambridge University Press.

Coates, B. E. and Rawstron, E. M. (1971), *Regional Variations in Britain*, London, Batsford.

Cochrane, A. L. (1972), *Effectiveness and Efficiency: Random Reflections on the Health Service*, London, Nuffield Provincial Hospitals Trust.

Conrad, P. and Kern, C. (1981), *The Sociology of Health and Illness: Critical Perspectives*, New York, St Martin's Press.

Conrad, P. and Schneider, J. W. (1980), *Deviance and Medicalisation*, St Louis, Mosby.

Cox, C. (1983), *Sociology: An Introduction for Nurses, Midwives and Health Visitors*, London, Butterworths.

Day, S. (1980), 'Is obstetric technology depressing?', *Radical Science Journal*, vol. 4, pp. 17–45.

Dijksterhuis, E. J. (1961), *The Mechanisation of the World Picture*, London, Oxford University Press.

Dorner, G. (1976), *Hormones and Brain Differentiation*, Amsterdam, Elsevier.

Doyal, L. (1979), *The Political Economy of Health*, London, Pluto Press.

Doyal, L. and Epstein, S. S. (1983), *Cancer in Britain: The Politics of Prevention*, London, Pluto Press.

Dubos, R. (1959), *Mirage of Health*, New York, Harper & Row.

Ehrenreich, J. (ed.) (1978), *The Cultural Crisis of Modern Medicine*, New York, Monthly Review Press.

Eyles, J. and Woods, K. J. (1983), *The Social Geography of Health and Medicine*, London, Croom Helm.

Ferguson, T. and MacPhail, A. N. (1954), *Hospital and Community*, London, Oxford University Press.

Friedson, E. (1970), *Professional Dominance: The Social Structure of Medical Care*, New York, Aldine.

Gilbert, E. W. (1958), 'Pioneer maps of health and disease in England', *Geographical Journal*, vol. 124, pp. 172–83.

Girt, J. L. (1972), 'Simple chronic bronchitis and urban ecological structure', in N. D. McGlashan (ed.), *Medical Geography: Techniques and Field Studies*, London, Methuen.

Girt, J. L. (1973), 'Distance to general medical practice and its effect on revealed ill-health in a rural environment', *Canadian Geographer*, vol. 17, pp. 154–66.

Gold, R. (1973), 'Stop it, you're making me sick!', *American Journal of Psychiatry*, vol. 130, pp. 1211–12.

Harris, R. J. C. (1970), 'Cancer and the environment', *International Journal of the Environment*, vol. 1, pp. 59–65.

Hart, N. (1982), 'Is capitalism bad for your health?', *British Journal of Sociology*, vol. 33, pp. 435–43.

Hart, N. (1985), *The Sociology of Health and Medicine*, Ormskirk, Lancs., Causeway Press.

Hill, J. D., Hampton, J. R. and Mitchell, J. R. A. (1978), 'A randomised trial of home-versus-hospital management for patients with suspected myocardial infarction', *Lancet*, vol. 1, pp. 837–41.

Howe, G. M. (1963), *National Atlas of Disease Mortality in the United Kingdom*, London, Nelson.

Illich, I. (1975), *Medical Nemesis*, New York, Pantheon.

Inglis, B. (1981), *The Diseases of Civilisation*, London, Hodder & Stoughton.

Jewson, N. D. (1976), 'The disappearance of the sick man from medical cosmology 1770–1870', *Sociology*, vol. 10, pp. 225–44.

Kelman, S. (1971), 'Towards a political economy of health', *Inquiry*, vol. 8, pp. 30–8.

Kennedy, I. (1983), *The Unmasking of Medicine*, London, Granada.

Knox, P. L. (1975), *Social Well-Being: A Spatial Perspective*, London, Oxford University Press.

Knox, P. L. (1982), *Urban Social Geography: An Introduction*, London, Longman.

McKeown, T. (1979), *The Role of Medicine: Dream, Mirage or Nemesis?*, Oxford, Blackwell.

McKinlay, J. B. (1984), *Issues in the Political Economy of Health Care*, New York, Tavistock.

McKinlay, J. B. and McKinlay, S. (1977), 'The questionable contribution of medical measures to the decline of mortality in the United States in the twentieth century', *Millbank Memorial Fund Quarterly*, vol. 56, pp. 405–28.

Mather, A. G. (1971), 'Acute myocardial infarction: home and hospital treatments', *British Medical Journal*, vol. 3, pp. 334–6.

Melville, A. and Johnson, C. (1983), *Cured to Death: The Effects of Prescription Drugs*, New York, Stein & Day.

Miller, N. F. (1945), 'Hysterectomy: therapeutic necessity or surgical racket?', *American Journal of Obstetrics and Gynecology*, vol. 41, pp. 804–10.

Mishler, E. G. (1981a), 'The social construction of illness', in E. G. Mishler, L. R. Amarasingham, S. D. Osherson, S. T. Hauser, N. E. Waxler and R. Liem (eds), *Social Contexts of Health, Illness and Patient Care*, pp. 141–68, Cambridge, Cambridge University Press.

Mishler, E. G. (1981b), 'Viewpoint: critical perspectives on the biomedical model' in E. G. Mishler, L. R. Amarasingham, S. D.

Osherson, S. T. Hauser, N. E. Waxler and R. Liem (eds), *Social Contexts of Health, Illness and Patient Care*, pp. 21–3, Cambridge, Cambridge University Press.

Mohan, J. and Woods, K. J. (1985), 'Restructuring health care?: the social geography of public and private health care', *International Journal of Health Services*, vol. 15, pp. 197–215.

Navarro, V. (1975), 'The industrialisation of fetishism or the fetishism of industrialisation: a critique of Ivan Illich', *International Journal of Health Services*, vol. 5, pp. 351–71.

Navarro, V. (1984), 'Medical history as justification rather than explanation: a critique of Starr's "The Social Transformation of American Medicine" ', *International Journal of Health Services*, vol. 14, pp. 511–28.

Open University (1985), *Medical Knowledge: Doubt and Certainty*, Course U205, Milton Keynes, Open University Press.

Osherson, S. and Amara, S. L. (1981), 'The machine metaphor in medicine' in E. G. Mishler, L. R. Amarasingham, S. D. Osherson, S. T. Hauser, N. E. Waxler and R. Liem (eds), *Social Contexts of Health, Illness and Patient Care*, pp. 218–49, Cambridge, Cambridge University Press.

Patrick, D. L. and Scrambler, G. (1982), *Sociology as Applied to Medicine*, London, Baillière Tindall.

Phillips, D. R. (1981), *Contemporary Issues in the Geography of Health Care*, Norwich, Geo Books.

Pinchuck, T. and Clark, R. (1984), *Medicine for Beginners*, London, Writers and Readers Cooperative.

Posner, T. (1984), 'Magical elements in orthodox medicine: diabetes as a medical thought system' in N. Black (ed.), *Health and Disease: A Reader*, Milton Keynes, Open University Press.

Powles, J. (1973), 'On the limitations of modern medicine', *Social Science and Medicine*, vol. 1, pp. 1–30.

Pyle, G. (1979), *Applied Medical Geography*, New York, Wiley.

Rose, S., Kamin, L. J. and Lewontin, R. C. (1984), *Not in Our Genes: Biology, Ideology and Human Nature*, Harmondsworth, Penguin.

Salmon, J. W. (1984), *Alternative Medicines: Popular and Policy Perspectives*, New York, Tavistock.

Shannon, G. W. and Dever, G. E. A. (1974), *Health Care Delivery: Spatial Perspectives*, New York, McGraw-Hill.

Smith, D. M. (1977), *Human Geography: A Welfare Approach*, London, Edward Arnold.

Stark, E. (1982), 'Doctors in spite of themselves: the limits of radical health criticism', *International Journal of Health Services*, vol. 33, pp. 435–43.

Starr, P. (1982), *The Social Transformation of American Medicine*, New York, Basic Books.

Sweet, W. H., Mark, V. H. and Ervin, F. R. (1967), 'Role of brain disease in riots and urban violence', *Journal of the American Medical Association*, vol. 201, pp. 895.

Tesh, S. (1982), 'Political ideology and public health in the nineteenth century', *International Journal of Health Services*, vol. 12, pp. 321–41.

Thomas, L. (1979), *The Medusa and the Snail*, New York, Viking.

Thunhurst, C. (1982), *It Makes You Sick: The Politics of the NHS*, London, Pluto Press.

Waddington, I. (1973), 'The role of hospitals in the development of modern medicine: a sociological analysis', *Sociology*, vol. 7, pp. 211–24.

Waitzkin, H. (1981), 'A marxist analysis of the health-care systems of advanced capitalist societies', in L. Eisenberg and A. Kleinman (eds), *The Relevance of Social Science for Medicine*, Dordrecht, Reidel.

Waxler, N. E. (1981), 'Learning to be a leper: a case study in the social construction of illness', in E. G. Mishler, L. R. Amarasingham, S. D. Osherson, S. T. Hauser, N. E. Waxler and R. Liem (eds), *Social Contexts of Health, Illness and Patient Care*, pp. 169–94, Cambridge, Cambridge University Press.

Wildavsky, A. (1980), 'Wealthier is healthier', *Regulation, AEI Journal on Government and Society*, January, pp. 10–12.

World Health Organisation (1948), *The Constitution*, Geneva, WHO.

Wright, P. and Treacher, A. (1982), *The Problem of Medical Knowledge: Examining the Social Construction of Medicine*, Edinburgh, Edinburgh University Press.

Yoxen, N. (1982), 'The social construction of genetic disease' in P. Wright and A. Treacher (eds), *The Problem of Medical Knowledge: Examining the Social Construction of Medicine*, Edinburgh, Edinburgh University Press.

Zola, I. K. (1972), 'Medicine as an institution of social control', *Sociological Review*, vol. 20, pp. 487–504.

CHAPTER 2

The collection of
epidemiological information

Geographers in the disease ecology tradition have contributed to the understanding of disease aetiology (or causation) by employing the methods and techniques of epidemiology. Epidemiology has been defined by MacMahon and Pugh (1970) as 'the study of the distribution and determinants of disease frequency in man', and geographers have concentrated on the spatial variations of disease frequency in an attempt to provide clues to disease causation. Much epidemiological and geographical research accepts the biological definition of disease but extends the biomedical model by considering explanations that involve social and environmental causes. This chapter is concerned with the availability and collection of data, while Chapter 3 deals with procedures for analysing such data in the search for causes, and provides an example of the application of epidemiological methods in discerning the links between water hardness and disease.

Two fundamentally different types of research design are commonly recognised in epidemiology: experiments in which the researcher changes and controls the variables, and observational studies in which the researcher does not deliberately intervene but passively observes the disease or illness and associated causal factors. Experiments are undoubtedly difficult to perform with human subjects (see Chapter 3), and the majority of epidemiological work is therefore based on observations of populations in their natural, real-world settings.

The data that are used by medical geographers in such observational studies are either collected on a routine basis (usually by government agencies) or by surveys which follow one of the three basic types of observational design: cross-sectional, case-comparison and cohort. The potential and problems of both routine and survey data are considered in this chapter while, at the same time, an attempt is made to illustrate some of the fascinating results that have been achieved by the application of epidemiological methods. A number of books present the findings of epidemiology in a more systematic manner, for example Morris (1975) and Alderson and Dowie (1979).

Before considering the nature of routine data, a fundamental distinction must be made between mortality and morbidity, and between communicable and non-infectious diseases. Communicable or infectious diseases are ones in which the disease spreads or is communicated from person to person, or from animal to human. Examples of such diseases are influenza, cholera, measles, polio and smallpox; they are considered in detail in Chapter 4. Non-infectious diseases are a diverse group and include the chronic degenerative diseases such as cancer (malignant neoplasms), strokes (cerebrovascular disease), heart disease and respiratory conditions such as bronchitis, as well as inherited and genetic diseases. Mortality statistics refer to information about death, while morbidity refers to information about illness. It is crucial to realise that there is not an equivalence between communicable disease and morbidity statistics nor between non-infectious diseases and mortality statistics. Firstly, a considerable number of deaths, even in an economically advanced country, are not caused by chronic degenerative diseases; accidents and deaths associated with birth and pregnancy can take a considerable toll, particularly amongst the young. Secondly, many chronic illnesses are morbid conditions that do not normally end in death, arthritis for example. Finally, and most importantly, communicable diseases such as measles and whooping cough need not be temporary morbid conditions but major killers if the host is in poor general health due to, for example, malnutrition or lack of exposure and immunity.

Pyle (1979, p. 20) has recognised generalised cycles of infectious and chronic disease which he associates with the level of economic development (Figure 2.1). In the least economically advanced countries of the world, high fertility is often linked with high infant mortality which is produced by infectious and parasitic diseases and general malnutrition. In contrast, in the economically advanced countries of the world, chronic degenerative diseases (which Pyle associates with affluent, 'killing' lifestyles) are the major killers with the bulk of the deaths occurring to people aged over 60. Such global contrasts are further illustrated by a world map of life expectancy at birth, Figure 2.2.

While such differences can be recognised on a world scale, it is important to realise that remarkable contrasts can also be found

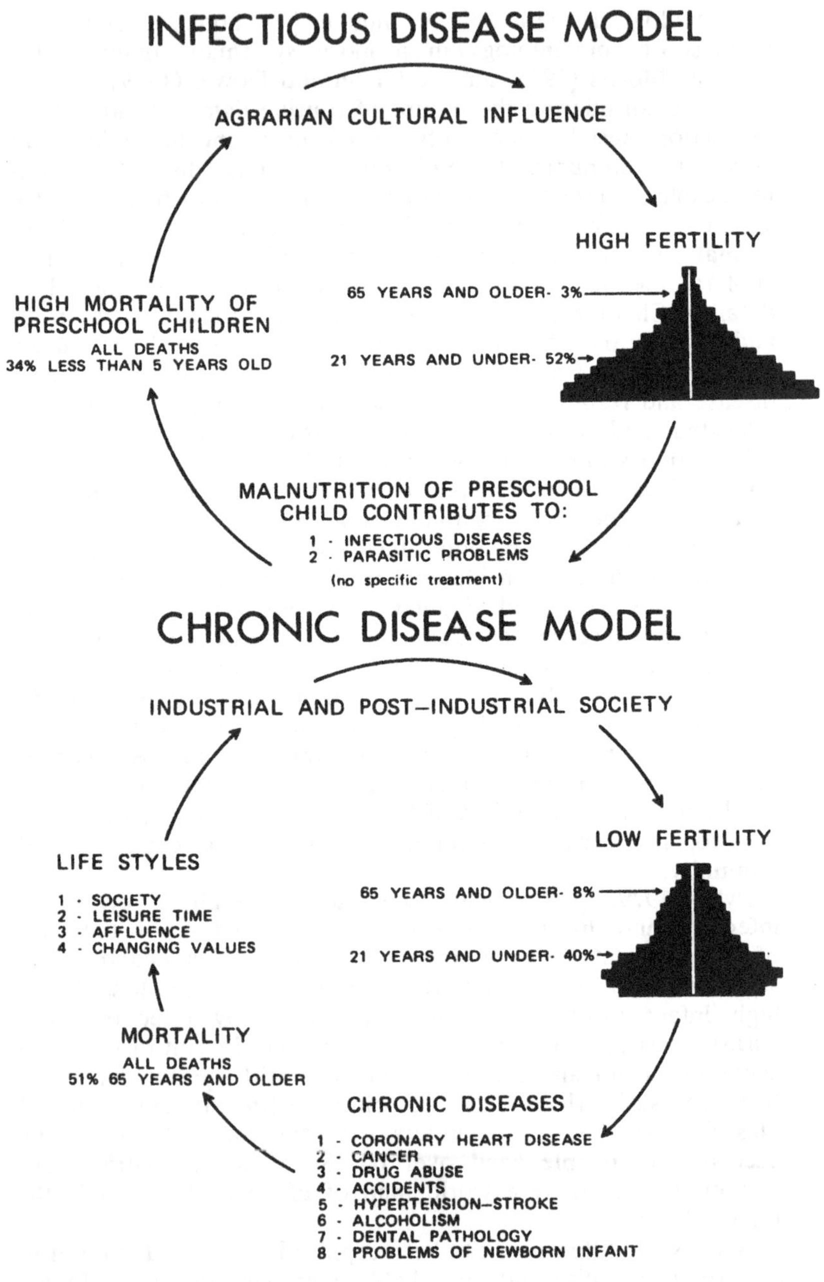

INFECTIOUS DISEASE MODEL

AGRARIAN CULTURAL INFLUENCE

HIGH FERTILITY

HIGH MORTALITY OF
PRESCHOOL CHILDREN
ALL DEATHS
34% LESS THAN 5 YEARS OLD

65 YEARS AND OLDER· 3%

21 YEARS AND UNDER· 52%

MALNUTRITION OF PRESCHOOL
CHILD CONTRIBUTES TO:
1 · INFECTIOUS DISEASES
2 · PARASITIC PROBLEMS
(no specific treatment)

CHRONIC DISEASE MODEL

INDUSTRIAL AND POST—INDUSTRIAL SOCIETY

LOW FERTILITY

LIFE STYLES
1 · SOCIETY
2 · LEISURE TIME
3 · AFFLUENCE
4 · CHANGING VALUES

65 YEARS AND OLDER· 8%

21 YEARS AND UNDER· 40%

MORTALITY
ALL DEATHS
51% 65 YEARS AND OLDER

CHRONIC DISEASES
1 · CORONARY HEART DISEASE
2 · CANCER
3 · DRUG ABUSE
4 · ACCIDENTS
5 · HYPERTENSION—STROKE
6 · ALCOHOLISM
7 · DENTAL PATHOLOGY
8 · PROBLEMS OF NEWBORN INFANT

Figure 2.1   Models of disease

Source: Pyle (1979)

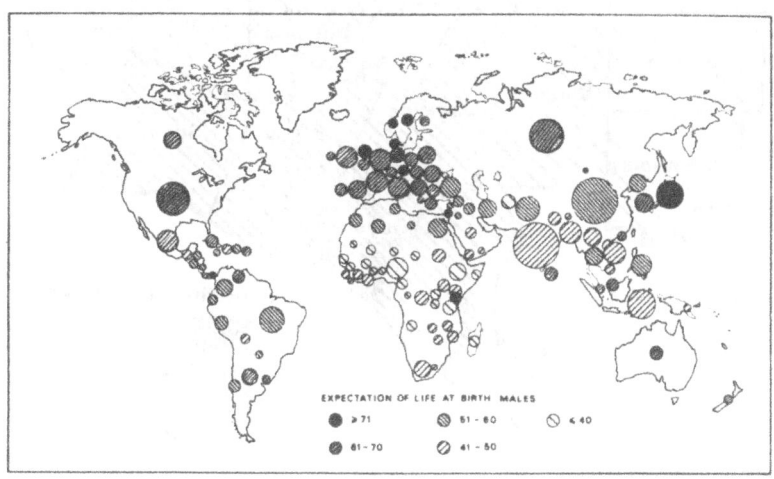

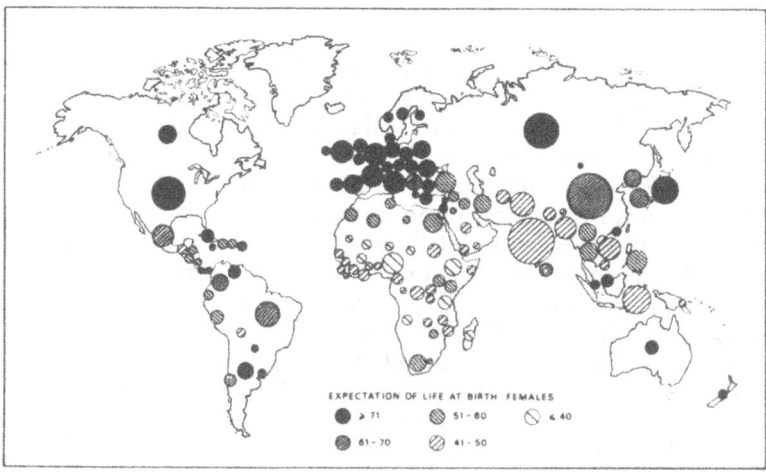

Figure 2.2 Global contrasts in life-expectancy at birth in the 1970s. The symbol for each country is proportional to the population
Source: Johnston (1984)

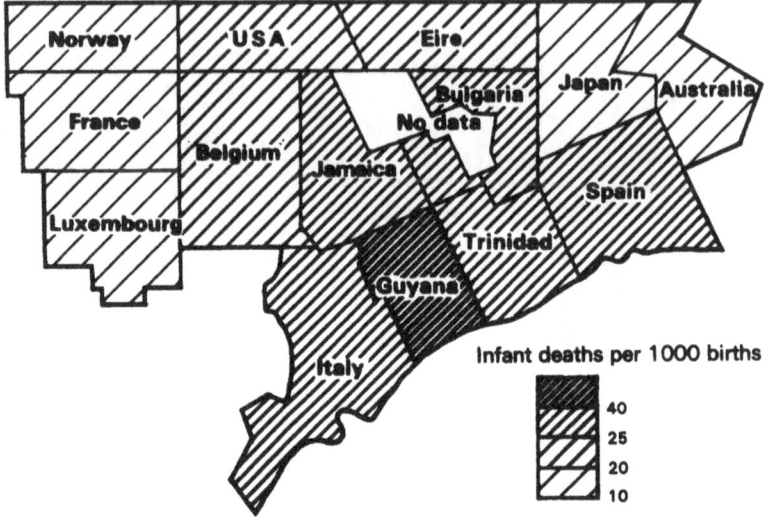

Figure 2.3   Infant mortality rates in various parts of the city of Detroit, 1969, compared with countries having similar rates at that time
Source: **Bunge and Bordessa (1975)**

within countries. Table 2.1 illustrates the profound differences between the mortality of whites, Africans and 'coloured' people (of mixed African/white or Malayan origin); the figures of Africans are based on better-off areas in which resources are sufficient for data collection. The blacks and coloureds live in a different world with concentrations of death at an early age from infectious diseases and nutritional disorders, while the white mortality is dominated by the chronic degenerative diseases. Moreover, by changing scale we can see the whole world in a microcosm; Figure 2.3 shows the 1969 infant mortality rate (deaths under age of 1 per 10,000 births) in Detroit as compared to rates for different parts of the world. While the suburbs have some of the lowest rates in the world, the inner-city areas are comparable with countries which are relatively economically backward. This chapter concentrates on data collection in the most economically advanced countries for it is here that routine collection is most developed and the observational designs are most commonly applied.

Routine sources

The chief advantage of using routine or archival data is that the cost of obtaining it is usually very low; it can literally be obtained

Table 2.1   Life and death in South Africa

(a) *Infant mortality rate (per 10,000)*   (b) *Life-expectancy (at birth)*

			Male	*Female*
White	227			
Coloured	1 394	White	66	72
African	1 357	Coloured	50	52
		African	45	47

(c) *Percentage of deaths according to age, 1970*

Age	White	Coloured	African
0–4	7	49	55
5–24	4	6	7
25–44	8	11	10
45–64	30	16	16
65+	52	18	12
Total	100	100	100

(d) *Main causes of death in South Africa, 1976*

Order of Importance	*White*	*Coloured*	*African*
1	Heart disease	Enteritis and other diseases	Pneumonia
2	Cerebrovascular diseases	Pneumonia	Enteritis and other diseases
3	Pneumonia	Cerebrovascular diseases	Homicide and injury by others
4	Motor vehicle accidents	Heart disease	Cerebrovascular diseases
5	Bronchitis	Homicide and injury by others	Tuberculosis

Sources: Van Rensburg and Mans (1982); Susser (1984).

off the shelf. With such data, however, the researcher has little control over what is collected and published and how it is obtained and processed. In particular, there may be problems of precision and accuracy (which may involve biased, under-reported or misclassified data) and problems of purpose, for the data are usually collected for legal and administrative needs and not research. Thus, no individual-level data may be available or the age-disaggregation may not be sufficient to allow detailed study. Despite these problems, routine sources are used extensively, particularly in descriptive studies in which relatively little is known about a disease, while aetiological or analytical studies designed to test *a priori* hypotheses are usually based on specially designed surveys. In descriptive, routine studies, disease frequency can be examined through time, over space, and in relation to any demographic and social variables that have been collected. But studies rarely stop at this stage and usually a whole series of hunches are generated which can be tested subsequently by more detailed data. Moreover, routine data are often used to monitor the public health as a sort of early-warning surveillance system to identify, for example, a sudden increase in birth abnormalities.

Mortality statistics

Mortality statistics are the most frequently used form of routine statistics and they are available for many countries. The system for collection and publication described below is for England and Wales and it has operated on a compulsory basis since 1874; complete coverage in the USA was only achieved some sixty years later when Texas began submitting returns. The local registrar in England and Wales will not issue a certificate authorising the disposal of the body until the medical practitioner attending the deceased has provided the cause of death, and until somebody (usually a relative or friend, but it could be the local social services department) has provided information on the deceased. If there are unusual circumstances or a sudden death, the death will be reported to the coroner who will attempt to ascertain the cause of death. The data collected by the registrar includes the following: date and place of birth, sex, marital status, usual address, occupation (father's for children, husband's for women), date and place of death. The medical practitioner or coroner has to describe the conditions leading directly to death, the underlying condition, and any other significant conditions contributing to death but unrelated to the condition actually causing death. Thus, a practitioner could write (Benjamin, 1968) 'peritonitis due to perforation of duodenum due to duodenal ulcer; epithelioma of skin of cheek also present'. It is the underlying cause, duodenal ulcer in this case, that is used to produce the routine statistics. The registrar sends all this information to the Office of Population Census and Surveys (OPCS) where it is coded, checked, verified and entered into (in recent years) a computer system; the address is coded by postcode and the cause of death is coded according to the International Classification of Diseases (ICD).

Table 2.2 The diseases and casualties in the week beginning 20 December 1664 for London

Abortive	2	Infants	3
Aged	24	Killed by a fall from a Scaffold	
Bedridden	1	at St. Martin in the Fields	1
Bruised	1	Lethargy	1
Cancer	1	Livergrown	1
Canker	1	Overlaid	1
Childbed	12	Palsie	1
Chrisomes	6	Plague	1421
Collick	2	Quinsie	1
Consumption	59	Rickets	8
Convulsion	25	Rising of the Lights	3
Dropsie	17	Rupture	1
Drowned in a Tub of VVash		Scowring	1
in a Brewhouse at St. Giles		Spotted Feaver	28
in the Fields	1	Stilborn	3
Feaver	82	Stopping of the Stomach	3
French-pox	1	Suddenly	1
Frighted	1	Surfeit	17
Grief	2	Teeth	41
Griping in the Guts	13	Tiffick	3
Jaundies	1	Winde	1
Imposthume	6	Wormes	8

$$\text{Christned} \begin{bmatrix} \text{Males} & 60 \\ \text{Females} & 44 \\ \text{In all} & 104 \end{bmatrix} \quad \text{Buried} \begin{bmatrix} \text{Males} & 951 \\ \text{Females} & 855 \\ \text{In all} & 1806 \end{bmatrix} \text{Plague } 1421$$

Decreased in the Burials this Week   1413
Parishes clear of the Plague   26   Parishes Infected   104

The Asize of Bread set forth by Order of the Lord Mayor and
Court of Aldermen
A penny Wheaten Loaf to contain Nine Ounces and a half, and three
half-penny White Loaves the like weight.

Source: Open University (1975).

Table 2.3 The basic categories of the International Classification of Diseases, Ninth Revision

Categories of disease	Examples
1. Infectious and parasitic diseases (001–139)	Malaria
2. Neoplasms (140–239)	Malignant neoplasm of larynx
3. Endocrine, nutritional and metabolic diseases and immunity disorders (240–279)	Diabetes mellitus
4. Diseases of blood and blood-forming organs (280–289)	Anaemia
5. Mental disorders (290–319)	Alcohol-dependence syndrome
6. Diseases of the nervous system and sense organs (320–389)	Epilepsy
7. Diseases of the circulatory system (390–459)	Hypertension
8. Diseases of the respiratory system (460–519)	Asbestosis
9. Diseases of the digestive system (520–579)	Appendicitis
10. Diseases of the genito-urinary system (580–629)	Acute renal failure
11. Complications of pregnancy, childbirth and the puerperium (630–676)	Postpartum haemorrhage
12. Diseases of the skin and subcutaneous tissue (677–709)	Dermatitis
13. Diseases of the musculoskeletal system and connective tissue (710–739)	Rheumatoid arthritis
14. Congenital anomalies (740–759)	Spina bifida

Table 2.3 cont'd

Categories of disease	Examples
15. Certain conditions originating in the perinatal period (760–779)	Slow fetal growth
16. Symptoms, signs and ill-defined conditions (780–799)	Sudden death cause unknown
17. Injury and poisoning (800–899)	Poisoning by psychotropic drugs
V Code. Supplementary Classification of Factors Influencing Health Status and Contact with Health Service (V01-V82)	
E Code. Supplementary Classification of External Causes of Injury and Poisonings (E800–E998)	

Source: WHO (1977).

One of the earliest examples of a classification of causes of death is given in Table 2.2 which is a Bill of Mortality for London in the week beginning 20 December 1664. Several features may intrigue us here, for example the people who died 'suddenly', 'frighted' and due to 'lethargy' and 'winde'! Moreover, several of these categories can be interpreted according to modern knowledge; thus 'chrisomes' is the death of children before christening, 'childbed' is the death of a mother during labour, 'consumption' and 'tiffick' are forms of tuberculosis, 'griping in the guts' is probably dysentery, 'livergrown' is possibly rickets, and 'rising of the lights' is possibly eclampsia. Two other features are noticeable, firstly the deaths are dominated by the plague, and secondly the causes are arranged in alphabetical order thereby emphasising the rudimentary nature of this early classification. The first modern international classification was published in 1893 and this has been revised at roughly ten-year intervals. The current Ninth Revision, published in 1975 and used from the beginning of 1979, has 1,778 categories which allow the classification of morbid as well as fatal conditions (Ashley, 1979). The nineteen major categories are given in Table 2.3, together with the three-digit code numbers in parentheses. In the OPCS publication *Mortality Statistics By Cause*, DH2, the numbers of

Table 2.4   The abbreviated mortality lists from the Ninth Revision of the International Classification of Diseases used to code cause of death

Ninth Revision (Adapted Mortality List)
All Causes (001–E999)
AM 1   Cholera (001)
AM 2   Typhoid fever (002.0)
AM 3   Other intestinal infectious diseases (Remainder of 001–009)
AM 4   Tuberculosis (010–018)
AM 5   Whooping cough (033)
AM 6   Meningococcal infection (036)
AM 7   Tetanus (037)
AM 8   Septicaemia (038)
AM 9   Smallpox (050)
AM10   Measles (055)
AM11   Malaria (084)
AM12   All other infectious and parasitic diseases
          (Remainder of 001–139)
AM13   Malignant neoplasm of stomach (151)
AM14   Malignant neoplasm of colon (153)
AM15   Malignant neoplasm of rectum, rectosigmoid junction and
          anus (154)
AM16   Malignant neoplasm of trachea, bronchus and lung (162)
AM17   Malignant neoplasm of female breast (174)
AM18   Malignant neoplasm of cervix uteri (180)
AM19   Leukaemia (204–208)
AM20   All other malignant neoplasma (Remainder of 140–208)
AM21   Diabetes mellitus (250)
AM22   Nutritional marasmus (261)
AM23   Other protein-calorie malnutrition (262, 263)
AM24   Anaemias (280–285)
AM25   Meningitis (320–322)
AM26   Acute rheumatic fever (390–392)
AM27   Chronic rheumatic heart disease (393–398)
AM28   Hypertensive disease (401–405)
AM29   Acute myocardial infarction (410)
AM30   Other ischaemic heart diseases (411–414)
AM31   Cerebrovascular disease (430–438)
AM32   Atherosclerosis (440)
AM33   Other diseases of circulatory system (Remainder of 390–459)
AM34   Pneumonia (480–486)
AM35   Influenza (487)
AM36   Bronchitis, emphysema and asthma (490–493)
AM37   Ulcer of stomach and duodenum (531–533)

AM38 Appendicitis (540–543)
AM39 Chronic liver disease and cirrhosis (571)
AM40 Nephritis, nephrotic syndrome and nephrosis (580–589)
AM41 Hyperplasia of prostate (600)
AM42 Abortion (630–639)
AM43 Direct obstetric causes (640–646, 651–676)
AM44 Indirect obstetric causes (647–648)
AM45 Congenital anomalies (740–759)
AM46 Birth trauma (767)
AM47 Other conditions originating in the perinatal period
(760–766, 768–779)
AM48 Signs, symptoms and ill-defined conditions (780–799)
AM49 All other diseases (Remainder of 001–799)
AM50 Motor vehicle traffic accidents (E810–E819)
AM51 Accidental falls (E880–E888)
AM52 All other accidents, and adverse effects
(Remainder of E800–E949)
AM53 Suicide and self-inflicted injury (E950–E959)
AM54 Homicide and injury purposely inflicted by other persons
(E960–E969)
AM55 Other violence (E970–E999)

Source: United Nations Demographic Yearbook (1980).

deaths in all three-digit categories are given as well as the number in all four-digit categories that are expected to attain 20 deaths annually. Thus, the code 157 represents deaths from malignant neoplasms (cancer) of the pancreas; 157.0, 157.1 and 157.2 are cancers of the head, body and tail of the pancreas respectively. Most analysis, however, is done on much less detailed classifications and Table 2.4 gives the abbreviated list of the Ninth Revision which is commonly used in epidemiological research.

There are several problems associated with the use of such classifications. Because of the perceived need to take account of the most recent medical knowledge, the classification is periodically revised and this may preclude the examination of changes over time. At each revision the OPCS undertake a 'bridge-coding' exercise and this facilitates the comparison of the old and new versions of the classification: in 1978 a quarter of the deaths were coded by both systems (OPCS monitor DH 81/2). Despite this and despite the WHO's contention that since 1948 the revised classifications have been characterised by continuity, any long-term trends must be examined with caution. In many countries, few of the deaths are attended by medically qualified persons, so that international comparisons may also have their problems. Even in England and Wales there can be difficulties with the scheme; for example, the physician attending the death may give an inaccurate diagnosis or OPCS may have difficulty in coding a certificate. For example, the perinatal morbidity rates (Figure 2.4) are widely used as an indicator of the quality of obstetric and early neo-natal care (Knox and Mallett, 1979), but they may be subject to some inaccuracy. Any foetus born dead after the twenty-eighth week is a stillbirth, and if it is alive but dies at once, it is a livebirth and a death. The mother may be unsure of her 'dates' and there is a tendency for doctors and midwives to classify an event as a miscarriage to spare parents the process of registering the death. This may vary according to local practice and the rates may reflect not quality of medical care but merely artifactual differences in certification.

A number of studies have been undertaken of the general problem of diagnosis and certification using a number of different methods (comparing clinical diagnosis against autopsy, reviewing complete case histories against death certificates, and duplicate coding of death certificates). Alderson (1983, pp. 21–2) reviews these studies and finds that there are errors, 20 per cent being minor, 5 per cent being major differences in disease classification, but he concludes that mortality statistics can still be used if they are interpreted with caution.

Other coding problems may also occur. For example, when the informants fail to give accurate details of occupation, believing this to be unimportant. Moreover, a problem can occur if people disabled by one job move on to a less demanding job, so that the registered occupation may not be the one that is causal. There may also be difficulties with the location of death; Gardner and Winter (1984) in their cancer atlas of England and Wales found that Stone Rural District, in Staffordshire, had a high rate of

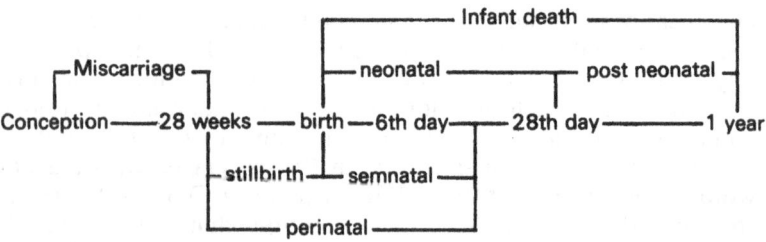

Figure 2.4 Definition of terms used in mortality statistics up to the age of one year   *Source*: Whitelegg (1982)

mortality for nineteen different cancers. By examining the abstracts of death certificates held on microfilm at OPCS, they found that a home for terminally ill cancer patients was located in the district. Since 1958, the OPCS rules state that if a death occurs within six months of admission to a hospital, the death is 'transferred' back to the person's previous area of residence. Following Gardner and Winter's findings, the rules are currently being reviewed. The recognition of these difficulties with mortality statistics should not prevent us from using the data, but should aid us in their interpretation.

Annual reports are published by the OPCS on mortality statistics in the DH Series (Table 2.5) and there is a decennial supplement on occupational mortality; prior to 1973 the data was published in the Registrar General's Statistical Review. There is also a considerable amount of data which has been processed and is available from OPCS but is not published. Davies and Chilvers (1980) discuss the detailed geographical data which can be obtained for 1,366 county boroughs, urban and rural districts for 1950 to 1978 (the so-called SD30s for 1950–62 which are not age-disaggregated and the SD25s for 1963–78 which give deaths by 10-year age-group); the 1950–67 data are classified according to

Table 2.5 Routine OPCS publications on mortality

Series DH   Deaths
DH1   Mortality statistics
DH2   Mortality statistics: cause
DH3   Mortality statistics: childhood; and Mortality statistics:
          perinatal and infant (social and biological factors)
DH4   Mortality statistics: accidents and violence
DH5   Mortality statistics: area

Series DS   Decennial Supplement on Occupation Mortality
Series LS   Longitudinal Survey

OPCS Monitors ('quick release of selected information as it becomes available')
Series DH   Deaths
DH1   Deaths, general: occasional
DH2   Deaths by cause: quarterly
DH3   Infant and perinatal statistics: occasional
DH4   Deaths from accidents and violence: quarterly

Source: HMSO (1984), *Government Publications Sectional List 56.*

the ICD Sixth and Seventh Revisions, after 1968 the Eighth Revision was used. They argue that this source is of high potential but has been grossly under-used. Since 1979 the data have been processed on the post-1974 reorganised local government areas, and information is therefore available on 369 'county' districts and 36 'metropolitan' districts. In addition to these SD25 tables for local authorities, data are available at the level of the electoral ward, and there are even 'primary' computer tapes which give postcoded individual mortality data (Thunhurst, 1985, p. 49).

Broken down by age and sex

Mortality data are routinely available disaggregated by age and sex, and these two factors are the major determinants of disease frequency classified by cause. Women live longer than men in England and Wales (77 years for women, 70 years for men) but certain diseases have a much higher occurrence in women than men, cancer of the breast and strokes (cerebrovascular disease) for example; the reverse is true of other diseases such as lung cancer and heart disease. Table 2.6 shows that the death-rate is relatively high in the very young and then it drops to a low level in childhood; the rate then increases throughout adult life. Moreover, different causes of death are of differing importance at different ages; problems of birth are the main cause of death in the very young, accidents in the young, cancer for the middle-aged, and heart disease for the old. So important are these demographic factors that rates are usually standardised for age and sex before further analysis. One way of doing this is to calculate sex- and age-specific rates, but this results in a plethora of values. What is required is an overall index and Alderson (1981), after reviewing a variety of indices, concludes that the most suitable general technique is the indirectly calculated Standardised Mortality Ratio (SMR).

The crude death-rate demonstrates the importance of such age-standardisation. This rate is defined as the total number of deaths divided by the total population; for Bournemouth in 1971 this yields a rate for males of 18.7 per 1000, while the comparable figure for England and Wales is only 12.2. One explanation for this difference is that Bournemouth is an unhealthy place in which to live; a much more likely reason is that the crude rate is inflated by the large number of retired men living and then dying in the resort. The calculation of the male SMR for Bournemouth

Table 2.6   Main causes of death by age-group, 1981, England and Wales (death-rates per 10,000)

Age	All causes	1st	2nd
Under 1	111	Perinatal causes 42	Congenital anomalies 30
1–4	5	Accidents 1	Congenital anomalies 1
5–14	2	Accidents 1	Cancer 1
15–34	6	Accidents 2	Cancer 1
35–44	15	Cancer 5	Heart disease 4
45–54	49	Cancer 18	Heart disease 17
55–64	132	Heart disease 49	Cancer 47
65–74	334	Heart disease 124	Cancer 95
Over 75	1 001	Heart disease 339	Respiratory disease 195

Source: OPCS (1982).

is shown in Table 2.7; the product of columns (a) and (b) gives the expected deaths (c) if the England and Wales age-specific rates applied in Bournemouth; the actual deaths (d) are those which did occur in 1971.

The male SMR is calculated by dividing the actual number of deaths by the expected deaths and multiplying by 100. If the resultant SMR is 130, the place has a mortality experience 30 per cent worse than the England and Wales average; a value below 100 represents a favourable mortality experience. Bournemouth, with a male SMR of 90 is, in the national context, a healthy place in which to live.

Table 2.7 Calculation of standard mortality ratio for males in Bournemouth in 1971

Age	Male population Bournemouth (a) (000s)	Male death-rates England and Wales (b) (deaths per 1,000)	Expected Male deaths (c) = (a × b)	Actual Male deaths (d)
Under 1	0.74	19.78	15	
1–4	2.93	0.76	2	
5–14	8.38	0.40	3	
15–24	8.83	0.92	8	
25–44	13.41	1.62	22	
45–64	18.36	13.45	247	
65–74	8.23	51.82	426	
75 and over	4.64	137.42	638	
All ages	65.52		1 361	1 223

$$\text{SMR} = \frac{\text{Total actual male deaths}}{\text{Total expected male deaths}} \times \frac{100}{1}$$

$$\text{SMR} = \frac{1\ 223}{1\ 361} \times \frac{100}{1} = 90$$

Source: Open University (1975, p. 24).

Temporal trends

Examining death rates over time can give clues to causal explanations, but attention must be paid to artifactual explanations, adopting the maxim 'anything interesting is probably a mistake'. During this century both England and Wales and the USA have experienced considerable increases in lung-cancer death-rates (Figure 2.5) and before attempting to develop an explanation in terms of changing social and environmental conditions, artifactual explanations must be ruled out. Lung cancer is primarily a disease of old age and it could be that the increase is a result of changing age structure; age-standardisation reveals that this is not the case. Another possibility is that changing diagnostic techniques, particularly x-rays, are the real explanation but the mortality statistics reveal that the rate of increase is much higher for males than females, and it is unlikely that diagnostic changes would have been sex-specific. Another artifactual suggestion is that there has been mis-diagnosis in the early part of the century and true lung-cancer deaths were misclassified as tuberculosis. Gilliam (1955) observed that while lung cancer rates have increased, TB rates have decreased. He concluded, however, that the observed changes would require unreasonable assumptions about age and sex differences in the mis-diagnosis for the observed increase to be artifactual. All in all, the increase does not appear to be artificial and the search for explanation can therefore proceed to examine genuine causes. The evidence for England and Wales suggests that the rise is most closely associated with the increase of cigarette smoking with lag or latent period of some twenty years. Routine data are insufficient to test this hypothesis in detail; other study designs are needed.

Another example, albeit on a smaller scale, is Lloyd and Barclay's (1979) finding of a marked increase in respiratory cancer in an industrialised community in Scotland. They discovered a low incidence (SMR, 68) prior to 1967 but a consistently high rate after that date (SMR, 178); over the period 1961 to 1967 only 16 deaths occurred but, in an equal period of time, 1968–74, the number leapt to 55. These changes were not due to changes in population or age-structure, nor due to ICD changes which should have decreased the rate as observed in the rest of the country. Moreover, there had been no alterations in medical practice or diagnostic facilities during this period. They concluded that the abrupt increase was genuine and suggested that a possible cause was the introduction of an environmental carcinogen into the town; they thought that this was associated with the local steel industry. Detailed survey data would be required to confirm or refute this hypothesis.

Geographical variations

We have already seen that mortality experience varies from place to place; this section attempts to show how spatial variaitons, at international, intra-national and intra-regional scales, can be used to aid the development of causal explanations. Providing care is taken that the observed differences are not in fact spurious, one of the most powerful uses of international mortality statistics is so-called 'migrant' studies which permit the assessment of the relative influence of genetic and environmental factors in determining disease patterns. The underlying rationale of such studies is that if genetic inheritance is the major determinant of disease frequency, the mortality rate of the migrants would be

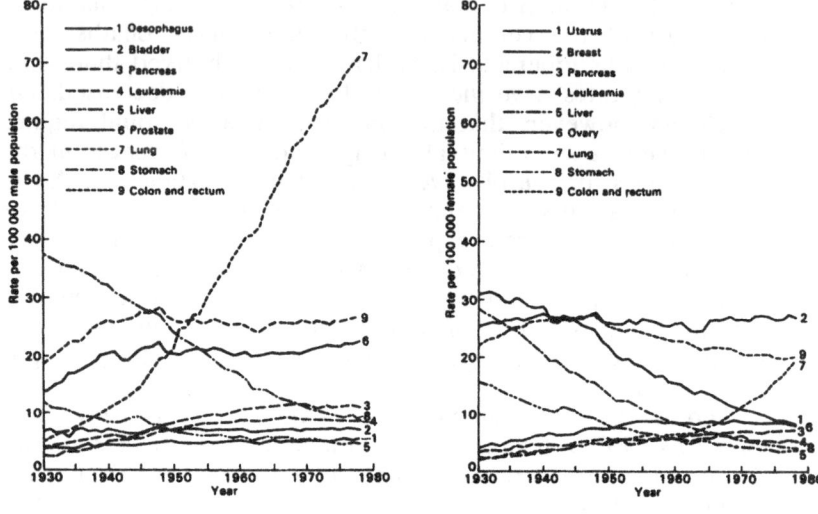

(a)

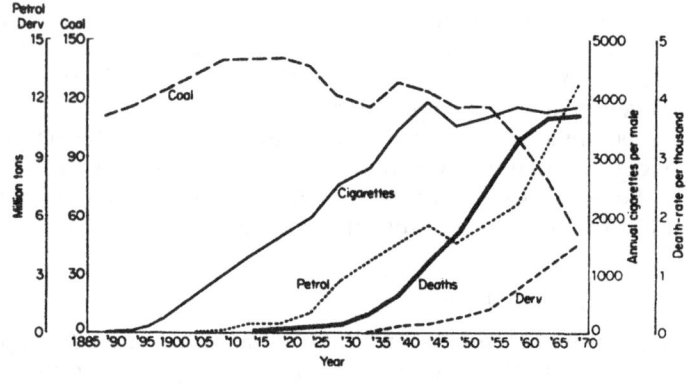

(b)

Figure 2.5 The increase in lung cancer in the USA and England and Wales (a) Cancer mortality in the USA, rates standardised for age on the 1940 population *Source: Mausner and Kramer (1985)* (b) Male lung cancer mortality in England and Wales, aged 60–4; smoking and atmospheric pollution
Source: Waller (1967)

roughly equivalent to the non-migrants in the country of origin. Conversely, if environmental factors are of prime importance, the rates for migrants would be similar to the host country, but different from the country of origin. Worth (1975) studied the coronary heart disease mortality rates of Japanese men living in Japan, Hawaii and California, USA. As Figure 2.6 shows, the rates for each age-group are consistent with the importance of environmental and social variables, for the highest rates are found for the San Franciscan Japanese who can be expected to live the least traditional lifestyle. The Japanese living in Hawaii are thought to have a lifestyle which combines some elements of traditional Japanese living with other elements of a more western life, and they have rates which are intermediate between USA migrants and Japan itself. Further studies have been undertaken of the variations of 'risk' factors for the same groups and it was found that serum cholesterol (Figure 2.6), diet and uric acid showed similar differences while blood pressure and glucose did not (Kagan *et al.*, 1974).

Several difficulties have to be overcome in a migrant study. It is obviously imperative that the ICD is being applied consistently in each country and that the origin and host countries have a contrasting mortality experience. Moreover, it is useful to have the ages of the migrants at the time of migration for this may

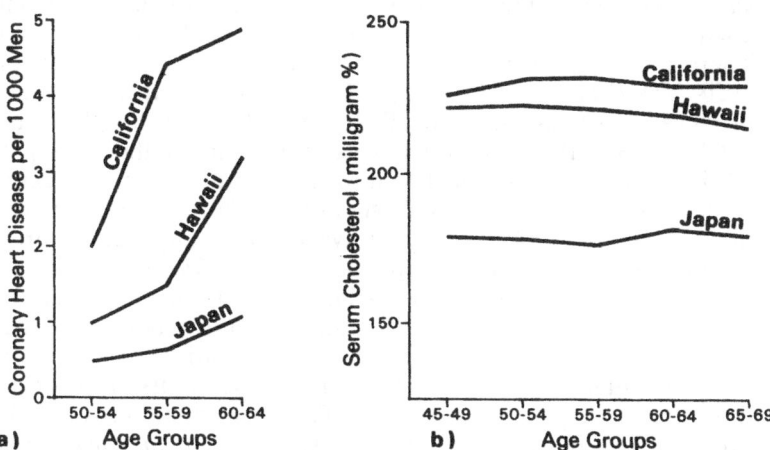

Figure 2.6 Mortality and risk factors for Japanese men living in Japan, Hawaii, and California (a) Coronary heart disease mortality
Source: Worth (1975)
(b) Mean values of serum cholesterol
Source: Kagan *et al.* (1974)

permit the consideration of the latent period of the disease as well as relative importance of pre- and post-migration environmental factors. Finally, the 'power' of the design is reduced when the migration has been selective, for example if only the fittest and healthiest migrate and the country of settlement only allows immigration after the migrant has passed a medical examination.

At the regional scale, a number of researchers have examined variations within particular countries. For example, Murray (1962) examines the all-causes death-rate, 1948–57, for more than 1,400 administrative units (county and municipal boroughs, urban and rural districts) in England and Wales. Cause-specific male rates, 1950–3, for bronchitis, pneumonia, respiratory tuberculosis, lung and stomach cancer, heart disease and strokes are also portrayed but these maps are less detailed, only showing variations among the county boroughs and counties. In a subsequent paper (1966) the all-cause death-rate for England and Wales is updated to 1958–62 and a similar map is given for the period 1950–9 for the 3,100 counties of the USA. Interestingly, he notes (p. 305) that 'US vital statistics for deaths from specific disease groups are hopelessly inadequate at this level'. Other researchers have produced extensive 'national atlases'. For example, Howe's (1970) atlas portrays the death-rates for specific causes for two time periods, 1954–8 and 1959–63, and the maps show variations in the United Kingdom for county boroughs, and amalgamations of urban and rural districts within each county. There is also an *Atlas of Cancer Mortality for the US Counties, 1950–1969* (Mason *et al.*, 1975), while Gardner *et al.* (1983) have produced a similar atlas for England and Wales for the years 1968–78; 23 of the 67 maps show variations among the 1,366 pre-1974 local authority areas. The most recent addition to this genre is another national atlas for England and Wales by Gardner and his colleagues (1985); this portrays a wide range of differing diseases by local authorities for the period 1968–78.

The observed geographical variations on these maps may be either genuine or artifactual. Genuine patterns may be produced by differing disease incidence because of variations in social and environmental factors, or variations in survivorship may be linked to inadequate medical care. The artificial patterns may be produced by differences in disease certification or they may be an outcome of chance variations. The latter is part of the general 'small-number problem' (Jones and Kirby, 1980) for the actual number of deaths in any one area may be quite low when disaggregated by sex, age and cause. Intuitively, we put greater faith in large numbers. For example, if we found 7,000 'heads' with 10,000 tosses of a coin we would conclude that the coin was biased in some way, but if 7 out of 10 tosses were 'heads' we would tend to dismiss the result as due to chance. Similarly, with the analysis of mortality rates, if there is a large number of people living in an area and consequently a large number of deaths, the SMR is reliable and can be confidently interpreted. However, if the areas have a small population and a small number of deaths, the SMR is generally unreliable and we can expect it to fluctuate from year to year. Imagine that the expected deaths from leukaemia are 2 per annum in a particular district with a low population, and the actual number of deaths for a four-year period are 4, 1, 0, 3. The SMR would vary wildly from 200 to 50, 0 and 150, while the absolute figures only vary from 0 to 4. The problem is compounded in the spatial display of such data because areas with low populations tend to be rural administrative areas with a large areal extent. Therefore, if the SMRs are shown on a choropleth map, the areas with the greatest visual impact are those with the most unreliable rates.

There are several ways of overcoming the problem: grouping several years together (but this may conceal temporal trends), aggregating areal units (but this may hide detailed geographical variations), drawing cartograms in which the areal extent of district is proportional to population size, or by statistically transforming the data. Statistical techniques can be used to calculate an index of whether the SMR differs sufficiently from 100 for a chance effect to be unlikely. This is given by the following formula:

$$\text{index of chance effect} = \frac{(\text{actual deaths} - \text{expected deaths})^2}{\text{expected deaths}}$$

Thus, if there are 20 actual and 15 expected deaths in area A, and 200 and 150 in area B, the SMRs will be 150 in both areas. But the index will be 1.66 in area A and 16.6 in area B. The index is a chi-squared statistic and with what is technically known as one degree of freedom, we can infer from statistical tables that a value as high as 3.84 would only occur 5 times out of 100 just by chance. Therefore, the SMR for area B may be judged significantly above the national average at the 5 per cent probability level, whereas there is insufficient evidence to decide whether or not area A is significantly different for the national rate; there are not enough deaths to calculate a reliable ratio. When the expected number of deaths is less than 5 we can use the Poisson generating function to calculate the probability of any

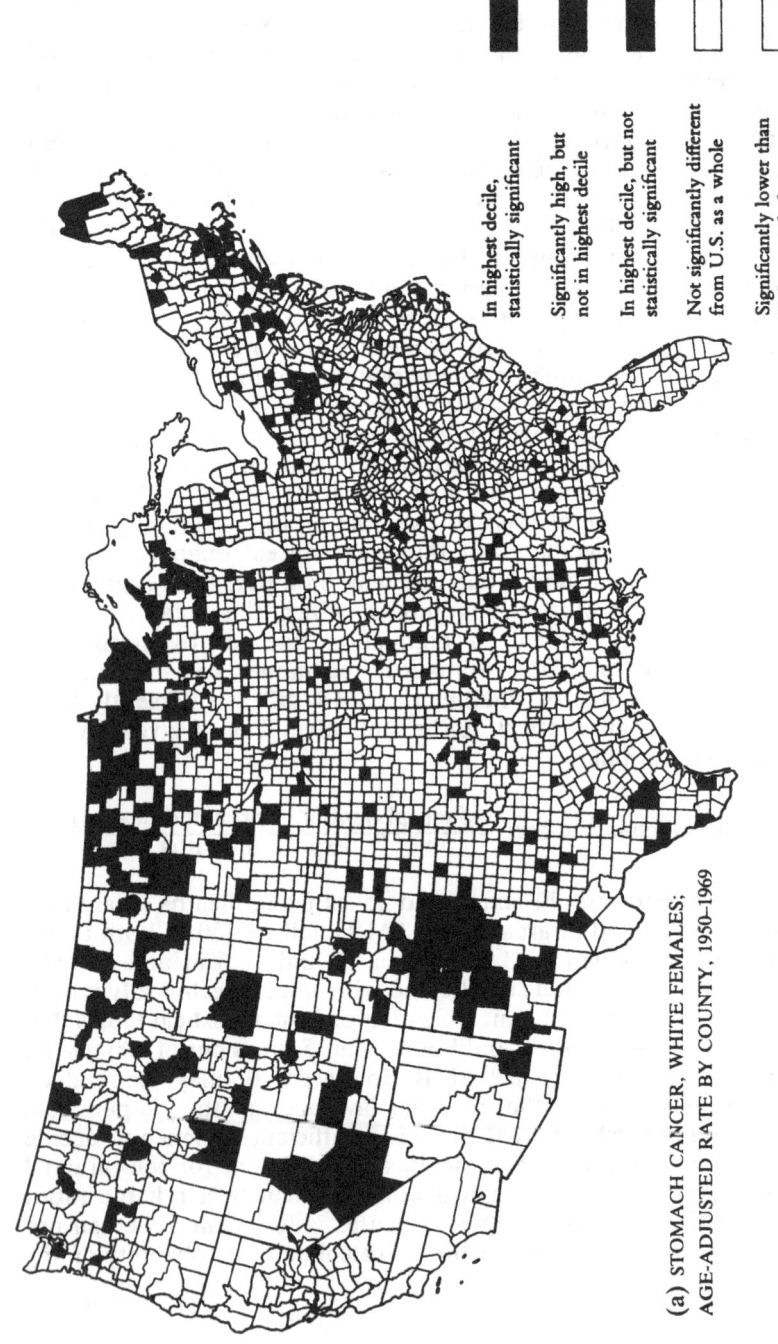

In highest decile, statistically significant

Significantly high, but not in highest decile

In highest decile, but not statistically significant

Not significantly different from U.S. as a whole

Significantly lower than U.S. as a whole

(a) STOMACH CANCER, WHITE FEMALES; AGE-ADJUSTED RATE BY COUNTY, 1950-1969

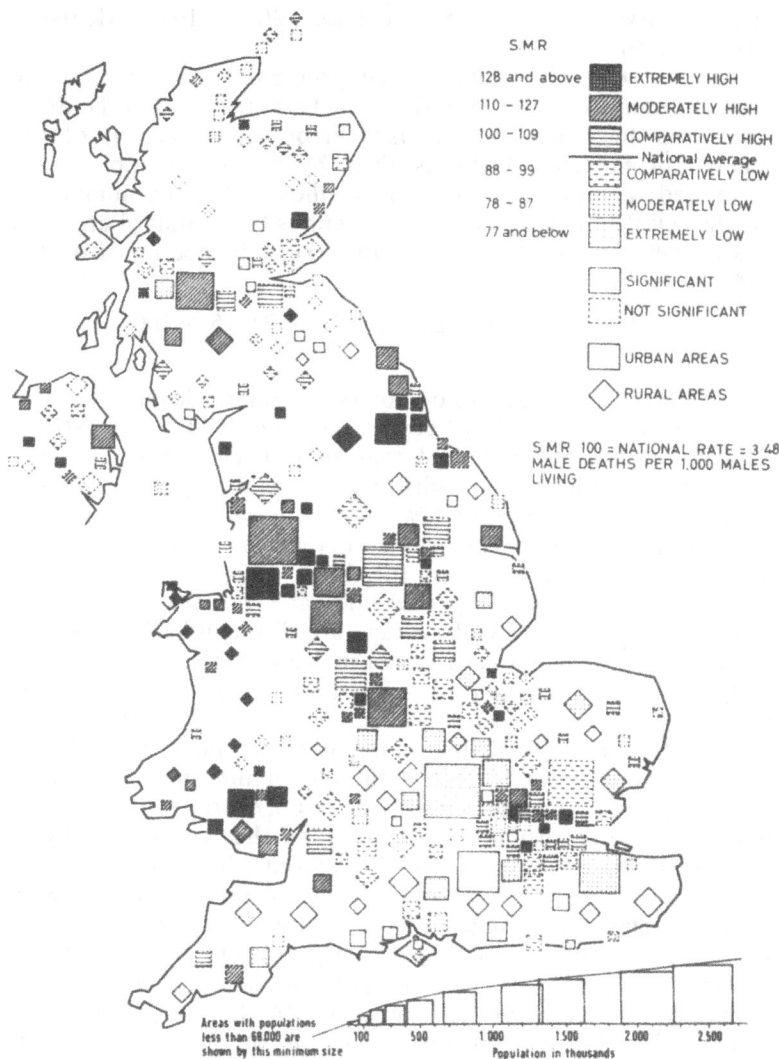

S.M.R

128 and above	EXTREMELY HIGH
110 – 127	MODERATELY HIGH
100 – 109	COMPARATIVELY HIGH
	National Average
88 – 99	COMPARATIVELY LOW
78 – 87	MODERATELY LOW
77 and below	EXTREMELY LOW

SIGNIFICANT

NOT SIGNIFICANT

URBAN AREAS

RURAL AREAS

S.M.R. 100 = NATIONAL RATE = 3.48
MALE DEATHS PER 1,000 MALES
LIVING

Areas with populations
less than 68,000 are
shown by this minimum size

Population in thousands

100   500   1,000   1,500   2,000   2,500

(b) CANCER OF THE STOMACH (MALES)

Figure 2.7   The geography of stomach cancer: intranational variations
(a) White females, age-adjusted stomach cancer rates, USA, 1950–69
Source: Mason *et al.* (1975)
(b) Males, aged 15–64 years, standardised mortality rates, stomach
cancer, United Kingdom, 1970–2
Source: Howe (1970)

given observation occurring by chance (White, 1972; Alderson, 1983, p. 310).

Another way of looking at the problem is to ascertain the expected range of variability that may be anticipated in the SMR from year to year. The standard error (or deviation) of the number of expected deaths is given by the square root of this value and we can be 95 per cent confident that the number of deaths will fluctuate within plus and minus two standard errors of this expected value. For area A the 95˙ confidence limits on the SMR range from

$$\frac{(20 - 2\sqrt{15})}{15} \times 100 \text{ to } \frac{20 + 2\sqrt{15}}{15} \times 100$$

that is from a SMR of 82 to one of 185. Similar calculations result in a confidence interval of between 117 and 150 for the SMR in area B. We can be confident that area B has a mortality rate above the average, while with area A it is impossible to tell whether the rate is high or low.

Figure 2.7 shows the variations in stomach cancer in the USA and in the British Isles. In the USA map, data have been aggregated for a 20-year period and rates that are significantly different from the national average are distinguished. There are some unusual high spots in northern New Mexico and southern Colorado, while the high rates for the north-central part of the country are thought to be associated with the consumption of smoked fish. For the British maps, data are plotted for deaths between the years of 15 and 64 thereby minimising the problem of mis-diagnoses; deaths in this age range are more likely to undergo a post-mortem examination and they are less likely to die of multiple causes than the elderly. The small-number problem is tackled by combining deaths for three years, by using data at a relatively aggregated scale (county boroughs, aggregates of urban and rural districts) and by drawing the symbols in proportion to the size of the population. The map reveals generally low rates in the south and east (except for some London boroughs), and high rates in the north and west. Intriguingly, there are strikingly high rates in some rural areas of north Wales and northern Scotland. Aird *et al.* (1953) examined these variations for an earlier period and noted that they were similar to the distribution of blood groups; there was a higher frequency of blood group O in persons living in the north. To test this hunch they examined the blood group of individual patients with stomach cancer in different parts of the country and found the opposite result: there were more patients with blood group A than with O! This is an illustration of a major difficulty with geographical studies, results found at the areal level may not necessarily apply at the individual level; this so-called ecological fallacy will be discussed further in Chapter 3.

An imaginative use of mortality data at the intra-national scale is Charlton *et al.*'s (1983) study of 'preventable' death in England and Wales. They collected information on deaths that should not occur, given adequate medical care, for the period 1974–8 for the

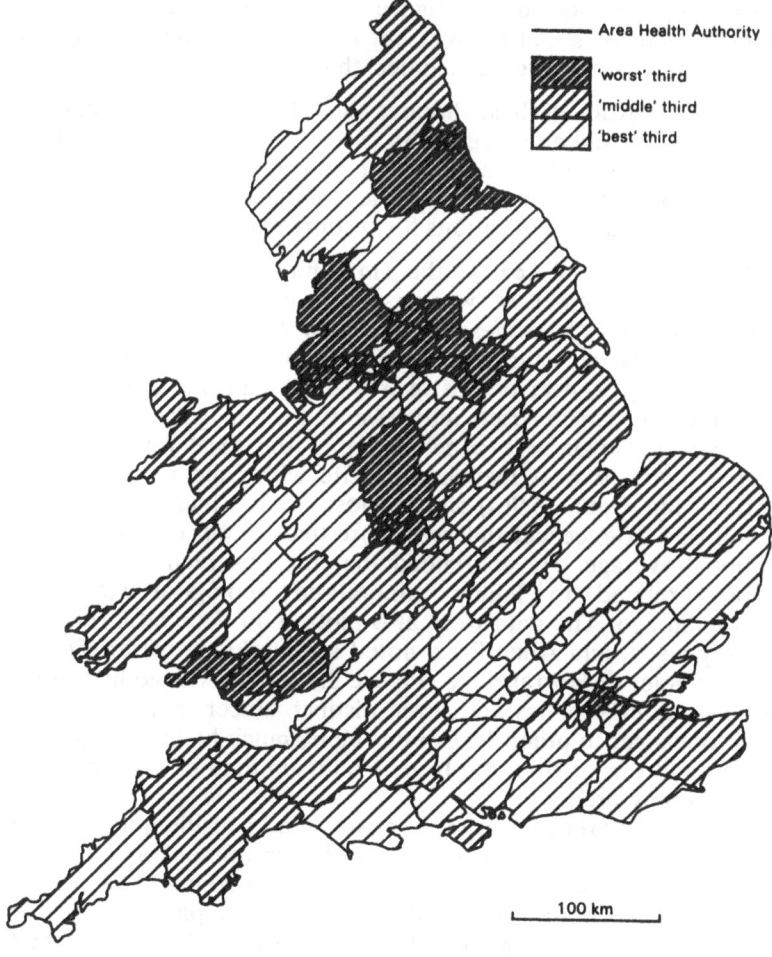

Figure 2.8 Variations in mortality from conditions amenable to medical treatment, England and Wales
Source: Charlton *et al.* (1983)

98 Area Health Authorities of England and Wales, provided that there were at least 200 deaths from that particular cause. After further excluding any cause that did not show significant spatial variation, the rates of twelve causes of death in the AHAs were divided into sextiles; a 1 was awarded if the area was in the 'best' group, a 6 was given if it was in the 'worst'. As Table 2.8 shows, some areas were consistently bad, others generally good. An overall index produced by summing the points awarded and the distribution is shown in Figure 2.8. The researchers then standardised for social factors, but the overall pattern remained roughly the same. They concluded that:

> Mortality from all these diseases should be largely avoidable, by efficient and effective health care. . . . If it is established that there are indeed large variations in the quality of health-care delivery from one part of the country to another, this will have important implications for resource allocation. (p. 696)

At intra-regional scales, the small-number problem can be acute but a number of workers have used mortality data to suggest and test hypotheses. In what a *Lancet* (1981) editorial called a 'classic' study, Allen-Price (1960) investigated the distribution of cancer deaths over a 20-year period (1939–58) in Tavistock urban district and twenty-five neighbouring rural parishes in Devon. He concluded that the local folklore, that certain parts of west Devon were rife with cancer, was indeed correct and Figure 2.9(a) shows those areas which had a crude death-rate from cancer that was more than 50 per cent above the national average. He contended that the observed pattern was closely related to highly mineralised geological strata of Devonian age. The data were also plotted on detailed Ordnance Survey maps, and Allen-Price noted a remarkable pattern in the small hamlet of Horrabridge (Figure 2.10). The village had three distinct water supplies and three distinct cancer mortalities, the ratio of cancer to other death-rates being much higher at 1 in 3 in the area to the north of the River Walkham than in the two other water-supply areas.

It has been argued that these results lend considerable support to the link between the geochemical environment and disease mortality, but the evidence should not be taken at face value. Despite using twenty years of data, many of the parishes had low actual numbers of cancer deaths. Two-thirds of the parishes had less than 20 deaths and six had less than 5. Chi-square statistics have been calculated for the urban district and all the parishes, and those areas with rates significantly above the national

Table 2.8 Sextile rankings of AHAs on deaths from 'preventable' mortality

Area		Hypertensive disease	Cancer cervix uteri	Pneumonia, Bronchitis	Tuberculosis	Asthma	Chronic rheumatic heart disease	Acute respiratory infection	Bacterial infection	Hodgkin's disease	Abdominal hernia	Maternal deaths	Anaemia	Overall score
'Worst' three	Walsall	5	5	6	6	3	6	6	6	5	6	4	3	61
	Bolton	5	5	4	6	2	6	6	6	4	6	5	6	61
	Sandwell	6	5	5	6	5	6	6	3	6	5	2	5	60
'Best' three	Gloucestershire	1	2	5	2	2	1	4	1	3	3	2	1	27
	Bromley	1	2	1	4	1	2	1	2	6	1	2	1	24
	Oxfordshire	1	1	1	1	1	2	2	1	4	4	3	1	22

Source: Charlton *et al.* (1983, p. 694).

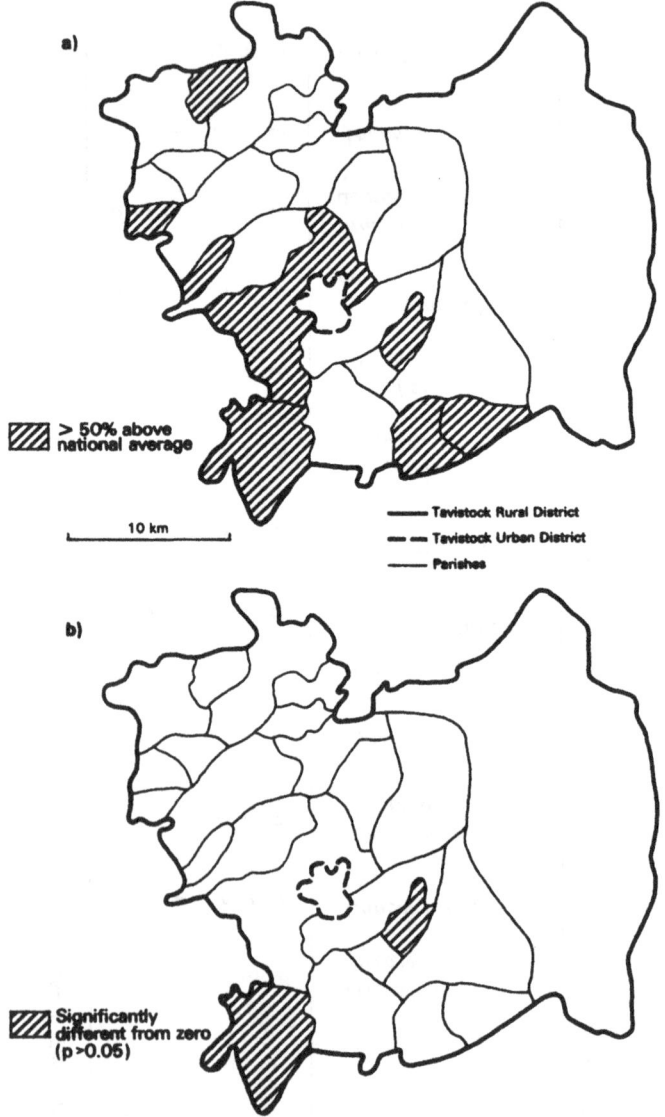

Figure 2.9 Cancer in West Devon, 1939–58 (a) Areas with death-rates 50 per cent above England and Wales average (b) Areas with death-rates significantly different from England and Wales average
Source: original data from Allen-Price (1960)

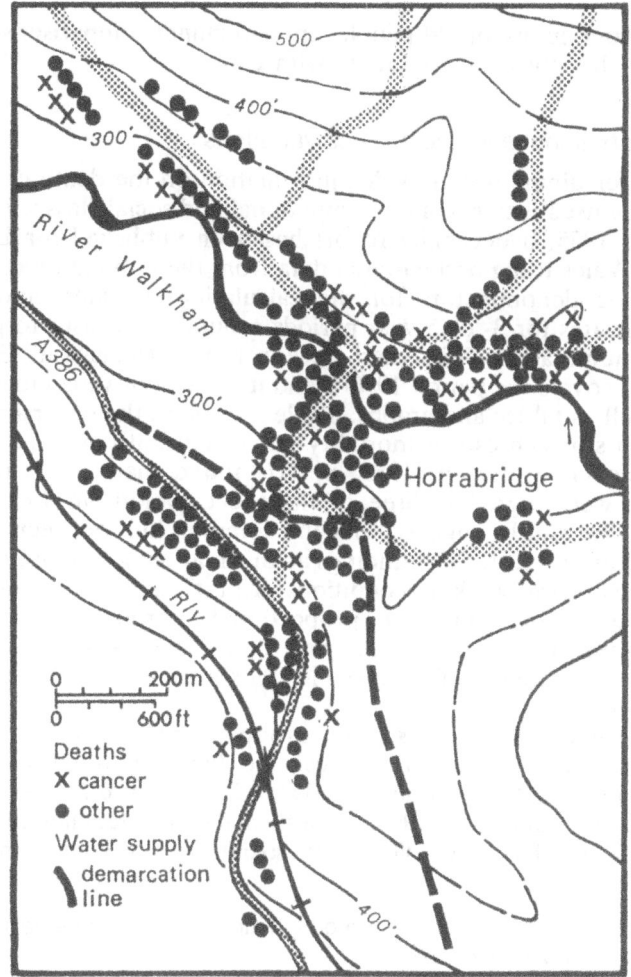

Figure 2.10 Distribution of deaths from cancer, Horrabridge, West Devon, 1939–58
Source: Allen-Price (1960)

average, at the 0.05 probability level, are indicated in Figure 2.9(b). The evidence for generally high rates in west Devon is now less compelling but even these results (as well as Figure 2.10) are based on very crude analyses which ignore age and sex differences in the local population. At this detailed scale of analysis, it is crucial that the researcher is not misled by very high, crude rates that could be based on a small number of deaths in a community of old people. Intra-regional comparisons can be done, but they must be done with care.

Occupational and social class variations

The remaining way in which information on the death certificate can be used is to examine occupational and social class variations. Since 1855, a decennial report has been published for England and Wales using occupational data from the population census to provide denominators for the calculation of rates, and death certificates for 3- or 5-year periods around the census to provide the numerators. The most recent report (Registrar General, 1978) covers the period 1970–2 and it has extensive commentaries as well as tables and graphs. Table 2.9 shows the occupations for which stomach cancer mortality was significantly high in 1970–2. Moreover, Alderson (1983, p. 287) found that cancer registrations were high for three of these occupations in 1966–7 and 1968–70, and he suggests that the common link between them is exposure to dust. Such findings as this may suggest more detailed studies of the working conditions of particular occupations. The figures in the table are proportional mortality ratios which measure whether the deaths from a particular cause are different from what one would expect in comparison to deaths from all causes.

Occupations can also be grouped into classes to form occupational or 'social' classes; the basic sixfold division for England and Wales is given in Table 2.10. Figure 2.11 shows the 'social pyramid of death'; for both males and females there is a strong social class gradient with social class V having the most

Table 2.9 Occupations showing high stomach-cancer mortality, England and Wales, 1970–2

Occupation	Proportional mortality ratio 15–64	65–74
Miners and quarrymen*	141	127
Glass and Ceramic makers	108	73
Labourers*	113	108
Warehousemen	107	108

Source: Registrar General (1978).

*Cancer registrations were also significantly high in 1966–7 and 1968–70.

Table 2.10 Registrar General's classification of social class

Social class	Description	Example	Percentage of population
I	Professional executives	Judges Doctors Accountants	6
II	'Semi-professionals' managers	Teachers Nurses	20
III N	Skilled non-manual	Clerks	15
III M	Skilled manual	Taxi drivers Plumbers	33
IV	Semi-skilled manual	Farmworkers Assembly line workers	19
V	Unskilled manual	Building labourers	8

Source: 1981 *Census of Population*.

favourable rate. Moreover, as Figure 2.12 shows, the gradient is found throughout life; Figure 2.13 shows that it is also found for many specific causes of death, while Figure 2.14 reveals that the pattern is maintained in sub-regions of England and Wales. A fascinating contrast is given in Table 2.11; in the USA it is the lower social classes that have lower rates of cirrhosis of the liver, while the exact opposite was observed in England and Wales for 1949–53. Terris (1967) suggests that the high duty on alcohol in England and Wales meant that 'only the well-to-do can really afford the luxury of dying from cirrhosis of the liver'. Since the 1950s, however, things have changed in Britain and as Blaxter (1981, p. 199) has argued, cirrhosis is now one of the 'new' diseases of poverty. With the 'old' diseases of poverty, social class V always had the highest mortality rates and, indeed, the relative gap has widened since the 1930s (Figure 2.15). With the 'new' diseases, social class I used to have the highest rates but through this century the position of social class V has worsened until they now have the most unfavourable mortality. The

Table 2.11   SMRs for cirrhosis of the liver, males aged 20–64, USA 1950, England and Wales 1949–53

Occupational level	USA	SMR England and Wales		Occupational class
Professional	90	207	I	Professional
Technical, administrative, managerial	88	152	II	Intermediate
Clerical, sales, skilled	105	84	III	Skilled
Semi-skilled	118	70	IV	Partly skilled
Labourers	148	96	IV	Unskilled

Source: Terris (1967).

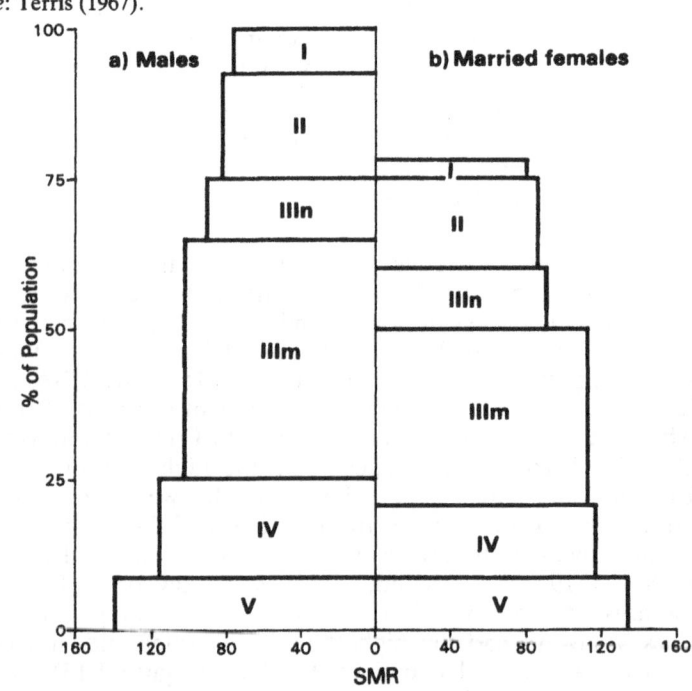

Figure 2.11   The social pyramid of death. Length of bar corresponds to the death-rate; height of bar corresponds to the number of people in each occupational class
Source: Mitchell, J. (1984)

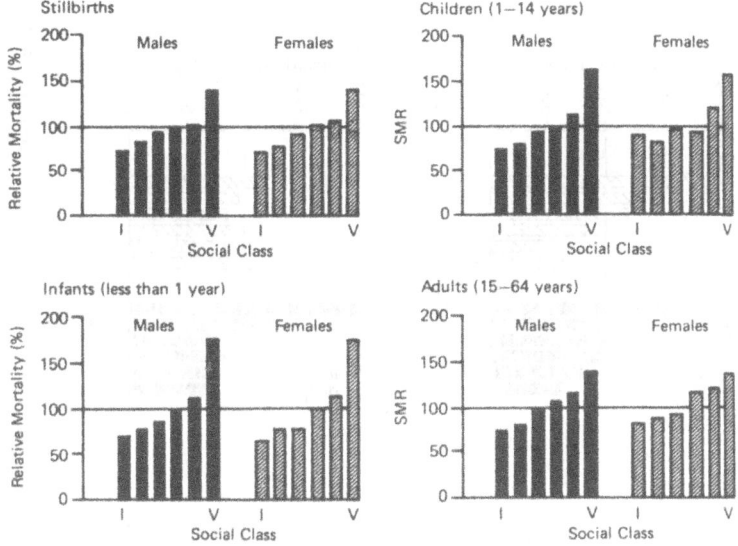

Figure 2.12   Social gradient throughout life: mortality by occupational class and age
Source: Registrar General (1978)

relationships between social class and ill-health have been extensively reviewed in the Black Report on *Inequalities in Health* (Townsend and Davidson, 1982); we will return to this topic in later chapters.

As with all data sources, we must be aware of potential pitfalls in interpretation. The small-number problem may occur when the number of workers in a particular occupation is small and we would certainly doubt the reliability of death-rates for single women in the construction industry. Other problems that may occur are selection bias (the deceased may have moved to the final less-demanding job due to ill-health) and confounding (it may be something in the home rather than in the work environment that is the true cause of the variations), and difficulties may occur because broad, aggregate occupation groups may hide very high rates for specific occupations. There is also a problem with the rates for married women, for these are generally based on their husband's occupation and not their own (McDowall, 1983). Despite all these difficulties, Alderson (1983) was able to conclude that routine data provides an extremely useful background to the study of occupational disease.

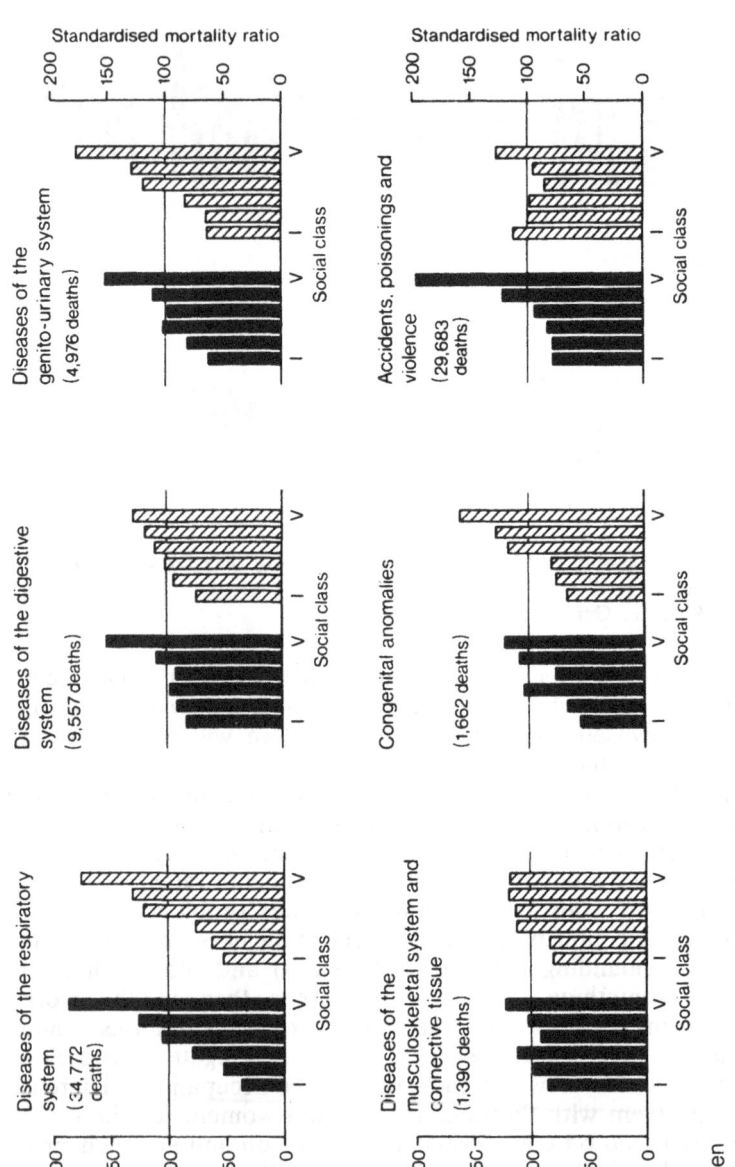

Figure 2.13 Social gradient and specific causes: mortality in adulthood, men and women by occupational class

Source: Registrar General (1978)

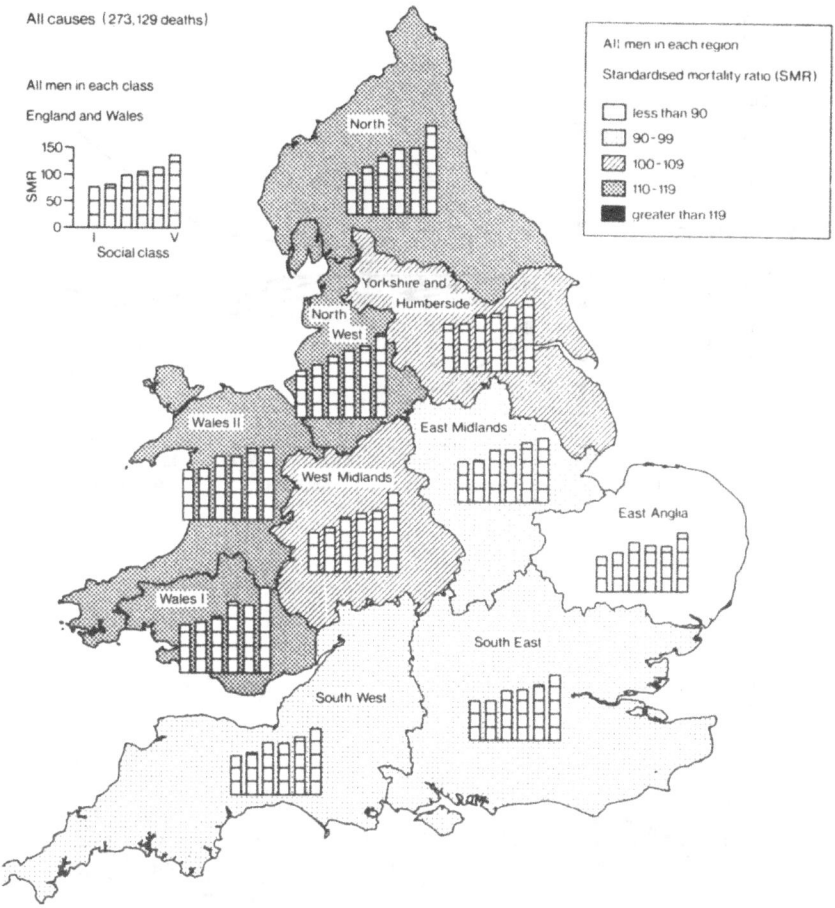

Figure 2.14 Mortality by occupational class and region, men aged 15–64
Source: Registrar General (1978)

Morbidity statistics

As a 1981 *Lancet* editorial put it, 'counting the dead is not enough' for mortality data are not applicable to all epidemiological problems. There are a large number of conditions such as arthritis and migraine which cause pain and discomfort but which are not fatal. Moreover, if a disease has a long latent period or a long duration period (Table 2.12) the routine data collected at death may bear little relation to the conditions prevailing at the time of the cause and onset of the disease. Consequently, there is

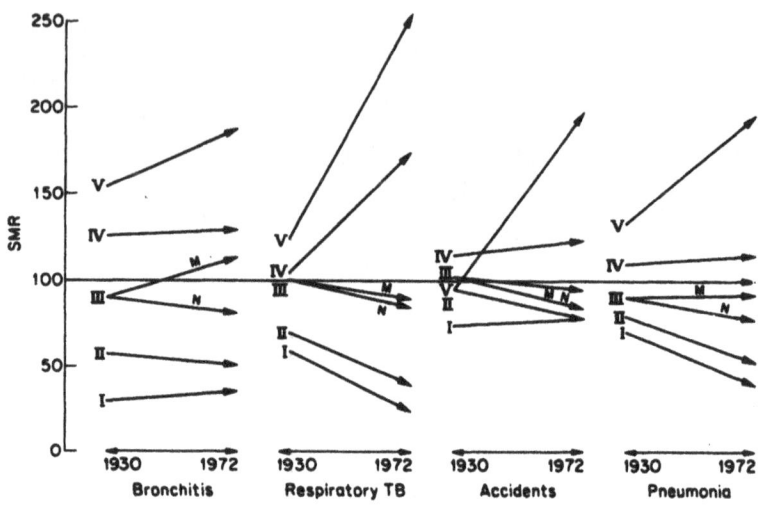

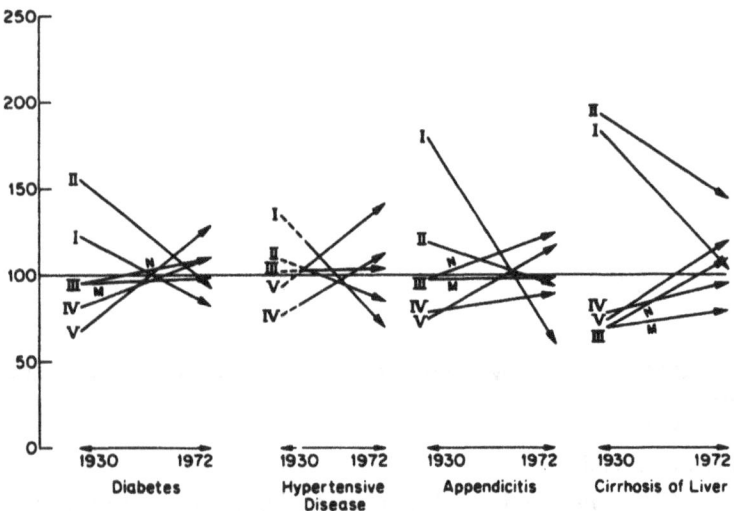

Figure 2.15 Trends in male adult (15–64) mortality rates by occupational class and cause, 1930–72 (a) 'Old' diseases of poverty (b) 'New' diseases of poverty
Source: Blaxter (1981)

a general need for information on the health and morbidity of the general population. This need is generally recognised in the USA and since 1956 the US National Health Survey has regularly undertaken a three-part study:

(i) the Health Interview Study is conducted continually with a sample of household interviews being completed each week;
(ii) the Health Examination Survey is a smaller sample targeted at specific age-groups and consists of physiological and psychological tests for specific diseases;
(iii) the Health Records Survey involves a sample of institutions and facilities providing care (Lawrence, 1977; McDowell, 1977).

Various types of national health surveys have been developed in European countries (Armitage, 1977) but in Britain nothing so extensive as a national epidemiological survey exists and so we have to rely on a wide range of different sources to provide only a partial picture. Indeed, in Britain there is an outstanding lack of information on health and illness. Here we will briefly review a number of sources that do provide some information, noting, however, that many of them involve the person who is ill contacting the health service. It is therefore very probable that the official statistics underestimate the true prevalence of a condition: the 'clinical iceberg'. For example, the Royal College of General Practitioners (1973) estimated that in a general practice of 2,500 there would be some 200 cases of anaemia and 150 bronchitics which were unknown to the GP. This review of routine morbidity sources is intended to be illustrative rather than exhaustive and for more detail see Alderson's (1983) review; Table 2.13 summarises the morbidity data routinely published by OPCS.

Table 2.12 The 'spectrum' of disease

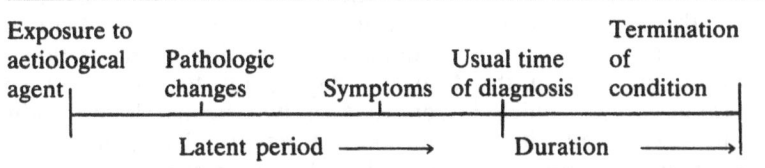

Source: Based on Lilienfeld and Lilienfeld (1980, p. 60).

Table 2.13 Routine OPCS publications on morbidity

Series MB   Morbidity
MB1   Cancer statistics
MB2   Communicable disease statistics
MB3   Congenital malformations
MB4   Hospital In-patient Enquiry

Series GHS   General Household Survey

Series AB   Abortion statistics

OPCS Monitors ('quick release of selected information as it becomes available')

Series MB   Morbidity
MB1   Cancer: annual
MB2   Infectious diseases: quarterly
MB3   Congenital malformations: occasional
MB4   Hospital In-patient Enquiry statistics: annual

Series WR   Weekly return (notifications of infectious diseases)

Series GHS   General Household Survey

Series AB   Abortion statistics

Source: HMSO (1984), *Government Publications, Sectional List 56.*

General Household Survey

The GHS is a continuous survey of a sample of non-institutional households in Great Britain which has been running since 1971; an annual report is published by OPCS. The survey covers not only health but also housing, employment and education; in 1982, 12,480 interviews were completed. The GHS provides information on self-reported illness in a number of forms: there are data on consultations with a GP during previous fortnight, visits to hospital out-patients departments in the last three months, and spells in hospital during the previous year. Data are also obtained on the prevalence of acute illness (restriction of normal activity in the fourteen days prior to interview) and chronic illness which is reported in two forms: long-standing illness with and without limitation of activity. This self-reported morbidity is presented in tables disaggregated by sex, age, economic activity and socio-economic group. Figure 2.16 illustrates the social gradient of illness and occupational class for the 1976 survey. In different years, different special topics of health interest are included so that the 1980 report provides data for tinnitus (sensation of noise in the head or ears). Table 2.14 gives the proportions who reported tinnitus by socio-economic group. Some 15 per cent of the adult population reported tinnitus that was not due to an external stimulus (such as swimming) and, indeed, 2 per cent of the adult population hear the noise continuously. Each of the annual reports contain the topics covered in previous years and a list of non-government research that has used the GHS; the individual data are available from the ESRC Survey Archive, University of Essex. From a geographical viewpoint, the source is a limited one permitting only gross regional comparisons to be made, while for the epidemiologist, the reliance on self-reporting and the lack of medical examination precludes any full analysis of a cause-specific nature. Despite these major shortcomings, as the Black Report (DHSS, 1980, p. 48) recognised: 'For the analysis of social class and occupational differences in morbidity, the only regular source of information provided by central government is the GHS.'

Notifiable disease statistics

Throughout this century in England and Wales, certain infectious diseases have required compulsory notification. These diseases range from mild, childhood infections such as measles to food

Table 2.14 Prevalence of adult tinnitus by sex and socio-economic group: age-standardised percentages

Socio-economic group	Males	Females	Total
Professional	10	11	10
Employers and managers	9	11	10
Intermediate and junior non-manual	12	16	15
Skilled manual and own account non-professional	14	18	16
Semi-skilled manual and personal service	16	20	18
Unskilled manual	16	20	18
All persons aged 16 and over	13	16	15

Source: General Household Survey (1981).

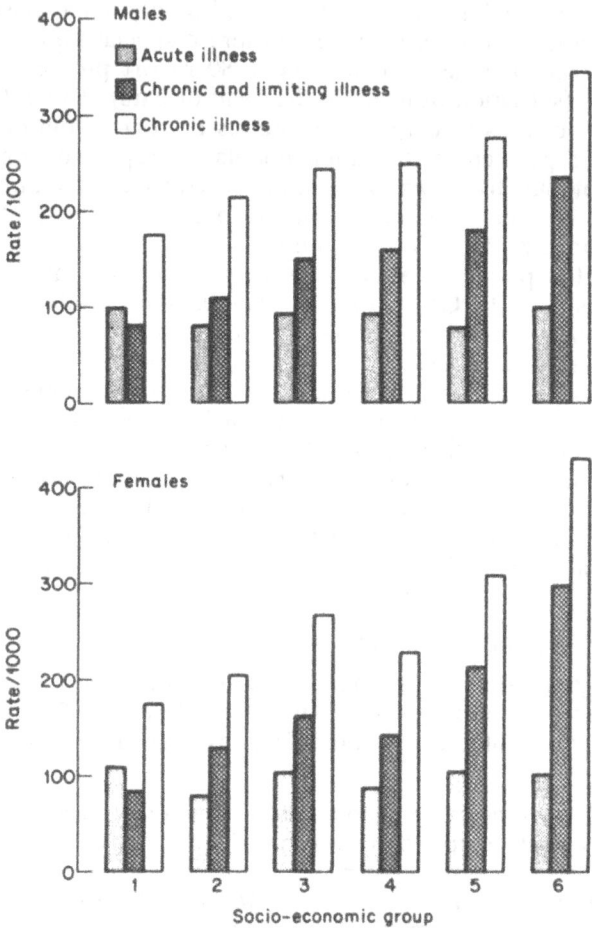

Figure 2.16 Rates of self-reported illness by socio-economic group
Source: General Household Survey (1976)

poisoning, dysentery and even exotics like Lassa fever and Marburg (Green Monkey) disease. The system of notification is as follows: the patient who is suffering contacts the doctor or a hospital and a preliminary diagnosis is made, a sample is frequently taken for laboratory analysis and OPCS and the Communicable Disease Surveillance Centre are informed. The notifications are available in printed form (disaggregated by cause and local authority district) at the end of the following week in the *Weekly Return*. This also contains newly diagnosed episodes of other non-notifiable, communicable and respiratory diseases such as chicken-pox and influenza for forty general practices. In the United States, physicians, clinics and hospitals are required to report specific communicable diseases ultimately to the Centre for Disease Control in Atlanta, Georgia, but Pyle (1979, p. 29) contends that, in practice, the procedure is 'voluntary' and is not fully adhered to. In both the United Kingdom and the USA, it is likely because of the system of notification that the reported figures, particularly for the milder conditions, are considerable underestimates of the true prevalence.

Although infectious diseases only account for a small proportion of the morbidity in economically advanced countries, the notification system can play a vital intelligence role, monitoring the spread of an epidemic and facilitating such arrangements as quarantine and vaccination. An impressive example of this surveillance role is the work of Gill *et al.* (1983). In 1982, 269 patients were found to have *Salmonella Napoli* in their faeces in comparison to only 15 reported cases in the previous thirty years. This was a major and dangerous epidemic of food poisoning and a detailed examination of the records revealed that the majority were children and that over 20 per cent had been admitted to hospital. On interview, a common link was found: the sufferers had eaten chocolate manufactured by an Italian company. Some 32 tons of chocolate were recalled from shops and warehouses and it was estimated that 200 hospital cases and several thousand cases of food poisoning were avoided by this swift action.

Young (1974), as part of a general study of the medical geography of County Durham, noticed something odd about the food-poisoning notifications in Consett Urban District during 1964–5. He found high rates for suspected cases but laboratory analysis did not reveal a bacteriological cause for 95 out of 137 cases (Figure 2.17). Plotting these data revealed that non-bacteriological cases were clustered in an area around the steelworks. Young argued, following a statistical analysis of temporal variations, that the pollution and 'red dust' produced by large, open, spoil heaps was a strong enough irritant to cause symptoms of diarrhoea, abdominal pain and nausea. Since the closure of the steelworks in 1981, Young (1982) concluded that despite the social problems of unemployment, the town has become a healthier place to live in. Other examples of the analysis of notifiable disease will be given in Chapter 4, for geographers have made extensive use of such data to test epidemic diffusion models.

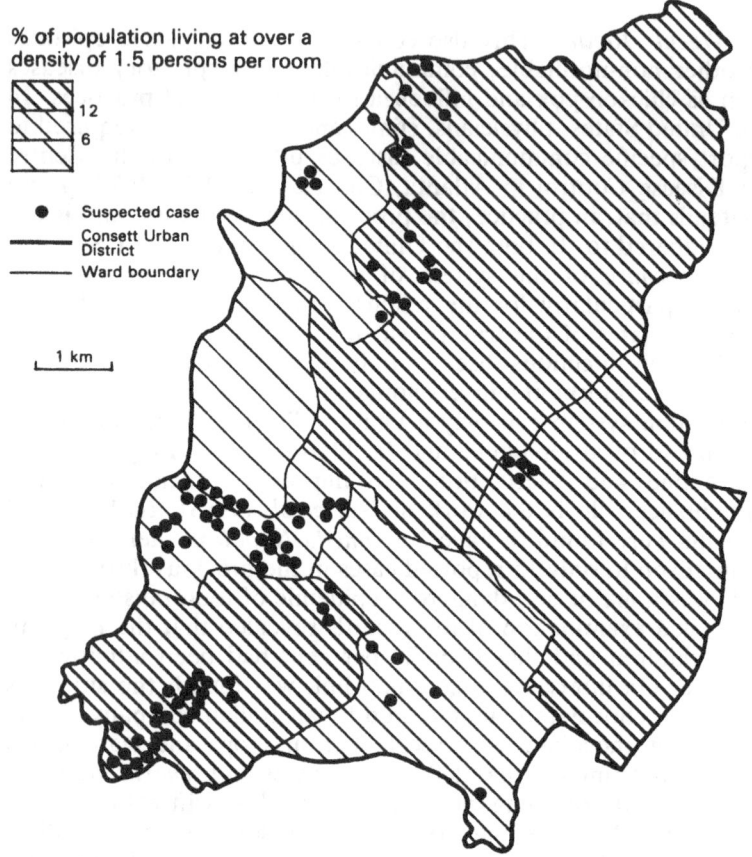

Figure 2.17 Suspected food poisoning in Consett Urban District, 1964–5
Source: Young (1974, 1982)

Cancer registries

Following Denmark's example in 1942, many countries have set up a national cancer registry. Each register attempts to collect as much information as possible on newly diagnosed cases of malignant disease, and this often requires the collection of data from several sources including general practitioners, hospitals, pathology laboratories and even death certificates. The usefulness of the register depends critically on the quality of the data collection and Magnus (1982) has examined methods of maintaining an accurate register as well as its potential uses. As an example of the geographical use of such a register, we will consider Craft and Openshaw's (1985) submission to Black's (1984) investigation of cancer incidence in West Cumbria. A television programme, transmitted in 1983, caused great concern about the rate of childhood leukaemia in the village of Seascale which is close to the Sellafield nuclear processing plant. Four cases had been observed when the expected number was considerably less than one. Craft and Openshaw tried to assess whether this high value could have occurred by chance, and whether there were any other comparable 'high spots' along the Cumbrian coast. Their source of data was the Northern Region's Childhood Malignant Disease Registry which seeks to include the diagnosis, sex, age and location of all children aged under 15 who have cancer. Registration is mainly by direct notification from clinicians and pathologists but it is supplemented by information from hospital-activity analyses and the national cancer registration system. The diagnoses are checked by pathological review and by an examination of the clinical notes; the data are thought to have high accuracy and coverage (Birch *et al.*, 1980).

Each of the 1,113 cancer sufferers identified in the period 1968 to 1982 were allocated to their appropriate ward of residence and using the population aged under 15 in 1981, incidence rates were calculated for all 675 wards. The analysis of these data represents the small-number problem in its severest form. Pomiankowski

Table 2.15   Five wards in order of descending Poisson probability, lymphoid malignancy, 1968–82, Northern Region

	Cases	Children aged under 15	Poisson probability	Rates per 1,000	Ward
1	4	411	0.0001	9.73	Seascale, Cumbria
2	4	976	0.0032	4.09	Fairfield, Stockton
3	2	144	0.0036	13.88	Whittingham, Alnwick
4	4	1,207	0.0068	3.31	Sedgefield, Durham
5	4	1,353	0.0100	2.95	N. Ormesby, Middlesborough

Cumbrian rate: 0.61 per 1000

Source: Craft *et al.* (1984, p. 96).

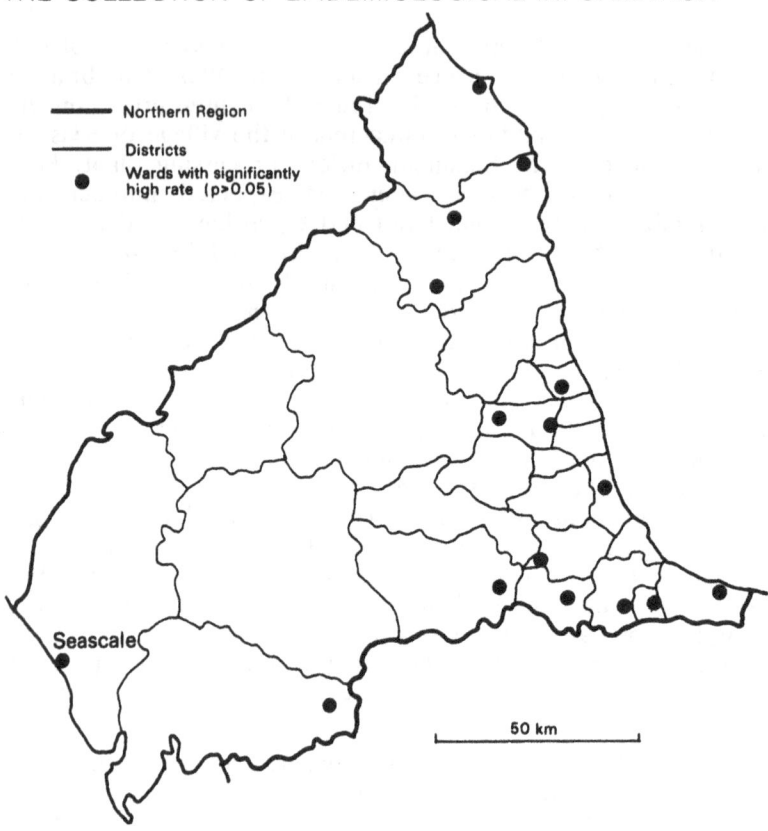

Figure 2.18   High incidence of lymphoid malignancy, Northern Region.

(1984) pointed out that as most wards have either 0 or 1 case, all that is being mapped are variations in the denominator which ranges from 100 to 2,000 children. Consequently, Craft and Openshaw (1985) also calculated the cumulative Poisson probability index for each ward in comparison with an overall Northern figure. As Table 2.15 reveals, Seascale has a high rate of incidence of lymphoid malignancy and a low Poisson probability of this occurring by chance. But the map of Figure 2.18 reveals a number of similar, improbably high rates in the region and these are not clustered around Sellafield nor along the Cumbrian coast. In fact, the pattern is simply inexplicable given our present knowledge of childhood leukaemia. Moreover, in a series of simulation experiments with a completely random generating process, Craft and Openshaw (1985) found that it was inevitable that some clustering occurred in areas with small populations. They concluded that the occurrence of four leukaemias, even with very low Poisson probability, could still occur in a random series just by chance with a modest degree of regularity. Thus the evidence for a Sellafield effect is not overwhelming, but they also point out that the Seascale rate is exceptionally high, particularly when the two further cases which have appeared since 1982 are taken into account. Black (1984) recommended that further work on this problem was required and this is currently being undertaken.

Hospital discharge data

The final source of routine morbidity statistics to be discussed is the hospital in-patient enquiry (HIPE) which is processed for every tenth patient discharged from a general hospital in England and Wales. Data are collected on the characteristics of the patient (for example, sex and age, but nothing is collected on race and social class), administrative particulars (for instance, date of admission and discharge, number of visits to the operating theatre) and medical details (diagnosis and type of operation if performed). This source is frankly a very limited one for epidemiological research and is primarily used for assessing the utilisation of facilities and workload of the health service.

Routine sources: a summary

High-quality mortality statistics are routinely available for most economically advanced countries and because they are usually disaggregated by place, by sex, by age and by occupation, it is possible to use such sources to suggest causal hypotheses. In Britain only very limited data are available on morbidity in the population: the GHS provides survey data on the prevalence of acute and chronic illness classified by social class, but there is inadequate geographical disaggregation and little disease-specific information. In contrast there is detailed weekly information on notifiable diseases for every local authority, but such illnesses represent only a small part of overall morbidity. Cancer registries can provide details of this particular disease but very little associated data is collected; HIPE data are of very limited epidemiological value for this information is collected primarily to monitor efficiency of hospital health care. For a wide range of diseases, including arthritis, migraine, heart disease, bronchitis, there is minimal routine morbidity data. The major reorganisation that is being currently undertaken of the collection of information in the British National Health Service is unlikely to substantially improve this situation with regard to epidemiological data (Mason and Morrison, 1985). Consequently, epidemiologies will continue to design and undertake their own surveys.

Epidemiological survey methods

It is generally recognised that while routine data (both mortality and morbidity statistics) are useful for descriptive work and suggesting hypotheses, analytical research involving the testing of hypotheses requires that epidemiologists collect their own information. There has therefore been a major effort to produce systematic methods for data collection in relation to the design, conduct and interpretation of observational studies. In essence the survey method is simple: a group of people are selected in some way and information is obtained on disease status (usually presence, absence and type), potential causal factors (such as occupation, cigarette consumption), risk factors (blood pressure and cholesterol levels, for instance) and background variables (for example, age and sex). The information may be obtained by physical measurements and examination or by questionnaires for it has proved possible to develop questions that provide a reliable diagnosis of disease. Here is not the place to detail aspects of scientific surveys for there are numerous books that cover such issues as sampling, non-response and the design of unambiguous questions. This section, however, attempts to outline the distinctive character of epidemiological surveys, to illustrate a few of the enormous number of uses, and to evaluate which of several alternative designs are appropriate for particular epidemiological problems. Before looking at these differing designs, we need to consider the general features of subject selection and the concepts of directionality and timing.

In any observational study we cannot include everybody alive in the world today and so we have to select our subjects. There are two basic forms of selection: sampling and restriction. Probability sampling (Stuart, 1968) is the recommended method and providing the sample is sufficiently large, the sample should be representative of the general population. Statistical significance tests are available for deciding whether observed relationships in the sample are genuine (and can be carried over to the population), or whether they are 'false' and a result of sampling error. Restriction is often done to reduce cost and to facilitate the selection and contact of the subjects. For example, studies have been used on particular occupational groups such as doctors (Doll and Peto, 1976), postmen (Reid and Fairbairn, 1958) and civil servants (Morris *et al.*, 1966). Other studies have chosen their subjects if they live in a particular area; for example, Dawber *et al.* (1963) selected 6,500 adults between the age of 30 and 62 from the total population of 28,000 living in Framingham, Massachusetts. These latter subjects were followed for twenty years to investigate the relationships between heart disease and certain risk factors: cholesterol, weight, blood pressure and smoking. Other forms of restriction are confining subjects to workers at a particular factory or to pupils attending particular schools and even to children born in a specific week. With restriction it is crucial to consider whether the subjects are representative of the general population and to what extent bias or systematic distortion has been introduced.

Directionality is the temporal relationship between the observation of disease outcome or status and causal factors. In a non-directional design, both sets of variables are measured simultaneously; with backward directionality, research begins with the presence or absence of disease and then historical information about possible causes is obtained. In a forward study the research begins with the causal factors and people are followed over time to see what happens to their health status. Timing is another important element and refers to the temporal relationship betwen the most recent data collection and the occurrence of the causal factor and the disease. In a completely prospective study, the causal factors and the disease are measured after the start of the study while, in a completely retrospective study, all information comes from the past in terms of records and recollections. Finally, in an ambiperspective study, either the cause or disease is measured prospectively and the other retrospectively.

Although Kleinbaum *et al.* (1982) manage to list fifteen different types of design which accommodate a wide variety of different types of subject selection, directionality and timing, it is generally recognised that there are three basic designs which can be described in terms of Table 2.16.

Type A: If there is little prior knowledge at the start of the study on how individuals would be distributed in each cell of the table, we would have to study the

Table 2.16 Three basic types of survey design

		Disease			
		Present	*Absent*		
Causal	*Present*	a	b	a + b	Type C
factor	*Absent*	c	d	c + d	
		a + c	b + d	n	
		Type B			Type A

Source: Roht *et al.* (1982, p. 365).

entire sample (n) to identify the health problems at one point in time; *cross-sectional* designs provide such a snapshot of the prevalence or incidence of a disease.

Type B: With this design, persons with a disease (cases, a + c) and without (non-cases, b + d) are compared in a *case-comparison* study.

Type C: The essential feature of this *cohort* design is that disease-free persons exposed to a particular causal factor (a + b) are compared to persons not exposed (c + d) to determine the proportion of each group that develops the disease over time.

We will now consider and illustrate each of these basic designs.

Cross-sectional designs

Cross-sectional studies have either a non-directional or backward design and are often called survey or prevalence studies. Subjects are selected and then questioned or examined for disease, past and present causal factors and other relevant background variables. These studies are usually undertaken when relatively little is known about the aetiology and incidence of a disease.

In the early 1960s there was some evidence that cancer of the gullet (oesophagus) varied greatly throughout the world. For example, it was commonly seen in almost all the hospitals in Kenya, but neighbouring Uganda had a much lower frequency. There was also some evidence that there had been a considerable increase in the disease in some parts of Africa over the previous twenty or so years. But studies of the disease were hampered by the lack of medical records in the areas of possible high incidence. In 1965 a geographical survey was undertaken in central Africa in which 103 hospitals were visited on an 8,000-mile tour.

Straightforward questions were asked of the hospital staff of the form, 'How many patients with disease X do you see in a year?'; data were also collected on local customs and diet. The course of the disease is so distinctive, beginning with an inability to swallow solids, then liquids, usually leading to death by starvation within a year, that there is generally no need for x-rays and tests; a clinical diagnosis is usually sufficient. Figure 2.19(a) shows the distribution of hospitals with more than six cases per year and this pattern closely resembles the area of consumption of Kachasu or Malawi Gin (Figure 2.19(b)). This is an indigenous spirit distilled from maize and sugar on very crude equipment including bicycle frames and car exhausts. Samples of the spirit were obtained (McGlashan, 1972) and were found to contain zinc, copper and Dimethyl-N-nitrosamine (DMN). The latter is a known carcinogen with experimental animals but it was the first time it had been found in a human foodstuff. Moreover, these findings help explain some of the intriguing ideas discussed earlier. The post-war increase of the cancer could be due to the recent use of metal equipment in distilling, as opposed to clay utensils, and the abrupt change in prevalence between Uganda and Kenya can be explained by the former country's discouragement of maize growing (fearing soil erosion) and the latter's encouragement (Cook, 1977). One finding that is not explained by the consumption of distilled spirits is the high rate in northern Iran amongst devout Muslims; here grape vinegar was found to be contaminated by DMN (McGlashan, 1972).

While McGlashan's basic unit of analysis was an individual hospital, it is much more common for an individual person to be the fundamental unit in a cross-sectional study, as in Lambert and Reid's (1970) analysis of postal questionnaires sent to a sample of households in England, Scotland and Wales. The 10,000 respondents, aged between 35 and 69 years, provided information on age, current smoking habits, place of residence and respiratory symptoms. The answers to the place of residence were used to estimate exposure to air pollution (high or low), while on the basis of the respiratory questions, the presence or absence of chronic bronchitis was determined. Their basic results are presented in Figure 2.20. The prevalence of the disease increased with age and was generally lowest for non-smokers. There appeared to be a synergistic or interaction effect with the highest rates being found for people who lived at high pollution

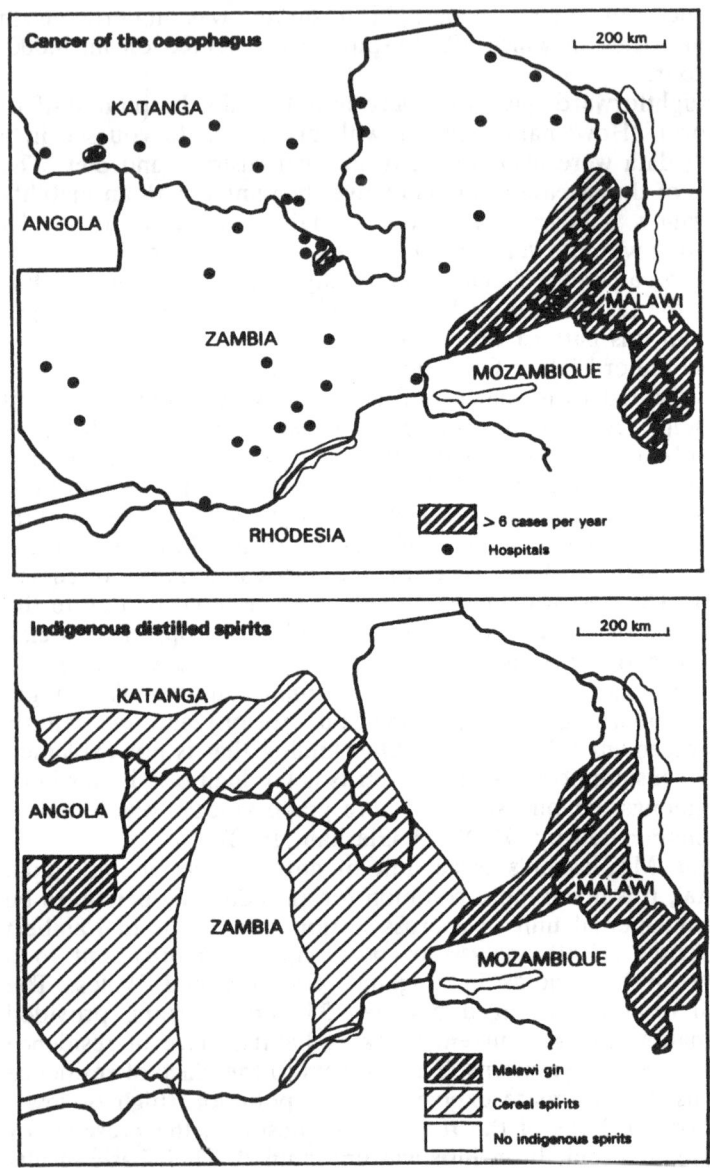

Figure 2.19 Cancer of the oesophagus (gullet) and indigenous distilled spirits
Source: McGlashan (1972)

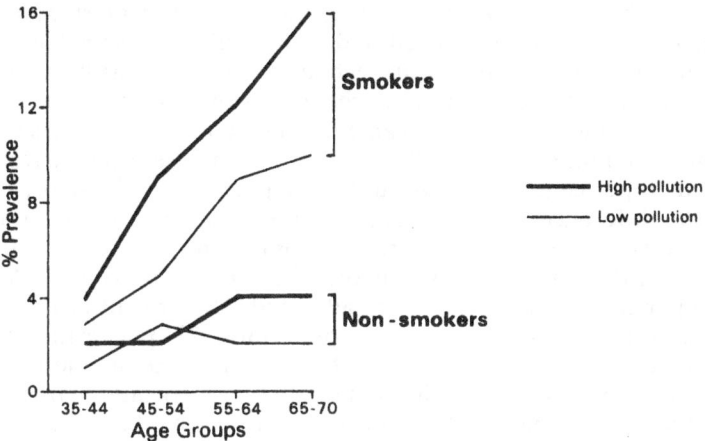

Figure 2.20 Prevalence of chronic bronchitis by age, smoking habit and exposure to pollution
Source: Lambert and Reid (1970)

levels and who smoked. The study can be criticised for not considering previous smoking habits and it is likely that current place of residence would give a fairly poor measure of a lifetime's exposure to pollution and especially pollution at work.

Case-comparison designs

Case-comparison studies have been in use for over a hundred years and are regarded as epidemiology's major contribution to research methodology. This type of design involves a backward or non-directional approach and, unlike cross-sectional surveys, the subjects are chosen from two populations, 'cases' with a disease and 'comparisons' without it. The primary purpose of this design is the testing of hypotheses by searching for differences in the previous or current exposure of the cases and comparisons to a range of possible causal factors. Information is obtained on the suspected past and present causal factors from routine data collection and by interviews with the subjects themselves or from their relatives, friends or physicians. The source of cases may be a cross-sectional prevalence survey but, more usually, cases make themselves known to the health service by attending a general practitioner, hospital or clinic, or by dying. Sources for the comparison groups range from surveys of the general population to the selection of friends and neighbours. A frequently used comparison group is hospital patients with other diseases.

As always in subject selection, the researcher must be aware that bias can be introduced and a particular problem known as Berkson's bias can occur when individuals with different characteristics have different chances of inclusion in the study. For example, if we have (Table 2.17) 6,000 cases of disease A in the community (cases), 6,000 people with disease B (comparisons) and 20 per cent of each group have a particular risk factor, there is no difference in the population on this possible causal variable. However, if there are differential admission rates so that 50 per cent of those with the risk factor, 10 per cent with disease A and 70 per cent with disease B are admitted, 58 per cent of the cases group have the risk factor, while only 23 per cent of the comparisons have it. Thus, no difference in the population can be turned into a major difference if the association rates are sufficiently variable. The reverse is also possible, differential admission rates concealing an association that exists in the population. The prudent researcher has to select comparison subjects with a great deal of care.

Table 2.17 Berkson's bias

(a) *Frequency of characteristics in the community*

	Number of individuals Disease A (cases)	Disease B (comparisons)
Risk factor		
with	1,200	1,200
without	4,800	4,800
Total	6,000	6,000

	Percentage of total Disease A (cases)	Disease B (comparisons)
Risk factor		
with	20	20
without	80	80
Total	100	100

The ideal comparison group is one which is exactly the same as the cases group expect for the characteristics which are to be studied. Consequently, it is common to reduce the differences between cases and comparisons by selection based on matching and this takes two general forms. In 'individual' matching, each individual in the comparison group is paired with a similar individual in the cases group. For example, we could have a male, 25-year-old civil servant in both groups, thereby matching for sex, age and occupation. This is often difficult to achieve exactly and 'group' or 'frequency' matching must then be used. Thus, if there are ten white-collar males aged between 30 and 40 in the cases group, there should be a similar number in the comparison group. Even with group matching, the selection of comparisons can be difficult without a large pool of subjects, and it is common to limit matching to only two or three factors. The best advice is to only match on a few background factors (usually sex and age) whose association with the disease is already known, for any variable that has been matched cannot be studied as a

---

(b) *Frequency of characteristics in the admitted group with differential rates of admission*

| | *Number of individuals* | |
	Disease A (cases)	*Disease B (comparisons)*
Risk factor		
with	660	1,020
without	480	3,360
Total	1,140	4,380

| | *Percentage of total* | |
	Disease A (cases)	*Disease B (comparisons)*
Risk factor		
with	58	23
without	42	77
Total	100	100

---

causal factor. For example, if matching is done on occupation, comparison of the two groups will not reveal the differing effects of occupation but merely the successfulness of the matching.

As an illustration of a case-comparison design, we will examine a retrospective study of sudden infant deaths (SIDs) in Gosport (Powell *et al.*, 1983). Sudden infant death (or cot death) is when a baby dies unexpectedly and a thorough post-mortem examination fails to find an adequate cause. Gosport, a part of the Portsmouth city region, had experienced a very high rate of SID in the period 1977–81, and the town's rate (at 7.3 per thousand) was more than three times the national average. Over half the births in the area are to service families, for there is an extensive naval estate of nearly 4,000 married quarters in the Rowner area of Gosport, and such births tend to be to young mothers who have only recently moved into the area. The average length of stay is eighteen months, husbands can be away on nine-month tours of duty and the mothers usually lack the support of an extended family. Previous research had found a high level of child abuse and a major aim of this case-comparison study was to discover whether the SID rates were higher for naval families as compared to civilian.

The cases were forty-nine SIDs that had occurred in the period 1977–81 (two deaths could not be used owing to lack of records) and the 'comparisons' were selected by choosing the infant entered immediately before and after the SID entry in the health visitor's casebook. Data were abstracted from the birth and death certificates, coroner's report, obstetric records and health visitor's records. Analysis revealed that the cases were more likely to be male, of lower birthweight, and a twin, while the mother tended to be younger and was planning to bottle-feed. No differences were found between the service and civilian families. However, there are problems with the design of this study. As no distinction was made between the ranks of the services, lower SID rates for officers may be 'suppressing' (Chapter 3) the overall rate for the services. Moreover, by selecting the comparisons by their position in the casebook, the design has explicitly matched for geographical area and implicitly for service or civilian families. Thus, if a case is a service family and lives in Rowner then the comparison is also likely to live in Rowner and be a service family. The finding of no difference could therefore merely reflect the success of matching, and the main aim of the research may have been thwarted by the study design.

Cohort designs

In a concurrent cohort study, information about causal factors is collected at the outset of a study and the selected subjects are followed into the future to measure disease incidence. The subjects' disease status is either measured directly by periodic re-examination and re-interview, or indirectly by such surveillance methods as hospital records or death certificates. The cohort can either be fixed, in that no new entries to the group are allowed after the onset of the study, or dynamic. The concurrent approach is a prospective design but it is possible to study a cohort retrospectively in a non-current design by tracing a group back into the past. This can only be done when good records are available for a well-defined cohort.

A classic cohort study is Doll and Hill's (1954) prospective study, which examined the association between tobacco consumption and lung cancer. Having already completed cross-sectional and case-comparison studies, they sent a very brief, self-administered questionnaire in 1951 to all UK medical practitioners aged over 35 concerning their smoking habits. Of the potential 60,000 subjects, 68 per cent returned the questionnaire. The researchers then waited for the doctors to die and all death certificates with this occupation were automatically sent to Doll and Hill; similarly the official medical register provided details of any doctor who had been removed from the list because of death. In the following twenty-nine months, 36 male doctors died and with even this small number the results were consistent with an increased risk of dying from lung cancer for smokers. In the following ten years, many more deaths occurred and they found a linear dose-response rate, a fourfold increase in cigarettes smoked resulting in a sixfold increase in lung cancer deaths (Doll and Hill, 1954). One must always question whether the cohort is representative of the general population, and by selecting only doctors an important causal variable, occupation, may have been inappropriately omitted from the design. Consequently, this study does not generalise to other occupations and it does not allow the assessment of working in a carcinogenic environment as a determinant of lung cancer. We return to this issue in Chapter 3.

Some of the most elaborate and comprehensive British cohort studies are the three national longitudinal studies based on births during one week in 1946, 1958 and 1970. In the earliest study, 5,362 children were selected and re-contact was made at intervals of two years or less until the age of 26, and subsequently at five-year intervals. Twenty-five years after the start of the study, re-contact was made with 4,041 of them (Douglas *et al.*, 1977) and indeed the children of this cohort are now being studied as the Second Generation Study. Initially, the study concentrated on social and economic aspects of pregnancy but a wide range of epidemiological issues have now been considered, for example the relationship between air pollution and respiratory illness in children and early adulthood (Colley *et al.*, 1973). While this earliest cohort was fixed, the other two are dynamic, with immigrants to Britain born during the appropriate week being identified and included. The second cohort was based on the Perinatal Mortality Study, a cross-sectional study of 17,000 births in the week beginning 3 March 1958. Re-contact was made at 7, 11, 16, and 23 (Fogelman, 1983). Various sub-groups of the children have been reported on; for example, at the age of 11, children who were defined as 'deprived' on the basis of their environment and family circumstances had been absent from school for over three months during the previous year, five times more often than non-disadvantaged children (Wedge and Prosser, 1973). The third cohort also began as a study of pregnancy, childbirth and the first week of life, and examined over 16,000 births. The whole cohort was re-contacted at 5 years of age and various sub-samples had been contacted at 22 months and 3½ years.

Interestingly, the idea of a fourth birth cohort was discussed for 1982 but as Cartwright (1983, p. 38) notes, there was some disappointment that the 1970 study had made little impact on birth practice and had yielded few aetiological insights, with the major exception of the elucidation of the effects of maternal smoking. Despite this, the previous three cohorts remain a rich store of information, particularly as the groups move into middle and old age and disease prevalence rises.

A comparison of three basic designs

No one type of study design is uniformly superior to the others in every situation, and any appropriate choice depends on several factors including the rarity of the disease, its latency and duration and how much is already known of the aetiology of the disease. Each of the designs has drawbacks and advantages and these are listed in Table 2.18. Before making some recommendations, we will consider four issues in a little more detail: temporal relations, recall bias, rarity and cost.

Table 2.18 Advantages (A) and disadvantages (D) of three basic study designs

	Cross-sectional	Case-comparison	Cohort
Many hypotheses to test		A	D
Rare diseases	D	A	D
Selective recall of important events		D	A
Attrition (death, migration, loss of participation)	A	A	D
Non-response	D	A	D
Time needed to complete study	A	A	D
Cost		A	D
Inference to population	A	D	A
Temporal relation of aetiologic factor and disease onset	D	D	A
Establishing directness of association	D	D	A
Selection of controls leading to bias		D	A

Source: Based on Roht *et al.* (1982, p. 384).

An important element in the choice of a design is the temporal relationship between the possible causal factors and the disease. When the causal factor and disease are measured simultaneously, it is impossible to discover whether the possible cause actually preceded the occurrence of the disease. For example, if it is found in a cross-sectional study that people who have suffered a stroke have a high cholesterol level, the researcher cannot ascertain whether this high value preceded or followed the stroke. Thus, a forward design is best for testing hypotheses, non-directional least.

Cross-sectional and case-comparison studies involving retrospective designs may involve memory distortion or recall bias; the presence of disease may heighten or lower the recall of certain events. For example, the patient with cirrhosis of the liver may underestimate the amount of alcohol consumed over the past thirty years, while the mother whose child has leukaemia may overestimate the number of x-rays taken during pregnancy. This problem should be worse for subjective topics and least for factual events, but Lilienfeld and Lilienfeld (1980, p. 214) report that a third of males on interview could not correctly state whether or not they were circumcised in comparison to a physical examination! In a concurrent cohort study, the information collected cannot be biased by the outcome of diseases for it has not yet happened but, conversely, data collection may inadvertently influence behaviour.

With rare diseases, cross-sectional and cohort designs are difficult to perform and a very large sample followed over a long period may be needed to uncover just a few cases. For example, the Royal College of General Practitioners have estimated that it would require some 125,000 person years of use to identify significantly different rates of pulmonary embolism in oral contraceptive users and non-users. Similarly, it has been estimated that 100,000 individuals would be needed in a cohort study with a prevalence of 1 per 1,000 for the unexposed group and 2 per 1,000 for the exposed (Lilienfeld and Lilienfeld, 1980, p. 248). For very rare conditions, the only feasible type of study is the case-comparison. This issue of data efficiency is closely related to the cost and speed of the study: case-comparisons are usually the quickest and cheapest to do while concurrent cohort studies are the most expensive and are subject to long time delays.

In the light of these comments and the information of Table 2.18, it is possible to make some general recommendations. The cross-sectional approach is best suited to discovering the extent of a problem and exploratory work when little is known about the aetiology of a disease. Such studies form a valuable and essential supplement to routine data, and they need not necessarily be time consuming or expensive. Case-comparison studies are useful for the rapid testing of hypotheses, particularly for rare diseases, but care must be taken of recall and selection bias. They permit the testing of several hypotheses at once and provide preliminary support for tentative explanations. The cohort study is not good at generating ideas and is unsuitable for rare diseases. But providing there is no participation or attrition bias, this design allows rigorous testing of causal hypotheses derived from other studies. It is most useful when specific hypotheses are well developed and supported by previous studies and final, additional evidence is required. Clearly, all three designs must be part of the repertoire of the medical geographer involved in epidemiology and disease ecology.

Record linkage

So far in this chapter we have described methods of data collection that are in widespread use. In this final section, we describe briefly an approach that is likely to be of much greater importance in the future, that of record linkage. In many ways this approach is an amalgam of routine data collection and cohort survey methods.

Table 2.19   Routine data collection throughout life

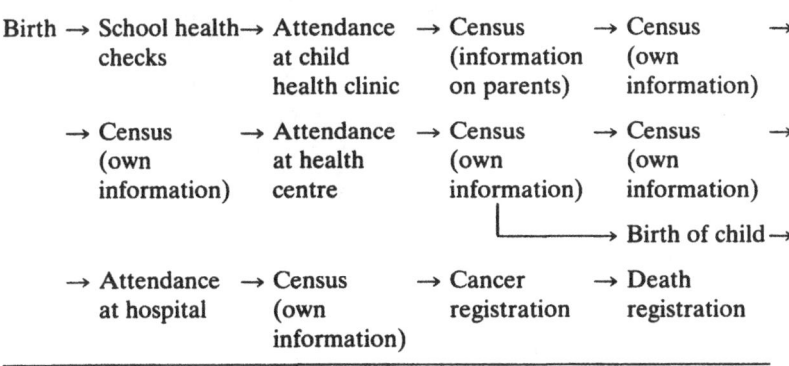

Table 2.20   Perinatal and infant mortality rates (per 1,000) by mother's age, 1982, England and Wales

	Perinatal	*Infant*
under 20	13.7	16.1
20–24	11.3	11.7
25–29	9.7	9.3
30–34	10.8	9.0
over 35	15.6	10.3
All ages	11.2	10.6

Source: OPCS Monitor DH 3 84(7).

Throughout life a large amount of data is collected on every person; from birth to death (Table 2.19) routine information is gathered but in an *ad hoc* fashion, and information in one system is rarely passed on to another. Record linkage aims to bring these disparate data together so that individuals can be followed to produce a cumulative life history without the need for any special survey. Much of the necessary records are already collected and current information technology is capable of dealing with such

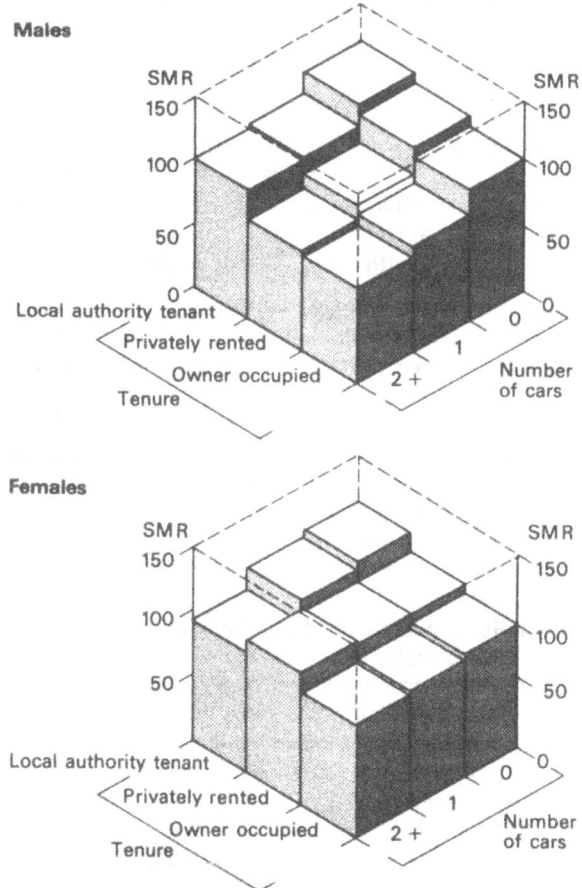

Figure 2.21 Standardised mortality ratios in 1971–5 of private households by tenure and access to cars
Source: Fox and Goldblatt (1982)

vast amounts of data. Researchers could formulate a research hypothesis and test it directly on the linked data, providing confidentiality restrictions are met. While no such all-embracing system is currently in operation, there are a number of examples of record linkage. In 1962 (Acheson, 1967) a pilot scheme was begun in Oxford to link data about birth, each hospital discharge and each death, and since 1975 OPCS has routinely linked infant death records for England and Wales with the corresponding birth record to obtain information on, for example, age of mother, social class, twin or singleton. Table 2.20 shows the type of information that can be obtained from the system: the highest death-rates are found for 'young' and 'old' mothers while the lowest rates are mothers in their late twenties. But the most extensive system of record linkage currently in operation in Britain began with the 1971 census: the OPCS Longitudinal Study (Brown and Fox, 1984). A 1 per cent sample of all people living in England and Wales was selected from the 1971 census by nominating four birthdays; immigrants with these birthdays have also been included as a dynamic cohort. A number of vital events have been linked with the census so that of the 530,000 initial subjects, 60,000 have died during 1971 to 1981 and 9,000 registered as cancer sufferers from 1971 to 1975. The records are currently being linked with 1981 census data and therefore will be data on changing housing and occupational characteristics. Although researchers do not have access to any identifiable individual records, a considerable amount of analysis has been undertaken, for example the relationships between socio-economic circumstances and mortality in different parts of the country, bereavement and cancer, and unemployment and health. Over the 1971–5 period it was found that if you have two cars and live in an owner-occupied house you have almost half the risk of dying before you retire than someone who does not have a car and lives in a council house (Figure 2.21). Noticeably, this longitudinal study still contains very little information about morbidity apart from cancer registration and the census identification of the permanently sick, but this may become more available as general practitioners convert their records to a computerised form.

Conclusions

Basic data collection of an unbiased and representative kind is said to be vital to any epidemiological study. Medical geographers have been generally too content in their reliance on routine data and we must become aware of the opportunities offered by survey methods. Moreover, we must also recognise that all types of observational studies have little control over the causal variable in comparison to experimental designs. Consequently, the researcher has to pay particular attention at the analysis stage to the potential sources of error that could invalidate the findings; it is to this analysis that we turn in Chapter 3. Finally, it must be noted that some commentators require that social researchers divorce themselves from the 'scientific' methods of epidemiology discussed in the present chapter. Arguing that epidemiology is too closely tied to the biomedical model, they contend that data collection should be more 'intensive' and ethnographic in character, and that the biological definition of disease should not just be accepted; we will explore these ideas in the final chapter as part of a 'critical' approach to epidemiology.

Guided reading

A number of epidemiological textbooks cover the material in this chapter. For example, Alderson (1983) provides detailed discussion which is appropriate for a British audience, while Lilienfeld and Lilienfeld (1980) and Mausner and Kramer (1985) have a more American flavour. For British mortality data the essential reference is Alderson (1974), while the same author (1981) has provided a valuable compendium of international mortality statistics. A critical commentary on British routine statistics is provided by Doyal (1979) and an 'alternative' guide has been produced by the Radical Statistics Health Group (1980).

General guides to survey analysis are Abramson (1979), Cartwright (1983), Moser and Kalton (1971), and McKinlay (1975); the first is a cookbook guide to actually performing such studies. Two essential overviews of case-comparison studies are Breslow and Day (1980) and Schlesselman (1982), while Goldstein (1979) provides an extended discussion of cohort studies. A well-indexed and copiously referenced guide to the results and findings of survey designs in the British context is Alderson and Dowie (1979); this can usefully be complemented by Blaxter's (1981) book which, despite the title, considers epidemiological findings for both children and adults.

Kleinbaum *et al.* (1982) is a clear (if at times highly technical) consideration of the relative merits of different designs. A large number of authors have argued over the competing claims of these different approaches; the spirit of these arguments may be appreciated from reading the 1979 editorial in the *British Medical*

Journal and associated letters (Mann *et al.*, 1979), and a whole issue of *Journal of Chronic Disease* (Ibrahim, 1979).

A review of record linkage is given by Brown and Fox (1984).

For those wishing a self-teaching guide to the material of this chapter, Roht *et al.* (1982) can be recommended.

Useful journals include: *American Journal of Epidemiology, Archives of Environmental Health, Community Medicine, International Journal of Epidemiology, Journal of Chronic Disease, Journal of Epidemiology and Community Health, Preventive Medicine, Scandinavian Journal of Social Medicine, Social Science and Medicine.*

References

Abramson, J. H. (1979), *Survey Methods in Community Medicine*, Edinburgh, Churchill Livingstone.

Acheson, E. D. (1967), *Medical Record Linkage*, London, Oxford University Press.

Aird, I., Bentall, H. H. and Fraser Roberts, J. A. (1953), 'A causal relationship between cancer of the stomach and the ABO blood groups', *British Medical Journal*, vol. 1, pp. 799–801.

Alderson, M. R. (1974), 'Central government routine health statistics', in W. F. Maunder (ed.), *Review of the UK Statistical Sources*, vol. 2, pp. 1–145, London, Heinemann.

Alderson, M. R. (1981), *International Mortality Statistics*, London, Macmillan.

Alderson, M. R. (1983), *An Introduction to Epidemiology*, 2nd edn, London, Macmillan.

Alderson, M. R. and Dowie, R. (1979), *Health Surveys and Related Studies*, Oxford, Pergamon.

Allen-Price, E. D. (1960), 'Uneven cancer distribution in West Devon', *Lancet*, vol. 1, pp. 1235–8.

Armitage, P. (1977), *National Health Survey Systems in the EEC*, Luxembourgh, Commission of the European Communities.

Ashley, J. S. A. (1979), 'The ninth revision of the International Classification of Disease', *Community Medicine*, vol. 1, pp. 106–14.

Benjamin, B. (1968), *Health and Vital Statistics*, London, Allen & Unwin.

Birch, J. M., Marsden, H. B. and Swindell, R. (1980), 'Incidence of malignant disease in childhood: a 25 year review of the Manchester Children's Tumour Registry', *British Journal of Cancer*, vol. 42, pp. 215–23.

Black, D. (1984), *Investigation of the Possible Increased Incidence of Cancer in West Cumbria*, London, HMSO.

Blaxter, M. (1981), *The Health of Children*, London, Heinemann.

Breslow, N. E. and Day, N. E. (1980), *The Analysis of Case-Control Studies*, Lyons, International Agency for Research on Cancer.

Brown, A. and Fox, J. (1984), 'OPCS Longitudinal Study: ten years on', *Population Trends*, vol. 37, pp. 20–2.

Bunge, W. and Bordessa, R. (1975), *The Canadian Alternative*, Downsview, Ontario, York University Geographical Monographs.

Cartwright, A. (1983), *Health Surveys in Practice and in Potential: A Critical Review of their Scope and Methods*, London, King Edward's Hospital Fund.

Charlton, J. R. H., Hartley, R. M., Silver, R. and Holland, W. W. (1983), 'Geographical variations in mortality from conditions amenable to medical intervention in England and Wales', *Lancet*, vol. 1, pp. 691–6.

Colley, J. R. T., Douglas, J. W. B. and Reid, D. D. (1973), 'Respiratory disease in young adults', *British Medical Journal*, vol. 3, pp. 195–8.

Cook, P. J. (1977), 'Geographical clues to the causes of cancer', paper given at the British Association, Aston Meeting.

Craft, A., Openshaw, S. and Birch, J. M. (1984), 'Apparent clusters of childhood lymphoid malignancy in northern England', *Lancet*, vol. 2, pp. 96–7.

Craft, A. and Openshaw, S. (1985), 'A geographical analysis of cancer among children in the Northern Region', paper given at IBG Meeting, Leeds.

Davies, J. M. and Chilvers, C. (1980), 'The study of mortality variations in small administrative areas of England and Wales, with special reference to cancer', *Journal of Epidemiology and Community Health*, vol. 34, pp. 87–92.

Dawber, T. R., Kannel, W. P. and Lyell, L. P. (1963), 'An approach to longitudinal studies in the community', *Annals of the New York Academy of Science*, vol. 107, pp. 539–56.

Department of Health and Social Security (1980), *Inequalities in Health*, Report of a Research Working Group chaired by Sir Douglas Black, London, DHSS.

Doll, R. and Hill, A. B. (1954), 'The mortality of doctors in relation to their smoking habits', *British Medical Journal*, vol. 1, pp. 1451–5.

Doll, R. and Peto, R. (1976), 'Mortality in relation to smoking: 20 years of observations on male doctors', *British Medical Journal*, vol. 2, pp. 1525–36.

Douglas, J. W. B., Kiernan, K. E. and Wadsworth, M. E. J. (1977), 'A longitudinal study of health and behaviour', *Proceedings of the Royal Society of Medicine*, vol. 70, pp. 530–2.

Doyal, L. (1979), 'A matter of life and death: medicine, health and statistics', in J. Irvine, I. Miles and J. Evans (eds), *Demystifying Social Statistics*, pp. 237–56, London, Pluto Press.

Editorial (1979), 'The case-control study', *British Medical Journal*, vol. 2, pp. 884–6.

Fogelman, K. (1983), *Growing up in Great Britain: Papers from the Child Development Study*, London, Macmillan.

Fox, J. and Goldblatt, P. (1982), 'Socio-demographic differences in mortality', *Population Trends*, vol. 27, pp. 9–13.

Gardner, M. J. and Winter, P. D. (1984), 'Mapping small area cancer mortality: a residential coding story', *Journal of Epidemiology and Community Health*, vol. 38, pp. 81–4.

Gardner, M. J., Winter, P. D. and Barker, D. J. P. (1985), *Atlas of Mortality from Selected Diseases in England and Wales, 1968–1978*, Chichester, Wiley.

Gardner, M. J., Winter, P. D., Taylor, C. P. and Acheson, E. D. (1983), *Atlas of Cancer Mortality in England and Wales, 1968–1978*, Chichester, Wiley.

Gill, O. N., Sockett, P. N., Bartlett, C. R., Vaille, M. S. B., Rowe, B., Gilbert, R. J., Dulake, C., Murrell, H. C. and Salmaso, H. (1983), 'Outbreak of Salmonella Napoli infection caused by contaminated chocolate bars', *Lancet*, vol. 1, pp. 574–7.

Gilliam, A. G. (1955), 'Trends of mortality attributed to carcinoma of the lung', *Cancer*, vol. 8, pp. 1130–6.

Goldstein, H. (1979), *The Design and Analysis of Longitudinal Studies: Their Role in the Measurement of Change*, London, Academic Press.

Howe, G. M. (1970), *National Atlas of Disease Mortality in the United Kingdom*, London, Nelson.

Howe, G. M. (1979), 'Mortality from selected malignant neoplasms in the British Isles: the spatial perspective', *Geographical Journal*, vol. 145, pp. 401–6.

Ibrahim, M. A. (ed.) (1979), 'The case-control study: consensus and controversy', *Journal of Chronic Disease*, vol. 32, pp. 1–90.

Johnston, R. J. (1984), 'The world is our oyster', *Transactions of the Institute of British Geographers*, vol. 9, pp. 443–59.

Jones, K. and Kirby, A. (1980), 'The use of a chi-square map in the analysis of census data', *Geoforum*, vol. 11, pp. 409–17.

Kagan, A., Harris, B. R. and Winkelstein, W. (1974), 'Epidemiologic studies of coronary heart disease and stroke in Japan, Hawaii, and California', *Journal of Chronic Disease*, vol. 127, pp. 345–64.

Kleinbaum, D. G., Kupper, L. L. and Morgenstern, H. (1982), *Epidemiological Research: Principles and Quantitative Methods*, California, Wadsworth.

Knox, E. G. and Mallett, R. (1979), 'Standardised perinatal mortality ratios: technique, utility and interpretation', *Community Medicine*, vol. 1, pp. 6–13.

Lambert, P. M. and Reid, D. D. (1970), 'Smoking, air pollution and bronchitis in Britain', *Lancet*, vol. 1, pp. 853–7.

Lawrence, P. S. (1977), 'Methods in the US National Health Interview Survey', in P. Armitage (ed.), *National Health Survey Systems in the EEC*, pp. 80–8, Luxembourg, Commission of the European Communities.

Lilienfeld, A. M. and Lilienfeld, D. E. (1980), *Foundations of Epidemiology*, 2nd edn, New York, Oxford University Press.

Lloyd, O. L. and Barclay, R. (1979), 'A short latent period for respiratory cancer in a susceptible population', *Community Medicine*, vol. 1, pp. 210–21.

McDowall, M. (1983), 'Measuring women's occupational mortality',

Population Trends, vol. 34, pp. 25–9.

McDowell, A. J. (1977), 'Health examination surveys in theory and application', in P. Armitage (ed.), *National Health Survey Systems in the EEC*, pp. 106–18, Luxembourg, Commission of the European Communities.

McGlashan, N. D. (1972), 'Food contaminants and oseophagal cancer', in N. D. McGlashan (ed.), *Medical Geography*, London, Methuen.

McKinlay, S. M. (1975), 'The design and analysis of the observational study – a review', *Journal of the American Statistical Association*, vol. 70, pp. 503–23.

MacMahon, B. and Pugh, T. (1970), *Epidemiologic Methods*, Boston, Little, Brown.

Magnus, K. (1982), *Trends in Cancer Incidence*, Washington DC, Hemisphere.

Mann, J. I., Vessey, M. P., Jones, R. and Young, D. (1979), 'The case-control study and retrospective controls', *British Medical Journal*, vol. 2, pp. 1507–8.

Mason, A. and Morrison, V. (1985), *Walk, Don't Run: A Collection of Essays on Information Issues*, London, King Edward's Hospital Fund.

Mason, T. J., McKay, F. W., Hoover, R., Blot, W. J. and Fraumen, J. F. (1975), *Atlas of Cancer Mortality for the US Counties, 1950–1969*, Washington DC, Public Health Service.

Mausner, J. S. and Kramer, S. (1985), *Epidemiology: An Introductory Text*, Philadelphia, Saunders.

Mitchell, J. (1984), *What is to be Done About Illness and Health*, Harmondsworth, Penguin.

Morris, J. N. (1975), *Uses of Epidemiology*, Edinburgh, Churchill Livingstone.

Morris, J. N., Kagan, A., Pattison, D. C., Gardner, M. J. and Raffle, P. A. B. (1966), 'Incidence and prediction of ischaemic heart disease in London busmen', *Lancet*, vol. 2, pp. 553–9.

Moser, C. A. and Kalton, G. (1971), *Survey Methods in Social Investigation*, London, Heinemann.

Murray, M. A. (1962), 'The geography of death in England and Wales', *Annals of the Association of American Geographers*, vol. 52, pp. 130–49.

Murray, M. A. (1966), 'The geography of death in the United States and the United Kingdom', *Annals of the Association of American Geographers*, vol. 57, pp. 301–14.

Open University (1975), *Vital Statistics Unit 2*, Milton Keynes, Open University Press.

Pomiankowski, A. (1984), 'Cancer incidence at Sellafield', *Nature*, vol. 311, p. 100.

Powell, J., Machin, D. and Kershaw, C. R. (1983), 'Unexpected sudden deaths in Gosport – some comparisons between service and civilian families', *Journal of the Royal Naval Medical Service*, vol. 69, pp. 141–50.

Pyle, G. F. (1979), *Applied Medical Geography*, Washington DC, Holt Rinehart & Winston.

Radical Statistics Health Group (1980), *The Unofficial Guide to Official Health Statistics*, London, BSSRS.

Registrar General (1978), *Decennial Supplement, England and Wales, 1971, Occupational Mortality*, London, HMSO.

Reid, D. D. and Fairbairn, A. S. (1958), 'The natural history of chronic bronchitis', *Lancet*, vol. 1, pp. 1147–52.

Roht, L. H., Selwyn, B. J., Holguin, A. H. and Christensen, B. L. (1982), *Principles of Epidemiology*, New York, Academic Press.

Royal College of General Practitioners (1973), *General Practice: Present State and Future Needs*, London, RCGP.

Schlesselman, J. J. (1982), *Case-control Studies: Design, Conduct and Analysis*, New York, Oxford University Press.

Stuart, A. (1968), *Basic Ideas of Scientific Sampling*, London, Griffin.

Susser, M. (1984), 'Apartheid and the causes of death', *Radical Community Medicine*, Winter, pp. 4–9.

Terris, M. (1967), 'Epidemiology of cirrhosis of the liver', *American Journal of Public Health*, vol. 57, pp. 2076–88.

Thunhurst, C. (1985), *Poverty and Health in the City of Sheffield*, Sheffield City Council.

Townsend, P. and Davidson, N. (1982), *Inequalities in Health: The Black Report*, Harmondsworth, Penguin.

Van Rensburg, H. C. J. and Mans, A. (1982), *Profile of Disease and Health Care in South Africa*, Pretoria, Academica.

Waller, R. E. (1967), 'Bronchitis, and air pollution' in R. W. Raven and F. J. C. Coe (eds), *The Prevention of Cancer*, London, Butterworths.

Wedge, P. and Prosser, H. (1973), *Born to Fail*, London, Arrow Books.

White, R. R. (1972), 'Probability maps of leukaemia mortalities in England and Wales', in N. McGlashan (ed.), *Medical Geography: Techniques and Field Studies*, pp. 173–85, London, Methuen.

Whitelegg, J. (1982), *Inequalities in Health Care: Problems of Access and Provision*, Retford, Notts., Straw Barnes.

Worth, R. M. (1975), 'Epidemiologic studies of coronary heart disease and stroke in Japanese men living in Japan, Hawaii, and California', *American Journal of Epidemiology*, vol. 102, pp. 481–91.

Young, J. C. (1974), *Suspected Food Poisoning in Consett, County Durham*, Durham, Dept. of Geography, University of Durham.

Young, J. C. (1982), 'The dust menace over Consett', *Geographical Magazine*, vol. 54, pp. 402–6.

CHAPTER 3

The causal analysis of epidemiological data

A fundamental research task of the disease ecologist is to investigate the aetiology or causes of disease by searching for associations between some aspects of the physical and social environment and the disease under study. Having found an association, the relationship has to be elaborated to discover why it is happening and under what conditions it occurs. Is the relationship true and real or somehow artificial and false? What is the nature of the relationship, is it positive or negative, weak or strong? Is it an entirely fortuitous result brought about by chance or does it have any substantive meaning? How consistent is the relationship and does it generalise to other situations? In short, does the 'explanatory' variable cause the dependent variable? Typically, a researcher enquires whether air pollution causes bronchitis or whether water hardness causes heart disease. Whereas Chapter 2 examined data collection, attention now focuses on the techniques for the analysis and interpretation of such data. In particular, this chapter considers the nature of cause-and-effect relationships and the relative merits of experiments and statistical analyses in discerning between true causal relationships and spurious ones. While statistical control is inherently weaker than experimental control, it is possible to strengthen the validity of this approach by paying particular attention to a number of criteria such as the strength, consistency and plausibility of the observed statistical association. The application of these criteria to the study of the relationship between water hardness and disease forms the final part of this chapter.

Cause and effect

While explanation is often considered to be synonymous with 'establishing the causes of events', there remains a fierce controversy over whether or not the real world actually behaves in such a manner (Bunge, 1979). Despite these arguments, many researchers find it convenient to view the world as if it operates on a cause-and-effect basis. While epidemiologists and medical geographers do not often use the word 'cause', they frequently imply causal explanations by using such terms as 'results in', 'accounts for', 'influenced by' and even 'explained by'. Continuing in this pragmatic vein, we can define causation as a relationship between two variables such that a change in the level of the explanatory or independent variable (X) produces a change in the level of the dependent variable (Y). If an increase in X produces an increase in Y, there is a positive causal relationship; if an increase in X produces a decrease in Y the relationship is a negative causal one. The concept of forcing or producing is inherent in this notion of cause and effect; night is followed by day but these events are merely associated and one is not the cause of another.

Unlike the biomedical approach, disease ecologists hold a particular causal view of the world that can be best described as probabilistic and multicausal. According to this probabilistic version of causality, no effect is completely determined by a cause. Thus, while not all smokers will die of lung cancer, the smoking of cigarettes can lead to an increased chance of dying from this disease. This is clearly related to the concept of multicausality: each effect has a number of causes and each cause can produce a number of effects. For example, consider the causal model in Figure 3.1 in which a single-headed arrow (→) signifies a possible causal relationship from cause to effect. Here there are multiple causes and multiple effects; and effects in one causal model can be causes in another.

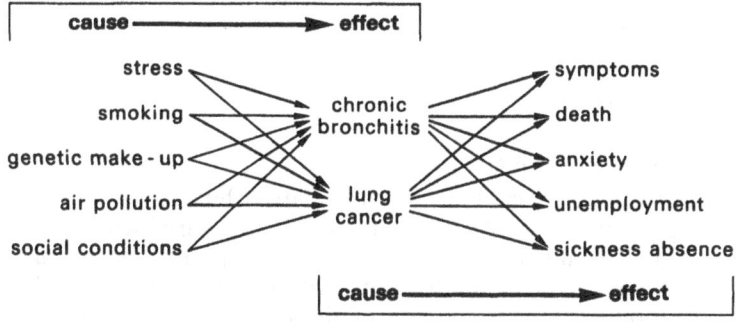

Figure 3.1 Multiple causality

How, then, do we know if an observed association is causal or not? Following J. S. Mill we can recognise three essential conditions for a causal relationship; if they are all met, then the relationship is deemed causal, but if any one of them is broken, then the relationship is not causal. The first condition is that of co-variation; that is X and Y are associated over space and time. Thus, if soft water is a cause of heart disease, areas of the country with soft water should have a high mortality from this disease. Similarly, if air pollution is a cause of bronchitis, an increase in bronchitis should be associated with an increase in air pollution. The second condition is that of temporal precedence; if X is causally prior to Y, changes in X produce a subsequent change in Y in the course of time. Unfortunately, the identification of temporal precedence is often plagued by time-lags and possible reciprocal causation (X causing Y, and Y causing X). For example, a person may be caused to be depressed by losing a job (X → Y) or the loss of a job may be caused by the inability to work effectively due to depression (Y → X), and this, of course, is why a long-term prospective study is so useful (Chapter 2). The third and final condition is that other possible explanations of the cause and effect have been eliminated. The researcher seeks other 'lurking' variables that are causally prior to both X and Y which, when controlled or removed, result in the disappearance of the original spurious association. No investigator can know when all possible important variables have been controlled and identified and, therefore, it is never possible to prove any relationship to be causal. However, in practice, researchers can be increasingly confident of their provisional causal interpretations as they fail to find them spurious after controlling several possible variables. Causal explanations are, thus, established by the elimination of plausible hypotheses by seeking to falsify them.

Causal models

Extraneous variables

Examining the concept of spurious relationships in more detail, let us begin with a simple model in which two variables, X and Y, have been found to occur in association. It may be postulated that X → Y, or overcrowding causes bronchitis. Such an association can occur for one of three reasons: there may be a true causal relationship, the association may occur simply by chance, or X may co-vary with Y because of some third variable

(Z). Taking air pollution as the variable Z, it is possible to recognise two fundamentally different types of relationships between the variables. Firstly, there is the possibility that:

overcrowding

air pollution ⟶ bronchitis

In this case, living at high densities and suffering air pollution are both independent causes of bronchitis. If such a relationship exists, the lowering of overcrowding and air pollution levels would result in lower death-rates, the lowest rates occurring when both crowding and air pollution are reduced. The alternative model is:

air pollution ⟶ bronchitis
overcrowding

in which only air pollution is truly a cause of bronchitis. The original association between overcrowding and bronchitis is, therefore, a spurious one produced by the association of air pollution and overcrowding. Sub-areas of a city, for example, may have both high air pollution and a great deal of overcrowding, but it may be only the former variable that is causally linked to chest disease. If overcrowding is reduced but air pollution remains at previous levels, this model predicts that bronchitis deaths would not decrease.

It is essential, therefore, that if there is deliberate intervention to lower death-rates, decisions are not made on the basis of such false, spurious relationships. To demonstrate the value of searching for spurious relationships, let us examine two research examples: first, cigarette smoking and disease and, second, the presumed increasing rate of mental illness. Reading the government warning that is printed on cigarette packets may lead one to believe that the simple causal relationship:

cigarette consumption ⟶ cancer

is undisputed. While the majority of researchers do believe in the essential validity of this model, others continue to argue that alternative models are more appropriate. While Doll and Peto (1981) have estimated that smoking accounts for some 30 per cent of all cancers, occupational cancers cause less than 5 per cent of all cancers and pollution only 2 per cent, Epstein (1979) argues that occupational carcinogens produce anything between 20 and 40 per cent of all cancers and the contribution of smoking has been overestimated. His fundamental model is that there is a synergistic interaction between smoking and occupation which can be conceptualised as follows:

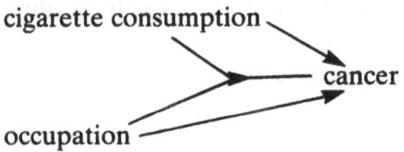

Both smoking and occupation are independent causes of cancer, but they also interact: the people most likely to develop cancer work in a carcinogenic environment and are heavy smokers. According to this view, studies that have considered smoking without controlling for occupation are bound to overestimate the smoking effect, for heavy smokers tend to be manual workers who work in high-risk occupations. Other researchers have contended that the relationship between smoking and cancer may have been misunderstood because the role of genetics and the behavioural environment have been ignored. Research has shown that smoking is related to personality factors (Eysenck, 1980) and so it is possible to suggest a whole series of alternative models. For example, in the following model:

genetic and behavioural factors are related to cancer through the intervening variables of extroversion and smoking; lower smoking would lower cancer. In contrast, according to the following version:

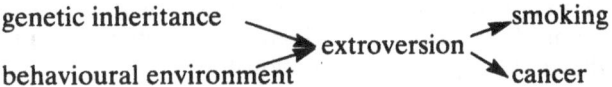

genetic and behavioural background variables are related to smoking and cancer through the intervening variable, extroversion. In this model, the link between smoking and cancer is spurious; lowering smoking levels would not alter the risks of getting cancer.

In order to evaluate such models, data on twins are required. Monosygous twins are derived from a single fertilised ovum and therefore should be genetically identical. If one could find such identical twins, one of whom smoked and the other who did not, it should be possible to evaluate the alternative models. Table 3.1

presents data on deaths among identical twins identified by the Swedish National twin registry; there does not appear to be a consistent pattern of excess deaths among moderate/heavy smokers in different age-groups. This evidence supports the argument that genetic factors may be more important than generally believed. The study, however, suffers from the limitation that the deaths are not cause-specific and a true relationship between smoking and cancer may be hidden within this gross picture based on a relatively small number of deaths. After thirty years of extensive research in this area, there are still fundamental differences of opinion about the basic form of a valid model.

Our second example considers the argument that the increasing pace, scale, mobility and anonymity of modern life have lead to increasing mental illness in developed societies. Goldhammer and Marshall (1949) examined the USA rates of hospitalisation for psychosis between 1845 and 1945 and they found a considerable increase. But they also showed that this was in many ways a false interpretation of the evidence, for an important extraneous variable, age, had been ignored. While, undoubtedly, the hospitalisation rates had increased, the bulk of this increase came from patients suffering senile dementia. There were many more such patients in 1945 than in 1845 simply because people were living longer and old people formed a greater proportion of the population. The increasing rate of illness was largely a reflection of changing age-distribution; the original association was revealed to be spurious when age was controlled.

Suppressor and distorter variables

Our discussion has considered only one type of spurious relationship, that is when an observed association between two

Table 3.1 Deaths among identical twins by smoking, and age

Age	No. of pairs	Male deaths		No. of pairs	Female deaths	
		Non-/ light smokers	*Moderate/ heavy smokers*		*Non-/ light smokers*	*Moderate/ heavy smokers*
45–59	177	6	11	262	10	9
60–69	69	12	7	64	3	5

Source: Friberg (1973).

variables is considerably reduced when a third variable is controlled. In the terminology of Rosenberg (1968) this third variable is called 'extraneous' and he recognises two other types of variable: 'suppressor' and 'distorter'. In the case of a suppressor variable, it is the absence of correlation between two variables that is spurious; the true strong relationship has been reduced or cancelled because the suppressor variable has not been controlled. Suppression usually occurs when only a relatively small part of the independent variable has an effect on the dependent variable. In any causal analysis, the research must, therefore, attempt to 'refine' the causal variable and to determine which aspects of the independent variable are truly causal. For example, the association between smoking and lung cancer deaths was considerably increased when, instead of just comparing those who did and did not smoke, the smokers were refined into different groups. Thus, cigarette smokers had a much greater risk of developing lung cancer than pipe and cigar smokers (Susser, 1973).

If a distorter variable is not controlled, the correct interpretation of the relationship is exactly the reverse of that suggested by the original bivariate relationship; observed positives are really negatives and vice versa. Durkheim (1951), in his analysis of suicide, had consistently argued that a major determinant of the decision to take one's own life was not stressful life events but the degree to which a person was integrated into society. An important test of these rival theories relies on the difference in suicide rates between married and unmarried people. If the married individuals with presumed greater life stress (!) had the higher rate, the evidence would be against Durkheim, but if unmarried people who presumably are less integrated into society had the highest rate, the evidence would support Durkheim's integration theory. Previous researchers had actually found that suicide rate was higher for married people, but Durkheim argued that this relationship had been distorted by age, for young people have both low marriage and suicide rates. He studied suicides in France for 1889 to 1891 and found that, for each five-year age group between 20 and 80 years old, married women and men had the lower rates, thereby supporting his theory (Table 3.2).

Any relationship between two variables (be it positive or negative, weak, strong or non-existent) should not be taken at face value. The presence of uncontrolled extraneous, suppressor or distorter variables may lead to a totally spurious interpretation. Techniques are required to control for such 'third' variables and, in essence, there are two distinct approaches: control is

Table 3.2   Suicides per 1,000,000 inhabitants, France, 1889–91

| | Men | | Women | |
Age	Unmarried	Married	Unmarried	Married
15–20	113	500	80	33
21–25	237	97	106	53
26–30	394	122	151	68
31–40	627	226	126	82
41–50	975	340	171	106
51–60	1,434	520	204	151
61–70	1,768	635	189	158
71–80	1,983	704	206	209
Above 81	1,571	770	176	110

Source: Durkheim (1951).

either achieved by design in experiments or by statistical analysis in observational studies.

Control by design

Experimental control

An ideal simple laboratory experiment to assess the influence of air pollution in producing lung cancer may be conducted as follows. One hundred male mice aged one week are placed in identical cages, given the same diet and are kept at the same temperature. Fifty of the mice in the experimental group receive a daily high dose of sulphur dioxide, while the remainder, the control group, breathe pure air. After fifteen weeks of such treatment, autopsies are performed on all the mice; if the mice who have breathed normal air have substantially fewer malignant neoplasms than the mice exposed to sulphur dioxide, there is a prima facie case for believing that this form of air pollution causes cancer. If we change the value of the independent variable with no other variations in experimental conditions and observe concomitant changes in the dependent variable, then these changes in Y must be due to X, that is X is the cause of Y.

The four essential elements of experimental design (manipulation, control, comparison and replication) are all evident in this example. The experimenter has not passively observed the relationship but has deliberately intervened by manipulating air pollution so that it is higher in the experimental group than in the control group. Control has been achieved by physical means (each mouse has been kept at the same temperature), by standardisation (they were all given identical diets) and by matching (the mice in the experimental group were matched with mice of the same age and sex in the control group). Such control allows for direct comparison between the two groups which should only differ in terms of air pollution. Replication can be seen in this experiment in two senses: firstly, it was not just one mouse that was placed in each group but fifty replicates. Secondly, this experiment could be replicated in another laboratory at another time and this should produce broadly comparable results; if this does not occur, there must be a searching investigation into why this is the case. It is because of these four elements that experiments are regarded by many as the only true and convincing way of testing a causal relationship.

So far, however, we have ignored the important assumption, that we know, a priori, all the extraneous factors that have to be controlled by experimental design. The classical solution to this problem was to emphasise the physical control of conditions – sterilisation, isolation, insulation, or sound-proofing, for instance. But in the 1920s, the statistician, Sir R. A. Fisher, proposed a brilliant alternative solution, by designing statistical experiments based on the key concept of randomisation. With randomisation, individual mice are allocated to either the experimental or control group by some random device so that every mouse has an equal probability of being allocated to either group. For example, using the tossing of an unbiased coin and taking each of the one hundred mice in turn, if the coin shows a 'head' it goes into the control group, while a 'tail' indicates allocation to the experimental group. While randomisation does not place a rigid control over any factor it does mean that we can expect that the average level of all variables (known and unknown, measured or unmeasured) will be approximately the same in both groups. For example, both groups will have roughly the same number of small mice, heavy mice, white mice, and so on. Control has been achieved by design so that the only way in which the two groups will differ is in terms of the experimental variable, pollution. Greater numbers in each group increase our confidence that the groups are equivalent before the experimental changes, and that the observed differences after the experiment are not a result of chance. Moreover, techniques of statistical inference allow us to specify exactly the degree of confidence we may have in the validity of the results. The incredible achievement of randomisation is that it controls all factors simultaneously to a known degree even without knowing what they are! Moreover, it is possible to design a statistical experiment so that more than one factor is varied at once, thereby allowing a number of generalisations to be made in a single study (Ehrenberg, 1975).

Clinical trials

The procedure in epidemiology that most closely approximates the laboratory experiment is the clinical trial in which the experimental group is given a new untried therapy, drug or vaccine, while the control group is given a treatment in current use or no treatment at all. Randomisation, in this context, has the particular advantage that no bias is introduced in the selection of patients in terms of the severity of the disease. With human beings, however, the mere fact they know that they are in an experiment may influence their behaviour. A classic example of this problem has been dubbed the 'Hawthorne' effect, after the experimental investigation into productivity at the Hawthorne Electrical Company (Roethlisberger and Dickson, 1939). A selected group of workers successively received changes that involved improved lighting, longer rests and higher incentives, with the result that, at each stage, their productivity increased. But when conditions were returned to the original ones of poor lighting, no rest and no incentives, productivity went up again! The explanation appears to be that productivity was improving not because of the specific changes but because of the extra management attention that was being given to this selected group. Similar problems are known to occur in clinical trials with patients receiving any treatment (even if it is truly ineffective) showing an improvement in their condition. This problem can be overcome by the single-blind experiment in which the patients do not know whether they are in an experimental group receiving treatment or in a control group receiving a placebo. In the double-blind experiment, the attending doctor, as well as the patient, does not know who is receiving the genuine, new treatment.

Validity of experimental research

While many researchers believe that experimental research is the most powerful form of scientific investigation, comparatively little human experimental work has been undertaken to investigate disease causation. Undoubtedly, the major reason for this is that it would be unethical to assign persons at random to do things that may cause suffering. For example, imagine taking a healthy group of children aged 10 and forcing half of them to live in overcrowded conditions, to smoke twenty-five cigarettes a day and to breathe air with high levels of pollution for the next thirty years. Such experiments would not be allowed by the ethical committees that grant permission for experimental research; they ensure that researchers adhere to the rule of *primum non nocere*, or 'first of all, do no harm'.

Moreover, there may be causal relationships of interest that are not amenable to manipulation and random allocation. Rosenberg (1968, Chapter 1) distinguishes between 'stimulus-response' relationships (which characterise the physical sciences and are amenable to manipulation), and 'disposition-response' relationships (which are found in the social sciences and which involve background characteristics), which are, for the individual, unchangeable. For example, sex, age, race and class are fundamentally unalterable however hard the experimenter may try. Importantly, societal, structural factors often cannot be randomised and controlled; for example, can you imagine any feasible experiment in which the causal variable is unemployment rate?

Another difficulty with experimental investigation is that, while the procedures may have high internal validity, they are poor in terms of external validity; their results do not easily generalise from an artificial situation to the real world. This is particularly true when animal experiments are performed. The particular problem here is that a disease may be species-specific. For example, the sweetener, saccharine, has been found to cause bladder cancer in rat trials but rat bladders are unusual. Rats concentrate their urine very highly before passing it and, as saccharine is not metabolised, it remains in the bladder for a comparatively long time. This situation is very unrepresentative of humans and it would be necessary to go outside the single-animal experiment to draw general conclusions.

An instructive example of how the artificial nature of an experiment may mislead investigators is to be found in the work of Pasteur, Koch and their colleagues, on the germ theory (Dubos, 1959). Their experiments apparently proved that it was sufficient to bring the germ or pathogen (be it cholera vibrios or tubercle bacillus) and animal host together to produce disease. When Koch's vibrios were described as the cause of cholera, another German, Pettenkoffer, tried to demonstrate that this was not so. Obtaining a culture, which Koch had taken from a fatal case, Pettenkoffer, and others, swallowed it and did not develop the disease. Why was this? One plausible explanation is that, by trial and error and by being superb experimenters, Pasteur and Koch had chosen the species of animal, the required dose of culture, the route of innoculation which produced, without fail, the disease. But these carefully selected, artificial experimental stations did not permit generalisation to humans nor did they reveal the importance of the social condition and general environment of the human host in determining the onset and course of such infectious diseases. We return to this example in Chapter 4 when we discuss the human host in relation to communicable diseases.

In conclusion, while much of the success of science has been its ability to explore hypotheses by creating artificial experimental situations, it may be improper, impossible or unrealistic to do so in aetiological research. When this is the case, the researcher has to use the inherently weaker form of analysis based on statistical control. There are two major interrelated methods of statistical control: using cross-tabulations to examine sub-groups of the independent and dependent variables, and using multiple regression (and associated methods of partial correlation and causal models) to examine the variability of the dependent variable in relation to the variability of the independent variables.

Control by analysis

Cross-tabulation

Beginning with a two-variable relationship, we may postulate that pollution causes bronchitis. A questionnaire has been administered to 100 respondents and they have been classified or 'cross-tabulated' into bronchitics, non-bronchitics, high and low pollution (above and below an annual average sulphur dioxide levels of 100 $\mu g/m^3$). Following the calculations in Table 3.3(a), 50 out of the 100 respondents live in high-pollution areas and, in the model of 3.3(c), this value is regarded as a constant that represents the factors outside the model that determine the probability (0.5) of living in a high-pollution area. For the two-by-two cross-tabulation of 3.3(b), the probability of being a bronchitic if one lives in high-pollution areas, is 0.4 (that is 20/50), and this compares with a probability of 0.2 if one lives in a low-pollution area. The remaining coefficient of the model is the constant term for bronchitis, which represents the effect of variables not included in the model which determine whether or not a person is bronchitic. This is derived as the proportion of

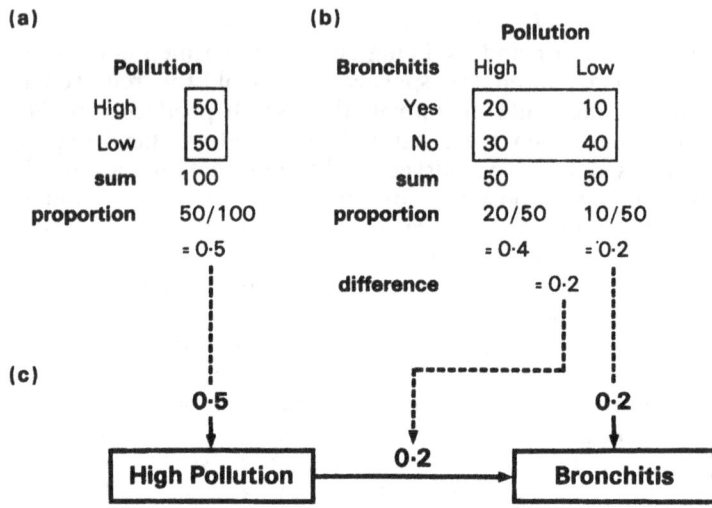

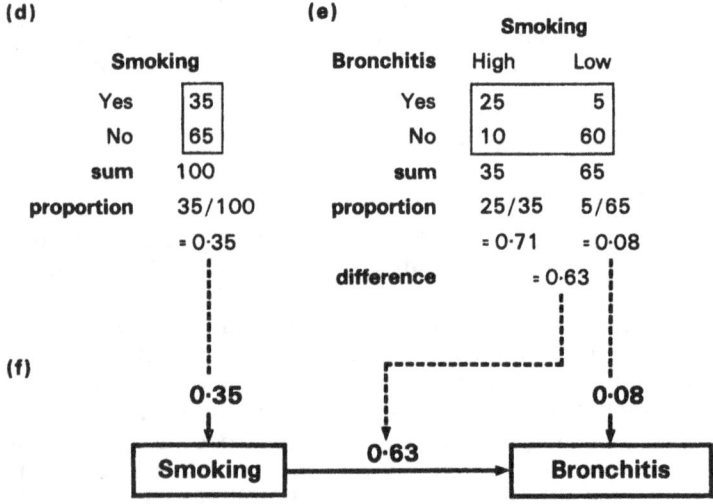

Table 3.3 Two-variable cross-tabulation

bronchitics who live at low population and equals 0.2 (10/50). Thus, the data reveal the probability of being bronchitic is 0.20 in low-pollution areas, but increases by 0.20 if the respondent lives in high pollution; there appears to be some relationship, if not particularly strong, between pollution and bronchitis.

The crucial next step is to test whether the relationship strengthens, reduces, disappears or remains the same when a third variable is introduced. One such additional variable could be cigarette smoking, and Table 3.3(d), (e), (f) reveals that the probability of being a smoker is 0.35, the probability of being bronchitic if you do not smoke is 0.08, but this is increased by 0.63 to 0.71, if you do smoke. This is evidence of a much stronger relationship, but the relationship between all three variables must be examined through the three-variable model and table (Table 3.4). In this model the probability of living in high pollution is 0.5, the probability of being a smoker is 0.35, and the probability of being bronchitic and a non-smoker living in low pollution is 0.08. Table 3.4(d) also shows the estimation of the partial effect of smoking holding pollution statistically constant. The probability of being bronchitic if one lives in high pollution and smokes is 0.78 (18/23), while, if one lives in high pollution and does not smoke, the probability is only 0.07 (2/27). The effect of smoking when pollution is held statistically constant at the high level, is to increase the probability of being bronchitic by 0.71. Somewhat contradictory evidence comes from comparing the probability of being bronchitic, if one smokes and lives at low pollution, 0.58, with the probability of being bronchitic if one lives in low pollution and does not smoke, 0.08. Thus, there are two partial effects for smoking when controlling for pollution: at the high-pollution level the effect is 0.71, while, at the low level, the effect is 0.50. To derive a single measure of effect, we can simply average the two values, (0.71 + 0.50)/2 to give a mean effect of 0.61. The unweighted average can be used in this case because the control groups are of exactly the same size with 50 respondents in both the high- and low-pollution categories (Hellevik, 1984, p. 14).

A similar analysis (Table 3.4(d)) produces two estimates of the effect of pollution when controlling for smoking: 0.20 for the smoker (0.78 − 0.58) and −0.01 for the non-smoker (0.07 − 0.08). Here, however, we cannot use a simple average to determine the overall partial effect because there are different numbers in each control group; there are 35 smokers and 65 non-smokers. Following Hellevik (1984, p. 15), we can derive an overall partial effect for pollution by weighting both partial effects by the

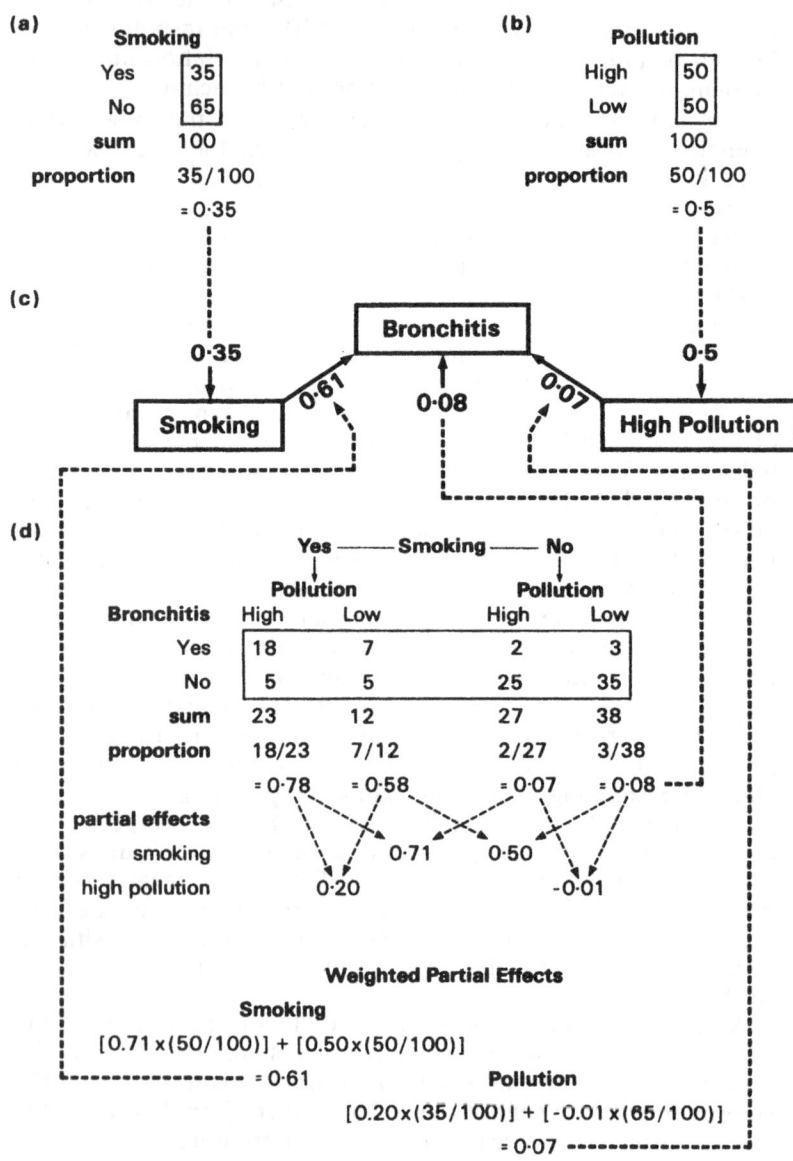

Table 3.4   Three-variable cross-tabulation

relative proportion of all respondents belonging to their respective control group. Therefore, the overall partial effect for pollution is: $[0.20 \times (35/100)] + [-0.01 \times (65/100)]$ which equals 0.08.

The completed causal model, Table 3.4(c), reveals that if one does not smoke and lives in a low-pollution area, the chances of becoming bronchitic are comparatively low at 0.08; in other words, causes producing 8 out of 100 bronchitics have, therefore, not been included in the model. If a person smokes, no matter whether they live at high or low pollution, the probability of becoming bronchitic increases by 0.61; controlling for pollution has only slightly reduced the original bivariate probability of 0.63 and there appears to be a genuine, strong, causal effect between these two variables. In contrast, the effect of pollution in the original bivariate relationship was 0.2 and this decreases to 0.08 when smoking is controlled; the original relationship is, to some extent, a spurious artifact produced by the tendency of smokers to live in high-pollution areas.

It is important, at this stage, to summarise and extend the argument; our example is illustrative and not exhaustive. Firstly, the results may or may not be representative of the general population, and for this they depend on the quality of the questionnaire survey; a biased survey can only give biased results. Secondly, the observed relationships may be, to some extent, an outcome of chance associations; there were only a hundred people in the survey and it is unwise to generalise on the basis of such small samples. To guard against such inferential error, it is possible to evaluate the results by the use of 'significance' tests. Thirdly, there is a danger of falling prey to the temptations of 'fishing' for relationships by carrying out a large multivariable survey, cross-tabulating each variable against all the remaining variables and then concocting an argument that accounts for the 'significant' relationships (Kish, 1959, p. 336). There are, however, ways to mitigate the dangers of finding 'chance correlations' by this means – the simplest way being cross-validation which is achieved by splitting the data into two random sub-sets, exploring for relationships in the first and attempting to confirm their existence in the second (Mosteller and Tukey, 1977). Finally, and most importantly, the analysis would not end here. The observed, and apparently causal, relationship between bronchitis and smoking would have to be subjected to further tests by the introduction of more controls and variables. For example, is occupation related to both smoking and bronchitis, and does the relationship hold for different age-groups and both sexes? Moreover, the differing partial effects that were obtained for the different levels of the control variables suggest that our simplistic analysis of Table 3.4 is inadequate. Bronchitis seems to be produced not only by the independent effects of smoking and pollution but also by their interaction; the probability of being a bronchitic is highest for the smoker living at high pollution. An alternative causal model is, therefore, appropriate:

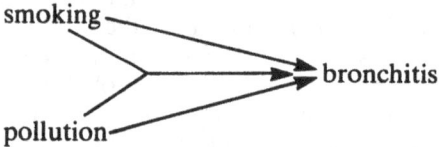

and such an interaction model (as well as models which require the simultaneous control of several variables) can be more effectively explored by a more advanced technique, 'log-linear' modelling (Gilbert, 1981; Wrigley, 1985).

Multiple regression

The second basic method of achieving statistical control is to use the multiple regression model and associated methods to ascertain to what extent the variations in the independent variables account for the variations in the dependent variables. For example, the verbal model:

heart disease is a function of (water hardness and air
pollution)

can be translated into the statistical linear equation of:

$$Y = \beta_0 + \beta_1 X_1 + \beta_2 X_2 + \varepsilon \qquad (3.1)$$

where Y could represent the spatial variations in heart-disease mortality (deaths per 10,000 population aged 15–64) among the states of the USA, X_1 and X_2 are, respectively, the variations in water hardness (ppm) and sulphur dioxide ($\mu g/m^3$) for the same areas, and ε symbolises the unsystematic variations produced in Y by innumerable, individually unimportant, ultimately self-cancelling causal variables that have not been included in the model. The intercept term, β_0, represents the value of Y when X_1 and X_2 are both zero; the key terms are the regression coefficients β_1 and β_2 which measure the change that would be produced in Y for a unit change in one independent variable while holding the other independent variable statistically constant. Thus, β_1 measures the change in the heart-disease death-rate if hardness was increased by one part per million while air pollution remained unaltered. The actual values of the coefficients are usually estimated by calibrating the model such that the sum of the squared differences, between the actual values of Y and $\hat{Y}$ values fitted by the model, are minimal; the so-called ordinary least-squares criterion of fit (Chatterjee and Price, 1977).

If the fitted model is as follows:

$$\hat{Y} = 850 - 2.0\,X_1 + 5.0\,X_2 \qquad\qquad (3.2)$$

the equation implies that, if there was no air pollution and water had no hardness, there would be 850 heart deaths per 10,000 population. Any increase in water hardness should lead to a decrease in this rate (as shown by the negative coefficient for X_1), while any increase in air pollution should lead to a higher death-rate. It is, therefore, possible to use this equation to make predictions of what would happen if we manipulated this statistical model. Thus, hardening an area's water supply by 100 ppm should reduce the death-rate by 200 deaths per 10,000; halving the sulphur dioxide levels from 100 to 50 $\mu g/m^3$ should result in 250 fewer deaths per 10,000 population. An important summary statistic that usually accompanies the estimated equation is the coefficient of determination, R^2, which measures the amount of variation in the dependent variable that has been accounted for by the combined variations of all the independent variables. It can range from 100 per cent, in which the independent variables completely determine the dependent variable, to 0 per cent when the independent variables are unrelated to the dependent variable. It would be unwise to make statistical predictions if the R^2 is nearer the latter than the former value.

The causal model behind the multiple regression model is as follows:

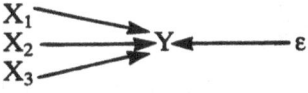

where there is only one Y, and each X is thought to be an independent cause of Y. Of course, as shown in Figure 3.1, postulated models may be much more elaborate than this, and if we wish to estimate these more complicated models, we must make use of sets of multiple regression equations according to the methods of Simon (1954) and Blalock (1964). This requires the research to postulate a set of alternative models, to produce predictions of the size of certain relationships, and to confront these predictions with the empirical evidence so as to eliminate models that are not consistent with the data.

To illustrate the use of these procedures we will, again, use the simplest possible case of three variables but will examine three alternative models:

A: low social class (X) →smoking (Z) →cancer (Y)

in which smoking is an intervening variable in the causal chain between class and cancer;

B: low social class (x)⟨ →smoking (z) →cancer (y)

with class being related to both smoking and cancer but the link between smoking and cancer being spurious;

C: smoking (z)⟍ low social class (x)⟋ →cancer (y)

in which cancer and class are both causes of cancer, that is, the usual multiple regression model. If we collected data on the varying rates of these variables for the counties of England and Wales, we could derive a correlation matrix such as Table 3.5(a). These correlation coefficients measure the linear association between pairs of variables; a value of zero signifies no relationship, a value of −1 and +1 represents a perfect negative and positive association respectively. The table reveals a positive association between all the variables; an area with a high rate of low social class will tend to have high rates of smoking and cancer. To implement the Simon-Blalock procedures, we must calculate the partial correlation between two variables (X, Y, for example) while controlling for a third (Z):

$$rXY.Z = \frac{rXY - (rXZ)(rZY)}{\sqrt{1 - rXZ^2}\sqrt{1 - rZY^2}} \tag{3.3}$$

where rXZ is the pairwise correlation between X and Z. This formula has an obvious intuitive meaning: in the numerator, the original correlation between Y and X (rXY) is 'reduced' by the combination of the correlation between the 'explanatory' and 'control' variables (rXZ) and the correlation between the dependent and control variables (rZY). The result is then standardised by two correction factors in the denominator, which

Table 3.5 Evaluating alternative causal models: the Simon-Blalock approach

(a) *Observed correlation matrix*

	X	Z	Y
X	1.0		
Z	.45	1.0	
Y	.32	.45	1.0

(b) *Predicted and empirical results for alternative models*

Model	Association	Prediction	Empirical results
A	rXY.z	0	0.14
B	rYZ.X	0	0.37
C	rXZ	0	0.45

summarise the variation explained in the explanatory and dependent variables by the control variable.

Returning to the three alternative models, model A implies that there should be no direct link between class and cancer; that is, rXY.Z should be zero. In model B, the observed association between smoking and cancer should be reduced to nothing when social class is controlled; that is, rYZ.X should equal zero. In the final model, both social class and smoking are independent causes of cancer and, therefore, there should be no association between the two causal variables; rXZ should be zero. Using the partial correlation formula and the correlation coefficients in Table 3.5, we can derive the predictions and empirical results and, on this basis, model A is preferred to models B and C, for this model has the smallest difference between these values. Model A is not necessarily a valid model but, given the empirical evidence, it has a better fit than the alternative models; it is tentatively accepted until an improved model is developed.

This discussion so far has only considered regression and correlation analysis in which the dependent variable is a continuous, quantitative variable, that is, the Y variable has been a rate of some kind. It is possible to extend the conceptual framework of such analyses to the case when Y is a qualitative variable as is so often found in survey analysis. For example, in order to evaluate the model:

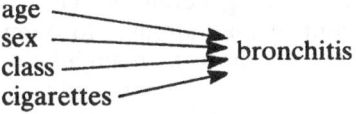

a sample survey could obtain the following data:

Y   bronchitis, a qualitative variable, a 1 indicating a bronchitic, 0 otherwise;

X_1 age, a continuous variable measured in years;

X_2 sex, a qualitative, dichotomous variable, a 1 if male, 0 otherwise;

X_3, X_4 a qualitative trichotomous concept (upper, middle, lower) represented by two variables: X_3 is a 1 if upper class, 0 otherwise, X_4 is a 1 if middle class, 0 otherwise; by definition lower class is represented by both X_3 and X_4 being 0;

X_5 cigarettes, a continuous variable measured by the number of cigarettes smoked daily.

The qualitative causal variables (so-called 'dummy' variables) can be accommodated within the usual multiple regression framework, but when the dependent variable is measured in a qualitative form (presence/absence), ordinary least-squares methods of fitting cannot be used. However, alternative procedures such as logistic models, weighted least-squares and maximum-likelihood methods of calibration, are available (Kleinbaum *et al.*, 1982; Wrigley, 1985), so that it is now possible to analyse both quantitative- and qualitative-dependent variables in a unified way (Wrigley, 1979). While these statistical techniques are generally suitable for the analysis of observational studies, they cannot be used without modification in the study of case-comparisons. This is because the 'matching' in the data collection has to be carried over into the analysis stage; the details are well described by Breslow and Day (1980).

Validity of statistical research

Although powerful techniques are now available for statistical analysis, it is generally accepted that the observational study is inherently weaker than the experimental approach. Experimental design permits all the operations that make causal inference possible: the control of extraneous factors (by design, matching or randomisation), the manipulation of the independent variables to determine the time sequence, and replication to discover under what conditions the result generalises. Observational studies based on statistical control may attempt to mimic these operations but typically they do not allow the manipulation of the explanatory variables nor the randomisation of subjects. As a result of these differences, statistical analysis is often beset with three key problems: specification, multicollinearity and the ecological fallacy.

The valid use of techniques for statistical control presumes that the researcher is able to specify exactly the correct form of the model that accounts for the variation in the dependent variable. For example, in the OLS regression model, it is assumed that all important explanatory variables have been included in the model and none has been inadvertently omitted. Moreover, the validity of any estimated coefficient depends on the whole causal model being correct; the coefficients of incorrectly specified models should not be interpreted. In experiments, randomisation guarantees that the control groups are equivalent within known statistical error and, thereby, all extraneous variables are removed by design. Unfortunately, with statistical control we can never be sure that important variables have been overlooked no matter how good the fit or how reasonable the result. While randomisation ensures the control of known and unknown factors, statistical analysis only permits predetermined factors to be controlled.

One of the assumptions of statistical control procedures is that causal variables are not highly interrelated, that is, they are not multicollinear. In the real world, however, everything tends to be related to everything else and social variables tend to be 'block-booked'. For example, people who work in hazardous environments are usually smokers who live in poor-quality housing in areas of high pollution. In experiments, the multicollinearity deadlock is broken by design; each causal variable is manipulated so that it becomes unrelated to any other variable, thereby ensuring that each independent variable is truly independent. While it may be conceptually possible to distinguish the effects of individual causal variables, statistical control may simply not be able to achieve this, in practice. It is as if common variation between the explanatory variables prevents the measurement of each variable's separate contribution to the variation of the dependent variable. This area is one of considerable statistical research (into so-called 'biased' estimation) but the fact remains, that as the association between two or more variables becomes stronger, the more difficult it is to tell one from another.

The final weakness of the statistical approach to be examined here, is a problem that occurs with the use of aggregate, routine data to make individual-level inferences. For example, areas of high unemployment may be coincident with areas of high heart-disease mortality in young males and there may be a very high correlation between the variables. This result, however, cannot be used to infer that unemployment causes heart disease for all; the deaths could be employed males who just happen to live in areas of high unemployment. An empirical example of this problem was given by Robinson (1950) in his classic paper on the ecological fallacy. When he correlated the variables 'foreign-born' and 'illiteracy' at the level of 93 million individual Americans, Robinson found a weak positive correlation (+0.12) but when the scale was changed to the nine aggregate census divisions of the USA, there was a moderate negative correlation (−0.62). Openshaw (1984) has shown that the ecological fallacy is only part of the more general Modifiable Area Unit Problem, for he recognises that there is not only a scale but also an aggregation problem. Results will differ not only as the number of observations is changed to a different scale but, at each scale, one can get incredibly varied results according to how the observations have been aggregated into groups. For example, with 99 original area units grouped into 24 new units, he found that the correlation between percentage Republican vote and percentage population aged 60 or over in Iowa could vary between −0.92 and +0.99 according to which particular aggregation was used! According to his arguments, areal correlation results depend as much on the arrangement of the areas on which routine data are published as on the underlying true causal relationship. The basic message of this work is that aggregate results can only be regarded as tentative, preliminary guidelines to further work, because the results may be an artifact produced by the particular scale and aggregation that has been used.

Making causal inferences

Our discussion has, so far, shown that while experiments are a powerful way of testing relationships, laboratory results cannot be easily generalised to the real world. In contrast, survey analysis can be highly representative, but it is plagued by problems of multicollinearity and unknown lurking variables that may remain uncontrolled. Two solutions are used in epidemiological research to overcome this dilemma. Firstly, researchers can use experiments in a 'natural' setting and, secondly, they can employ a set of criteria that allow the evaluation and combination of results produced by different methods of investigation.

Natural experiments

In a natural experiment, the researcher does not deliberately manipulate the supposed causal variable, but attempts to discover real-world situations in which the independent variable has varied and then to observe the effect of these changes on the dependent variable. The most famous of all such investigations is John Snow's study of cholera. At the time of his research in the 1850s, there were two rival theories of cholera causation: the miasmic theory proposed that the cholera victim had breathed foul air which had been in contact with putrefied bodies and decaying vegetation, while Snow believed that the disease was spread by drinking polluted water. Snow had observed, in the 1849 outbreak, that cholera death-rates were noticeably higher in those districts of London supplied by the Southwark-Vauxhall and Lambeth Water Companies (Table 3.6). Both companies took their water from the Thames at a point that appeared to be polluted by sewage. Shortly before the 1854 epidemic, the

Table 3.6   Cholera death-rates, London, 1849

Rank	District	Deaths per 10,000	Water-supply company
1	Rotherhithe	205	Southwark-Vauxhall, Kent, Tidal ditches
2	St Olave	181	Southwark-Vauxhall
3	St George	164	Southwark-Vauxhall, Lambeth
4	Bermondsey	161	Southwark-Vauxhall
5	St Saviour	153	Southwark-Vauxhall
6	Newington	144	Southwark-Vauxhall, Lambeth
7	Lambeth	120	Southwark-Vauxhall, Lambeth Pumpwells, Southwark-Vauxhall
8	Wandsworth	100	River Wandle
9	Camberwell	97	Southwark-Vauxhall, Lambeth
10	West London	96	New River
11	Bethnal Green	90	East London
12	Shoreditch	76	New River, East London
13	Greenwich	75	Kent
14	Poplar	71	East London
15	Westminster	68	Chelsea
16	Whitechapel	64	East London
17	St Giles	53	New River
18	Stepney	47	East London

Source: Snow (1855).

Lambeth Company moved their intake to a less polluted reach at Thames Ditton. Clearly, if polluted water was the cause of cholera, one would expect that those households supplied with this water would have a lower death rate. As Snow (1855) wrote:

> The experiment was on the grandest scale. No fewer than 300,000 people of both sexes, of every age and occupation, and of every rank and station, from gentlefolks down to the very poor were divided into two groups without their choice, and in most cases, without their knowledge; one group being supplied with water containing the sewage of London . . . , the other group having water quite free from such impurity.

While the householders were often unable to tell Snow which company supplied their water, he was able to distinguish the suppliers on the basis of a chemical test. In a detailed examination of the first seven weeks of the 1854 epidemic (Table 3.7), he showed that the mortality rate was over eight times higher in Southwark-Vauxhall-supplied houses than in the Lambeth-supplied houses. Many years before Robert Koch identified the cholera vibrio in 1883, Snow was able to conclude that there was a 'cholera poison' which was transmitted by water.

A second example of when researchers have been able to use 'natural' variations in the independent variables to strengthen their causal analysis is the work of Doll and Peto (1976). In the twenty-year period of the study, the male doctors who constituted their prospective cohort showed a substantial decline in cigarette consumption in relationship to all British men (Figure 3.2(a)). If smoking is a cause of early death, a concomitant decline in the death rate for doctors, in comparison with all males, should be observed. Moreover, if, as some researchers believe, smoking is specifically related to lung cancer, then only the doctors' lung-cancer rate should decline. As Figure 3.2(b) reveals, the doctors' lung-cancer rate has declined continuously

Table 3.7 Cholera deaths, first seven weeks of 1854 epidemic

Water-supply company	Number of houses	Cholera deaths	Deaths per 10,000 houses
Southwark-Vauxhall	40,046	1,263	315
Lambeth	26,107	98	37
Rest of London	256,423	1,422	59

Source: Snow (1855).

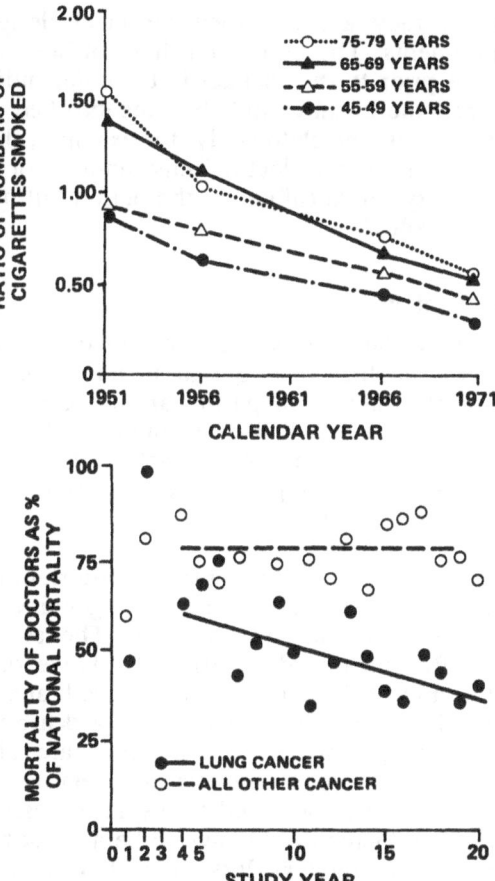

Figure 3.2 An example of a natural experiment: the case of doctors' smoking (a) Trends in the ratio of cigarettes smoked by male physicians to number smoked by British men (b) Trends in mortality of male doctors as percentage of national mortality
Source: Doll and Peto (1976)

over the twenty years, in relation to all males, while there has been no decline for all other causes when doctors are compared to the total male population. Thus, this evidence adds considerable support to the argument that the relationship between smoking and cancer is truly causal and is specific to cancer of the lungs. However, while the innovative researcher can use natural experiments to subject relationships to an exacting test, it must be remembered they are not such powerful designs as randomised experiments. The population has not been allocated to different groups prior to the changes in the independent variable and, therefore, there may still be uncontrolled extraneous variables. In the Doll and Peto study, for example, there may be some other variable in the doctors' environment or way of life that has changed systematically over the period but has not been observed or controlled.

Criteria for evaluation

Researchers in the late 1950s and early 1960s began to define operational criteria for making causal inferences that went beyond statistical analysis. In particular, the USA government published a report in 1964 that examined the link between smoking and ill-health according to a set of criteria for judging whether observed relationships are causal or not. Five of these criteria will be discussed here.

Strength of relationship   According to this criterion, the stronger the relationship, the more likely it is to be a genuine relationship and not the result of omitted variables. The strength of an association is often measured in terms of relative risk, which is defined as the ratio of the rates of the 'exposed' and 'unexposed groups'. Thus, Doll and Hill (1964) found a death-rate of 0.90 per thousand among smokers and a death-rate of only 0.07 amongst non-smokers, which suggests high relative risk of over twelve. For such a risk to be based on a spurious association, the unknown causative factor would have to be at least twelve times more frequent among smokers. It is argued that such a distinctive difference is unlikely to have been overlooked. However, the application of this criterion must not be too rigid; with a multifactorial disease and a number of causal factors, the association between any one of them and a disease will, in general, be relatively weak.

Dose-response relationship   If a relationship is causal then an increasing level and duration of exposure to the independent variable should lead to an increasing frequency of the disease. Thus, if cigarette smoking causes disease, there should be a consistent increasing relative risk as one examines people who smoke 0, 5, 10, 15 and 20 cigarettes a day. Again, one must be careful in the application of this criterion; dose-response relationships may fail to appear not because of spurious relationships but because threshold or saturation effects may be present. If this is the case, there will be a dose-response relationship only within a limited range of the causal variable.

Consistency Evidence for a causal relationship is considerably strengthened if an association in different population groups, different ages and different races is found by different researchers using different research designs. If the same results are found in laboratory beagles, vegetarian Trappist monks, cannibals and Welsh males, you know you are on to something, for it is unlikely that a spurious association will operate to the same degree in these disparate groups. As always, this criterion must not be strictly adhered to; in a multifactorial situation, the researcher is unlikely to find exact consistency.

Coherence with biological plausibility In essence, we need to know whether the observed relationship makes biological sense and whether it is possible to elaborate the biological mechanism between cause and effect. While a relationship between diet and stomach cancer seems intuitively plausible, an observed relationship between circumcision and lung cancer would be questionable. Two important caveats must, however, be noted. Firstly, any result can be made to appear plausible if the researcher is ingenious enough to invent or concoct the biological mechanism. Secondly, if we always insisted on this criterion, we would never discover anything new; coherence supports existing theory while incoherence potentially generates new theory. For example, Snow's theory of the water transmission of cholera was largely rejected at the time due to the dominance of the miasmic theory. Similarly, the observed association between oral contraceptives and circulatory disease was, at first, dismissed due to the lack of a known biological mechanism. What is biologically plausible can only depend on the knowledge of the day.

Specificity This final criterion suggests that there is an increased likelihood of truly causal relationships if the explanatory variable is only related to one disease or to one disease more strongly than others. In our view, this is a poor criterion on which to judge relationships, for it is based on the notion of specific aetiology. In a multicausal world, many factors can have multiple effects and all diseases can have multiple causes. Indeed, a more useful criteria may be *analogy*, when the researcher looks for other comparable examples of cause-and-effect relationships.

While each of these criteria may help in determining whether a relationship is causal, it must be stressed that none of these criteria is either essential or conclusive. Judgement is often not straightforward and the researcher usually has to balance conflicting evidence before deciding that the arguments are sufficient to permit and warrant action and intervention.

Water hardness: a health risk?

This particular case study was chosen to illustrate the designs and methods of disease ecology for a number of reasons. First, some researchers have argued that the effect of water hardness may be great; for example, Gardner (1976) calculated that the excess number of deaths for men aged 45–63 in England and Wales alone was of the order of 10,000 per year. Second, the relationship has been extensively researched in the last thirty years, and a wide variety of different methods and approaches has been applied. Many of the studies have a distinct 'geographical' flavour. Third, there is continued doubt about the validity of the relationship; the majority of studies have found a negative relationship, but some have found a positive one and others, none at all. Before outlining the story of the search for a relationship, we need to define the variables for study. In relation to water supply, there are two main components that make it hard: calcium and magnesium. In addition to these bulk elements there are trace elements such as lead, cadmium, lithium and vanadium. Both trace and bulk elements have been associated with disease. In the main, it has been cardiovascular disease (CVD) of the heart and blood-supplying vessels that has been most often associated with water hardness. In particular, two subcategories have been the most heavily researched: they are ischaemic heart disease, in which there is inadequate reception of oxygen-rich blood to the heart muscle, and cerebrovascular disease in which there is a substantial reduction of blood flow to the brain or intra-cranial bleeding. To the layperson they are known as heart attacks and strokes, respectively.

The water story

The modern beginnings of the study of disease in relation to hardness did not come from medical research but were made by a Japanese agricultural chemist, Jun Kobayashi, in the late 1950s. He had plotted on a map of Japan the acidity of river water and deaths from stroke, and he visually noted a marked association. This study was taken further by the American, Henry Schroeder, who found a negative statistical association: in Japan, the harder the water the lower the death-rate from heart disease and strokes. In 1960 Schroeder, on return from Japan, found similar statistical relationships for the USA: there were significant negative relationships between hardness and deaths from CVD for men, using routine data collected on a state basis. When data for the 163 largest American metropolitan areas were analysed, there were again significant negative associations between hardness, magnesium, and calcium and heart-disease mortality for whites.

The search was now on and in 1961 the first results were presented for Britain by a team led by Lady Crawford. They had found broadly similar results to Schroeder's except that magnesium was not related, and the negative association for calcium was much stronger than in the USA. The north and west of England and Wales have high mortality and soft water. This study was based on routine mortality statistics collected for the County Boroughs of England and Wales which were analysed by correlation. This was followed by numerous studies from all over the world using areal-level routine data. The majority found negative relationships between some elements of water hardness and some element of CVD mortality. But there were also some conflicting results, for example, a positive association was found for the counties of Oklahoma and no association was found when using Irish cities.

The water story has been subsequently extended to examine the mechanism by which the link may operate; this has involved post-mortem findings and surveys of individuals. Before discussing these, we need to consider the process of dying from ischaemic heart disease. This starts with a narrowing of the arteries which deprives the heart muscle (myocardium) of an adequate supply of oxygen-rich blood, and a part of the heart muscle may die or infarct. This normally heals but death can occasionally occur due either to an arythmia, an electrical disturbance which is more likely to occur in the first few hours following infarction; or by myocardial degeneration, a much slower process. Researchers tried to examine each stage of the disease to elucidate the relationships with water hardness. Thus, in 1968 over 23,000 sets of arteries were collected from fourteen countries but no systematic differences in narrowness were found in relation to water hardness. Other researchers tried to examine the links with fatal arythmias. As this is so sudden that the person is likely to die before hospital, Crawford and her team in Britain examined the geographical variations in 'sudden death' among hard and soft water towns. They once again found negative associations but a similar study in Canada did not confirm these results. Another approach was to examine people before they died to see whether certain 'risk' factors varied systematically with water hardness. There were again conflicting results with, for example, cross-sectional surveys finding that in south Wales 240 men in hard water areas did not differ significantly from 360 men in soft water areas in terms of blood pressure and cholesterol, while a study of male executive civil servants in six hard and six soft water towns in England and Wales did differ systematically on these same variables.

Despite, and perhaps because of, these conflicting results, the water hypothesis remains a fertile area for research. In Britain, the most elaborate study design is that of the British Regional Heart Study. The first phase of this study analysed the geographical variations in CVD mortality for 245 urban areas in England, Wales and Scotland for the period 1969–73 by multiple regression; a negative association with water hardness was again found (Pocock, Cook, and Shaper, 1982). The second phase was a cross-sectional survey of 7,735 men aged 40–59 randomly selected from the general practices of twenty-four British towns. A lengthy questionnaire was given to each respondent, and data were collected on a wide range of physiological and biochemical variables; measurements of the tap water of 10 per cent of the interviewees were also made (Shaper *et al.*, 1981). The third phase is currently being undertaken and this consists of a prospective study with the 7,735 men as the cohort. Each man is tagged at the OPCS registry for mortality, and there is a system of general practitioner reports for morbidity. The cohort is being followed for at least five years to assess which of the many personal risk factors measured at phase 2 are most closely related to cardiovascular events.

Assessing the validity of the relationship

While research on the water hypothesis obviously continues, it is possible to make a critical evaluation of the past thirty years' work using the 'criteria of judgement' discussed earlier. This section attempts to answer the questions of whether or not there is a causal relationship between the variables, and why studies have reached conflicting conclusions.

Consistency From the above discussion, it is obvious that this criterion is not met. While the majority of national studies find a negative association, this is not the case when smaller geographical areas are used as the basic unit for analysis. For example, for all the Canadian municipalities there is a negative relationship, but this becomes positive when the data are analysed separately for British Colombia, and the Maritime and Prairie Provinces; a clear example of the modifiable areal-unit problem. Another commonly mentioned inconsistency is the case of Birmingham, UK. This is supplied with very soft water imported from Wales, but the CVD mortality rates are not substantially different from surrounding areas that do not receive such soft water.

Strength of relationship Comstock (1979) in a very perceptive article points out that the correlation coefficient is an inappropriate measure for disease-ecology studies because it measures association (scatter about the line) and not the strength of the relationship (the slope of the line) which should be assessed by the regression coefficient. Thus, it is possible to have a high correlation but only a very small relationship, and vice versa. Unfortunately, the vast majority of studies only report correlation coefficients. Using those studies that do report regression coefficients, Comstock calculates the highest relative risk to be 1.25, that is, for every 1 person who would die in hard water areas (200 parts per million), 1.25 persons would die in soft water areas (0 ppm). In comparison with relative risk ratios of over 10 for lung cancer between smokers and non-smokers, a value of 1.25 does not indicate a very strong relationship; another point against the water hypothesis.

Specificity As we have seen, several different types of cardiovascular disease have been found to be related to water hardness, but it is not only CVD. Many other conditions have been found to be significantly related to hardness; the list includes cirrhosis of the liver, peptic ulcer, bronchitis, infant mortality, congenital malformations, anencephalus, spina bifida, and even motor vehicle accidents! Taking one of these relationships, bronchitis, we must ask is it conceivable that water hardness can adversely affect the lungs? Or is it more likely that there are lurking variables that are related to both hardness and this disease? This extreme non-specificity certainly does not help the case for a causal relationship.

Coherence with biological plausibility There are two main hypotheses of how hardness could determine disease. The first is a direct relationship which postulates that the body needs so much calcium, and individuals in soft water areas do not get enough for a long life. The evidence for this 'bulk addition' is not strong; ask yourself how much of your daily fluid intake comes directly from the domestic tap where you live. It has been estimated that calcium from the tap contributes only one-tenth of total dietary intake. The second hypothesis is an indirect one, and it is postulated that calcium lowers the capacity of water to take up toxic trace elements from the soil, supply pipes and even cooking utensils. This is the hypothesis now favoured by Schroeder and he implicates the trace element cadmium. Unfortunately, studies have found no consistent relationships between trace elements and disease.

Natural experiments One of the most remarkable pieces of evidence in support of a truly causal relationship is the findings for Monroe County, Florida (Gardner, 1976). In 1941, the water supply was changed from very soft (0.5 parts per million) to hard (220 ppm) as deep well water was brought into use. If the water hypothesis is correct, there should be a decline in the CVD mortality rate given sufficient time. In fact, the rate appeared to drop dramatically, and within four years it was only a half of its 1941 level. But we must ask whether there were other changes that accompanied the change to well water, a question that has not been generally asked by the supporters of the water story. Monroe County includes Key West Naval Base, and during and after the war there were major population changes in the area. As Table 3.8 shows, the crude CVD mortality rate did drop by nearly 50 per cent over ten years, but when this is age-, sex- and race-adjusted, the decline is a much less impressive 16 per cent,

Table 3.8   CVD mortality, 1940–50, rates per 100,000

Rate		1940	1950	Percentage change
Monroe County:	crude	570	340	−47
	age-adjusted	702	572	−18
	age-, sex-, race-adjusted	684	575	−16
United States:	age-adjusted	508	440	−13

Source: Comstock (1979, p. 50).

which is similar to that experienced by the whole of the USA over the same period. The second natural experiment that appears to support the water hypothesis is provided by Gardner (1973). He examined changes in hardness for county boroughs in England and Wales, identifying 11 that had experienced changes of over 50 ppm in the period 1930–60. For the 72 boroughs without substantial changes, the CVD mortality for men aged 45–64 had increased by 11 per cent over the period 1951–61. Of the rest, 6 had become substantially softer and they had suffered a rise of 20 per cent in the CVD rate; while the increase in those 5 boroughs that had become harder was only 9 per cent. Superficially, this seems good evidence for the water hypothesis but as one of us has shown (Jones, 1980) there were much larger variations for individual boroughs that did not experience water supply changes. Indeed, of the 11 towns whose CVD ranking changed by over 20 positions over the period (for example, Merthyr was the fortieth highest in 1981 and the highest in 1961), only one of these boroughs, Burton, had experienced a changed water supply. Without controlling for the variables that had brought about these much larger changes, it is impossible to assess the importance of water hardness. There have also been a few true experiments using animals but they have been inconclusive and as Comstock (1979, p. 50) has commented 'the more rigorous the experimental design, the less the association between hardness and disease'.

On balance, therefore, there does not appear to be much support for a causal relationship, and even if it does operate, it is only a weak effect and there appears to be cases where it does not operate at all. Why then do we get these conflicting results and why is the association in Britain between calcium and CVD mortality so marked? One of us has previously argued (Jones, 1980) that the underlying problem is poor research design and inadequate statistical analysis. There have already been some examples of this in the discussion on natural experiments, and we will now consider this in more detail to reveal the need for rigorous epidemiological research.

The majority of analytical studies of the water hypothesis are either case-comparison studies based on individual data, or correlation studies based on routine areal data. The fundamental weakness of many of these case-comparison studies is that they have not been based on matched controls. A typical design selects six hard and six soft water towns and comparisons are made between individuals living in the towns. There is a clear scale problem here, for even if the towns are matched, there is no mechanism by which the individuals are matched. Consequently, systematic differences between towns may not represent the effects of water hardness but other uncontrolled variables that affect individuals who live in the towns. For example, Crawford and Crawford (1969) compared the lead content of bones for men who had died in soft water Glasgow and hard water London. They were testing the indirect hypothesis that hard water could prevent the uptake of lead from pipes, and a significance test did show a statistical difference between the cities in the lead content of bones. But this may not be caused by water hardness for there has been no control for occupations (for example, of plumbers) or of lead piping in houses; indeed, Glasgow is notorious for the amount of lead plumbing that still remains. The observed lead differences may be more readily attributed to the social environment (occupation and housing) than the physical environment (water hardness); with this design you simply cannot discern which is the valid explanation.

The fundamental problem with the correlation studies is the failure of much of the research even to consider the causal mechanism that links the independent variables to the disease outcome. For example, in what is an excellent analysis at the technical level, Gardner (1973) includes the variables latitude, longitude and rainfall in his multiple regression model to explain the variations in mortality among the County Boroughs of England and Wales (he is not alone in this, see Pocock *et al.*, 1982). These variables merely soak up the geographical variation of disease mortality without providing a convincing explanation. Good model fits and rigorous quantitative analysis are not enough to ensure valid explanations; where is the forcing or producing element that is the hallmark of a true causal relationship? Finally, it must be pointed out that many of the models that have been calibrated are socially blind in including only variables of the physical environment, often a large number of water quality elements. Even those better analyses that have included social variables have been troubled by the problems of multicollinearity. This may be the explanation for the relatively strong correlation found for calcium in England and Wales in that calcium may be a very good surrogate for social variables. The soft water areas of the north and west of the country were the areas of early industrialisation, and today they house a disproportionate percentage of the 'deprived' of this country. The methods of quantitative analysis may be unable to disentangle the separate effects of the social variables from those of water hardness, making it exceedingly difficult to develop a coherent model.

Conclusions

Epidemiologists and disease ecologists have rightly rejected the exclusive use of experiments to test and develop theories, for results must apply to people in the real world and not animals in the laboratory. Population methods and quantitative analysis, however, are plagued by a whole series of technical problems, and it is undoubtedly difficult to calibrate valid models. There is also a need for research on people as social beings and, in relation to the social context developed in Chapter 1, the epidemiological approach as presented in Chapters 2 and 3 is sorely lacking. In particular, epidemiology appears untouched by the social–constructionist viewpoint for it merely accepts the clinical, biomedical definition of what disease is. Moreover, much epidemiology concentrates on individual explanation while ignoring societal and structural explanations; indeed, it could be argued that the epidemiological methods considered here are intrinsically attached to this 'atomistic' viewpoint.

We continue to be surprised at the research time, expertise and money that have been expended on the case study we have used here. This is especially so when it is realised that other studies have found that the relative risk from CVD mortality between the top and bottom grades in one occupational group, civil servants, is over three times that found for water hardness (Rose and Marmot, 1981). Even when such social differences are researched, explanation is usually couched in terms of individual risk factors and lifestyles, and the causal chain is not extended back into society. Of course, finding that the cause lies in the physical environment of individuals is non-threatening to the current economic and social order, while situating explanations in society, and calling for social change, certainly is. At the end of the next chapter we begin to explore the links between societal organisation and communicable diseases. The final chapter is an attempt to develop a critical epidemiology that takes account of the criticisms voiced in this conclusion.

Guided reading

The essential and readable discussion of causal thinking in medical research is Susser (1973) which contains many examples; this can usefully be supplemented by the shorter accounts of Armstrong (1980, pp. 51–66), Abramson (1979, Chapter 27) and Curson (1984). Harvey (1969, Chapter 20) provides a geographical discussion of the methodological background to causal approaches; Hirschi and Selvin (1973) and Rosenberg (1968) are excellent guides to the application of this type of causal thinking in the social sciences.

The principles of experimental research in relation to statistical inference are clearly explained in Ehrenberg (1975, Chapter 19) and Moore (1979). Nachmias and Nachmias (1976, Chapters 3 and 11) consider the relative merits of experimental and population methods of analysis. Peto *et al.* (1976) give details on designing and analysing clinical trials.

Nachmias and Nachmias (1976) also provide a straightforward introduction to statistical control by cross-tabulation and by multiple regression. For more detail on cross-tabulation see Davidson (1976) and Davis (1971); Marsh (1982) applies this approach to developing an explanation of the social origins of depression. The classic account of control by regression and correlation is Blalock (1964). Pringle (1980) provides an introduction for geographers, while epidemiological examples are given by Golden and Dohrenwend (1981), and Goldsmith and Berglund (1973). Statistical problems of spatial analysis in epidemiology are discussed in general by King (1979); problems of scale and multicollinearity are considered by Cleek (1979) and Moriarty (1973) respectively.

Criteria of evaluation are elaborated in Alderson (1983, pp. 194–220), Hill (1971, pp. 309–23), Lillienfeld and Lillienfeld (1980, pp. 289–321) and USDHEW (1964). In relation to the water hardness case study, four readable reviews that contain extensive bibliographies are Comstock (1979), Neri *et al.* (1974), Gardner (1976), and Sharrett and Feinleib (1975).

References

Abramson, J. H. (1979), *Survey Methods in Community Medicine*, Edinburgh, Churchill Livingstone.

Alderson, M. (1983), *An Introduction to Epidemiology*, London, Macmillan.

Armstrong, D. (1980), *An Outline of Sociology as Applied to Medicine*, Bristol, Wright.

Blalock, H. M. (1964), *Causal Inferences in Non-Experimental Research*, Chapel Hill, University of North Carolina Press.

Breslow, N. E. and Day, N. E. (1980), *The Analysis of Case-Control Studies*, Lyons, International Agency for Research on Cancer.

Bunge, M. (1979), *Causality and Modern Science*, New York, Dover.

Chatterjee, S. and Price, B. (1977), *Regression Analysis by Example*, New York, Wiley.

Cleek, R. K. (1979), 'Cancers and the environment: the effects of scale',

Social Science and Medicine, vol. 13D, pp. 241–7.

Comstock, G. W. (1979), 'The association of water hardness and cardiovascular disease: an epidemiological review and criticism', in E. E. Angino (ed.), *Geochemistry of Water in Relation to Cardiovascular Disease*, pp. 48–68, Washington, DC, National Academy of Sciences.

Crawford, M. D. and Crawford, T. (1969), 'Lead content of bones in a soft-water and a hard-water area', *Lancet*, vol. 1, pp. 699–701.

Curson, P. (1984), 'Geography, epidemiology and human health', in J. I. Clarke (ed.), *Geography and Population*, pp. 72–93, Oxford, Pergamon Press.

Davidson, N. (1976), *Causal Inferences from Dichotomous Variables*, Norwich, Geo Abstracts.

Davis, J. A. (1971), *Elementary Survey Analysis*, Englewood Cliffs, NJ, Prentice-Hall.

Doll, R. and Hill, A. B. (1964), 'Mortality in relation to smoking: 10 years observation of British doctors', *British Medical Journal*, vol. 1, pp. 1399–1410 and 1460–67.

Doll, R. and Peto, R. (1976), 'Mortality in relation to smoking: 20 years of observation on male British doctors', *British Medical Journal*, vol. 2, pp. 739–48.

Doll, R. and Peto, R. (1981), *The Causes of Cancer*, London, Oxford University Press.

Dubos, R. (1959), *Mirage of Health*, New York, Harper.

Durkheim, E. (1951), *Suicide*, New York, Free Press.

Ehrenberg, A. S. C. (1975), *Data Reduction*, London, Wiley.

Epstein, S. S. (1979), *The Politics of Cancer*, New York, Anchor.

Eysenck, H. J. (1980), *The Causes and Effects of Smoking*, London, Temple-Smith.

Friberg, L. (1973), 'Mortality in twins in relation to smoking habits and alcohol problems', *Archives of Environmental Health*, vol. 27, pp. 294–300.

Gardner, M. J. (1973), 'Using the environment to explain and predict mortality', *Journal of the Royal Statistical Society*, Series A, vol. 136, pp. 421–40.

Gardner, M. J. (1976), 'Soft water and heart disease', in J. Lenihan and W. W. Fletcher (eds), *Environment and Man*, pp. 116–35, Glasgow, Blackie.

Gilbert, G. N. (1981), *Modelling Society: An Introduction to Log-linear Models for Social Researchers*, London, Allen & Unwin.

Golden, R. R. and Dohrenwend, B. S. (1981), 'A path analytical method for testing causal hypotheses about the life-stress process', in B. S. Dohrenwend and B. P. Dohrenwend (eds), *Stressful Life Events and their Contexts*, pp. 201–40, New York, Prodist.

Goldhammer, H. and Marshall, A. W. (1949), *Psychosis and Civilization*, Illinois, Free Press.

Goldsmith, J. R. and Berglund, K. (1973), 'Epidemiological approach to multiple factor interactions in pulmonary disease: the potential usefulness of path analysis', *Annals of the New York Academy of Sciences*, vol. 96, pp. 361–75.

Harvey, D. (1969), *Explanation in Geography*, London, Edward Arnold.

Hellevik, O. (1984), *Introduction to Causal Analysis*, London, Allen & Unwin.

Hill, A. B. (1971), *Principles of Medical Statistics*, New York, Oxford University Press.

Hirschi, T. and Selvin, H. C. (1973), *Principles of Survey Analysis: Delinquency Research*, New York, Free Press.

Jones, K. (1980), 'Geographical variations in mortality: an explanatory approach', unpublished PhD thesis, University of Southampton.

King, P. E. (1979), 'Problems of spatial analysis in geographical epidemiology', *Social Science and Medicine*, vol. 13D, pp. 249–52.

Kish, L. (1959), 'Some statistical problems in research design', *American Sociological Review*, vol. 24, pp. 328–38.

Kleinbaum, D. G., Kupper, L. L. and Morgenstern, H. (1982), *Epidemiological Research: Principles and Quantitative Methods*, California, Wadsworth.

Lillienfeld, A. M. and Lillienfeld, D. E. (1980), *Foundations of Epidemiology*, New York, Oxford University Press.

Marsh, C. (1982), *The Survey Method*, London, Allen & Unwin.

Moore, D. S. (1979), *Statistics: Concepts and Controversies*, San Francisco, Freedom.

Moriarty, B. M. (1973), 'Causal inference and the problems of non-orthogonal variables', *Geographical Analysis*, vol. 5, pp. 55–61.

Mosteller, F. and Tukey, J. W. (1977), *Data Analysis and Regression*, Reading, Mass., Addison-Wesley.

Nachmias, D. and Nachmias, C. (1976), *Research Methods in the Social Sciences*, London, Edward Arnold.

Neri, L. C., Hewitt, D. and Schreiber, G. B. (1974), 'Can epidemiology elucidate the water story?', *American Journal of Epidemiology*, vol. 99, pp. 75–88.

Openshaw, S. (1984), *The Modifiable Areal Unit Problem*, Norwich, Geo Abstracts.

Peto, R., Pike, M. C., Armitage, P., Breslow, N. E., Cox, D. R., Howard, S. V., Mantel, N., McPherson, K., Peto, J. and Smith, P. G. (1976), 'Design and analysis of randomised clinical trials requiring prolonged observation of each patient', *British Journal of Cancer*, vol. 34, pp. 585–612, and vol. 35, pp. 1–38.

Pocock, S. J., Cook, D. G. and Shaper, A. G. (1982), 'Analysing geographic variations in cardiovascular mortality', *Journal of the Royal Statistical Society*, Series A, vol. 145, pp. 313–41.

Pringle, D. G. (1980), *Causal Modelling: The Simon-Blalock Approach*, Norwich, Geo Abstracts.

Robinson, W. S. (1950), 'Ecological correlation and the behaviour of individuals', *American Sociological Review*, vol. 15, pp. 351–7.

Roethlisberger, F. J. and Dickson, W. J. (1939), *Management and the*

Worker, Cambridge, Mass., Harvard University Press.

Rose, G. and Marmot, M. G. (1981), 'Social class and coronary heart disease', *British Heart Journal*, vol. 45, pp. 13–19.

Rosenberg, M. (1968), *The Logic of Survey Analysis*, New York, Basic Books.

Shaper, A. G., Pocock, S. J., Walker, M., Cohen, N. M., Wale, C. J. and Thomson, A. G. (1981), 'British regional heart study: cardiovascular risk factors in middle-aged men in 24 towns', *British Medical Journal*, vol. 283, pp. 179–86.

Sharrett, A. R. and Feinleib, M. (1975), 'Water constituents and trace elements in relation to cardiovascular diseases', *Preventive Medicine*, vol. 4, pp. 20–36.

Simon, H. A. (1954), 'Spurious correlation: a causal interpretation', *Journal of the American Statistical Association*, vol. 49, pp. 467–9.

Snow, J. (1855), *On the Mode of Communication of Cholera*, New York, reprinted 1936 by Harvard University Press.

Susser, M. (1973), *Causal Thinking in the Health Sciences*, London, Oxford University Press.

USDHEW (1964), *Smoking and Health: Report of the Advisory Committee to the Surgeon General of the Public Health Service*, Washington DC, Government Printing Office.

Wrigley, N. (1979), 'Developments in the statistical analysis of categorical data', *Progress in Human Geography*, vol. 3, pp. 315–55.

Wrigley, N. (1985), *Categorical Data Analysis for Geographers and Environmental Scientists*, London, Longman.

CHAPTER 4

Communicable diseases

Introduction

This chapter is concerned with diseases that are capable of being transmitted from someone who is infected (the host) to someone who has not yet had the disease (the susceptible); they are variously known as contagious, infectious and communicable diseases. In the industrialised countries, despite the recent threat of AIDS, they are generally of low importance; even in childhood, communicable diseases are a minor cause of suffering and morbidity and now cause less than 5 per cent of all childhood deaths in Britain. Historically, they have been of much greater importance; only forty years ago tuberculosis was the disease of civilisation, and in the economically backward countries they remain the major killers today. Communicable diseases are given a separate chapter here because of this worldwide importance (two-thirds of all human illness is attributed to such causes), and because their fundamentally different disease processes, when compared to chronic and degenerative diseases, require different modes of analysis. The discussion begins with a brief introduction to the biology of the diseases and then the different research traditions of 'disease ecology' and 'diffusion' studies are outlined and illustrated. As the discussion proceeds we move away from the biomedical model towards a social, and arguably deeper, understanding of communicable diseases.

The biology of infectious diseases

Communicable diseases have a bewildering diversity and it is wise at the outset to offer a classification. It is possible to produce a classification on a number of criteria (for example, portal of entry and exit of infectious agent) but a classification based on the various types of organisms and their differing means of transmission is most appropriate for our purposes.

Organisms

There are six basic groups of organisms (Open University, 1985b).

Viruses are chains of nucleic acid surrounded by a protein coat; they cannot reproduce themselves without entering the cells of plants or animals. They produce such diseases as influenza, measles (rubella), rabies, herpes, smallpox, encephalitis, yellow fever, dengue fever and German measles.
Bacteria are true cells but have no nucleus. One example is tetanus which secretes a toxin which is taken up by the nerve endings and disrupts the nerves to the spinal cord and the brain, thereby producing an intense muscular spasm; hence the term 'lockjaw'. Other bacterial diseases include diphtheria, pneumonia, sinusitis, botulism, gonorrhoea, typhoid, cholera, whooping cough (pertussis), yaws, syphilis and dental caries.
Fungi produce diseases such as 'thrush' which results in an intense itching in the vagina, and ringworm, a skin disease which is most prevalent among children.
Protozoa are single-celled organisms which are able to change their shape and are generally bigger than bacteria. *Plasmodia* produce malaria, while *Trypanasoma* results in sleeping sickness which currently affects an estimated 40 million people.
Insects such as ticks and mites are multicellular organisms.
Helminths are also multicellular organisms and they often have a complicated lifecycle. Examples include hookworms, which attack the wall of the intestine and feed off blood, thereby producing anaemia in the host, and the *Ascaris* roundworm which can be present in such numbers as to mechanically block the gut.

Modes of transmission

There are five main modes of transmission:

By air Coughs and sneezes are said to spread diseases and this they can do if the host is infected. Diseases which spread in this way include mumps, measles, chicken pox, colds and influenza (viruses) and whooping cough, diphtheria, tuberculosis (bacteria). Such diseases are easily spread and while they tend to be fairly mild childhood diseases in advanced countries, they cause about a third of all deaths in the Third World countries.

By physical contact Diseases transmitted in this way include syphilis and gonorrhoea (bacteria), herpes (virus) and pubic lice (insects). Leprosy and yaws are two other diseases that fall in this category and they are estimated to affect some 10 and 40 million people respectively.

By food and water Examples of diseases that spread in this way include the cholera bacillus which is transmitted by sewage-infected water, and infectious hepatitis, a viral infection of the liver. Another such disease is hookworm which is estimated to infect some 450 million people.

By an insect Diseases in this category include the plague, which is produced when the *Versinia Pestis* is transferred to humans from the rat by blood-sucking fleas; and malaria, which results from the injection of the protozoa *Plasmodium* into the bloodstream by mosquitoes. It is estimated that half a billion people are at risk from this latter disease.

By being there already Many organisms may be living quite happily within the host as 'commensals' but changes (such as other illness or drug treatment) may lead them to multiply to produce disease. For example, certain types of fungus (yeasts) are normal commensals within the vagina of about a quarter of women, but if the host is taking antibiotics these may destroy commensal bacteria resulting in more 'food' for the yeasts which increase rapidly to produce candidiasis and intense itching.

The process of infection

A distinction is made between endemics, epidemics and pandemics. A disease is said to be endemic when it is consistently present in a community; when this is the case there tends to be sporadic outbreaks involving relatively few cases. In contrast, an epidemic is a paroxysmal increase in the number of cases while a pandemic is a massive epidemic that involves the whole world. At the individual level (and taking the simplest case with person-to-person transmission) we can recognise susceptibles who have not yet but can be infected, infectives who have the disease and can transmit it to others, and immunes who have recovered from the disease and are, temporarily or permanently, immune from further infection.

In more detail, an individual is infected at a particular moment in time and this is followed by an incubation period in which there are no signs of disease but the infected agent is multiplying and developing within the host. The incubation period can vary from a few hours in bacterial dysentry, to months for hepatitis induced by blood transfusion, to many years in the so-called slow-virus diseases. A period of a few days is most common; for example, the incubation period of influenza is about two days, polio about nine days and rubella eighteen days. A part of the incubation period is known as the latent period and this refers to time during which there are no signs of disease and the host is not infective. At some stage the latent period comes to an end, the infectious period begins and the host is capable of transmission, but the incubation period may not yet have ended and there may still be no signs of the disease. During this infectious period the host may come into contact with a susceptible and this individual may be infected. The time between the signs of symptoms in the first and second individuals is known as the serial period.

After some time the infectious period ends, the disease may acquiesce and if the host is not dead, the host becomes immune, that is resistant to the particular micro-organism that produced the infection. Some infections usually give immunity for life, while others like gonorrhoea provide no long-lasting immunity. Viruses in particular can change their 'form' over time and the body's immune system may fail to recognise that it has already been infected by this agent in another guise. The influenza virus, for example, appears to have many forms and it periodically undergoes a major 'shift'.

Control can be achieved by isolating the infectives (but this is difficult to do with transmission by air and of limited use if the individual is infective before incubation ends) and by preventive immunisation to protect the individual and to raise the herd immunity of the community artificially. Herd immunity refers to the lowered probability of an epidemic occurring because of the high level of immunity in the community. If a large proportion of the population is immune, there is a reduced chance of contact between the infectives and the immunes; the epidemic may fade out even though some susceptibles remain. Immunisation remains difficult or impossible for some diseases. For example, because of the variety of the forms of the influenza virus, no one vaccine is generally effective. For other diseases the vaccine may be in very limited supply because the organism can only be cultured in humans or in rare animals. For example, the micro-organism causing leprosy only thrives in humans and nine-banded armadillos. In such cases medical intervention is limited to treating those infected with the disease by drugs.

Disease ecology tradition

In 1950, a French physician, Jacques May, attempted to systematise the medical geography of communicable diseases believing that 'there is nothing in medical geography which is not based on well-known facts widely scattered throughout the various branches of medicine' (p. 9). He saw infectious disease as a complex interrelationship which only occurs when certain factors coincide in time and space. He recognised two major sets of factors: the pathogens or disease set and the geogens, the geographical factors. May distinguished between physical geogens (e.g. climate, altitude), human geogens (e.g. addictions, superstitions, sanitation) and biological geogens (e.g. plant-life, blood groups). The pathogens were subdivided into five categories:

> the **host** who gets the disease;
> the **agents** which are the direct cause of the disease; they are the different types of micro-organism (bacteria, virus, etc.) considered earlier;
> the **vectors** which are living organisms which are involved in the transmission of the disease, the commonest being insects (inanimate objects such as water involved in the transmission are known as vehicles);
> the **intermediate host** in which the agent develops or multiplies;
> the **reservoir** which maintains the agent in addition to the human host.

May's (1950) contribution was to examine and tabulate the variety of complexes involving the pathogens in relation to twenty-nine geographical factors (May (1960) provides a book-length account while Knight (1974) also gives a substantial table for a large number of diseases). May classified diseases according to the number of factors involved. The simplest diseases, the so-called non-vectored diseases only require two factors, that is the human host and the causal agent; examples include colds, influenza and measles where the transmission is from person to person (Girt, 1974). Three-factor complexes involve a host, a causal agent and a vector; an example is malaria which requires a human host, a protozoa for an agent and a mosquito as a vector. Four-factor complexes such as scrub typhus involve a human host, a rickettsia for an agent, a mite as a vector and a rodent as a reservoir. The disease ecologist attempts to understand the differing ecologies of each of these organisms in relation to geogens in order to develop an explanation of why a disease is occurring in a particular area.

Swimmer's itch

As an illustration of May's approach we shall consider schistasome dermatitis or swimmer's itch as it is more commonly known (Jarcho and Van Burkalow, 1952). In the late 1940s this disease was undergoing an apparent increase in the number of people infected and the disease was being reported over a wider area. Swimmers were reporting, a few minutes after leaving the water, an intense itching (and sometimes a rash) of those parts of the body that had been in contact with the water. Such sensations often disappeared within a hour but to return within ten hours and to last a week, this time with swelling of the affected parts.

The disease ecology approach begins with a consideration of the number and nature of the pathogens involved and then the lifecycle of these pathogens is studied in relation to the changing geographical environments which they inhabit. In May's terminology swimmer's itch is a four-factor complex. Humans are the infected hosts, the causal agent is a microscopic worm, the intermediate host is a freshwater snail and aquatic birds are usually the reservoir. We can begin the life cycle of the worm with the aquatic birds in which the worm lives, copulates, and lays its eggs. The eggs are then discharged in excrement, and some of them will find themselves in fresh water. On hatching the microscopic larvae have to penetrate certain types of water snail for it is only there that they can develop. When this development ends, the worms are discharged from the snail and they normally re-infect the waterfowl by penetrating the skin, passing by the bloodstream to the internal organs where they fully mature and lay eggs, and the cycle begins again. Humans are not part of this natural cycle but the worm larvae, on discharge from the snail, may accidentally encounter a human host, burrow into the skin, producing inflammation and itching, and eventually die because they cannot enter the human bloodstream.

The larvae enter the snail in the summer and develop until the next summer when they are discharged into the water. The snail itself will normally mature, breed and die by this second autumn and the new generation of snails will be infected in the summer when the birds begin discharging worm eggs within a fortnight of infection. The environment required for this natural cycle is freshwater lakes which are suitable breeding grounds for snails and wildfowl. The reason for the increase in the number of human infectives is that people are increasingly entering what was previously the 'silent zone', that is where all the factors except the human agent have existed naturally. The development of car-based commercial recreation is increasingly using lakes with inevitable results. In terms of prevention there is little that can be done except to avoid swimming on warm days in late August (when the discharge from the snail is highest) and when on-shore winds are concentrating larvae at the water's edge. Control would have to be concentrated on the static snail rather than on the migrating wildfowl, if there was to be no restriction on recreational swimming.

It is stressed in this ecological work that organisms often exist in a delicate balance in nature but that this can be easily disturbed by humans. Knight (1974) relates the unexpected results of malaria control in Borneo by insecticide. The insecticide not only killed the mosquitoes but accumulated in cockroaches which then poisoned the geckoes which fed on them. The drugged geckoes became an easy target for the village cats who, being at the end of the food chain, died. This caused a substantial increase in the rodent population, with the attendant risk of plague. It finally required cats to be parachuted into the remote villages to control the rodents!

One gets the feeling in the ecological work in the May tradition that there was natural balance and harmony until human intervention, and the work has been accused of environmental determinism (Phillips, 1981, p. 14). It is the physical and biological factors that are stressed as the major determinants; humans are cast as 'intruders' and the social, and particularly the societal factors, are not considered fully.

Regularities, forecasting and control

Although the disease ecology tradition continues in geography (for example Learmonth (1978) on malaria, Hunter (1966) on river blindness and Knight (1971) on sleeping sickness, although Knight (1974) argues that there has been little original contribution beyond synthesis and review), there was a move in the late 1960s towards a different approach to infectious diseases as technically minded geographers tried to use the methods of quantitative analysis to examine the nature and timing of epidemics over a geographical area. The original aim of these investigations was not so much to study the nature of disease but rather to reveal the general nature of spatial processes and the spatial structure of the area over which the disease was spreading. One group in particular, led by Peter Haggett and based in Bristol, was active in this area in the late 1960s and through the 1970s. They required abundant data disaggregated by time and space to develop and test their models of spatial processes, and they found that the weekly returns of measles for urban and rural districts in Britain were fine for their purposes (in the original investigation, Cliff *et al.* (1975) data on unemployment were also used).

Measles was chosen because of its straightforward characteristics; it appears that virtually everyone who comes into contact with an infective becomes diseased, it has a relatively short incubation period, and there is subsequent lifetime immunity. There is a major contrast with the disease ecology tradition for in trying to understand spatial processes and structure, the disease has been deliberately chosen to be a 'simple' one; in ecological terms it is a two-factor disease (person-to-person) and there is no concern for the intricacies of the differing ecologies of host, agent, vector and reservoir. There is a price to be paid for this simplicity in that their results may not generalize to other more complicated diseases. Indeed, as Bradley (1982) has argued: 'there is a tendency towards premature quantitative simplification to produce a spurious elegance in the analysis of messy and incompletely comprehended reality' (p. 321) and 'far more theoretical work has been done on the direct infections. Yet it is the simplicity of the transmission process that may limit the utility of the models so far devised' (p. 327). Eschewing such debate, the spatial work on measles will be discussed in due course, but it is easiest to begin with regularities over time and then to proceed to consider regularities over space.

Regularities over time

The graphs in Figure 4.1 show the number of cases of measles on a monthly basis for four countries over a 25-year timespan and they show a number of interesting features. Measles is clearly a recurrent epidemic but the frequency of recurrence differs from country to country with a major epidemic nearly every year in the USA, every two years in the UK, every three years in Denmark (at least in the latter half of the period), and roughly every three and a half years in Iceland. While the USA, UK and Denmark go into an endemic period between epidemics, Iceland has a complete fade-out and there are no reported cases in the inter-epidemic period. Another feature is the substantial reduction that

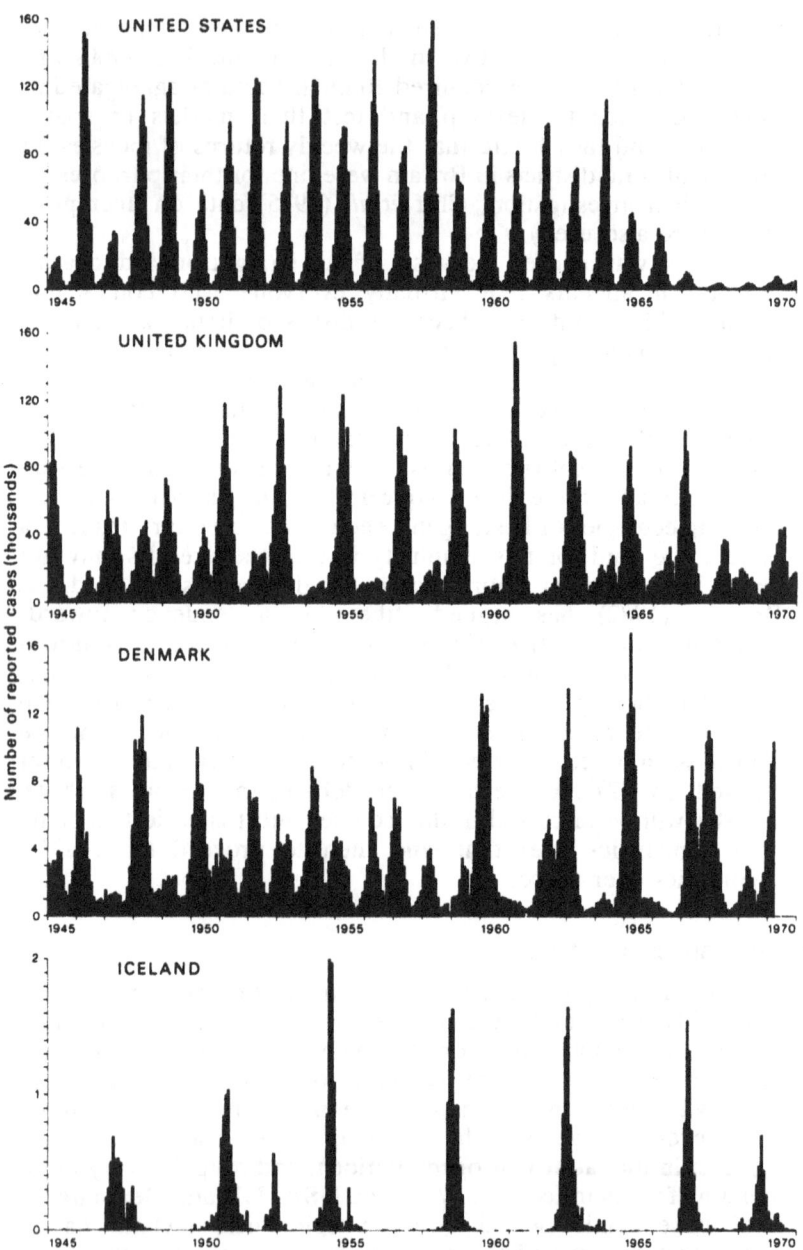

Figure 4.1 Number of reported cases of measles per month, 1945–70
Source: Cliff *et al.* (1981)

occurred in the USA after 1965, and to some extent in the UK after 1968. The latter feature is easily explained as the results of the introduction of widespread community vaccination programmes in these countries after those dates. The frequency of recurrence is clearly related to population size (USA, 210 million; UK, 56 million; Denmark, 5 million; Iceland, 0.2 million) and so too is the change from endemic to complete fadeout. Measles does not have a natural reservoir so it has to rely on host-to-host transmission to remain endemic in a community. If we take the British example, we can begin with the peak of the epidemic in which a large number of susceptibles are being converted into infecteds and then into immunes. At this point the stock of susceptibles is being reduced throughout the country and, because the number of new susceptibles (babies aged 6 months losing the immunity acquired from their mothers) is less than the number becoming immune, there must be a decrease in the number of new infectives.

The epidemic fades and retreats into large cities in which the number of new susceptibles is sufficient to permit some transmission and sporadic outbreaks. In the country as a whole the number of susceptibles continues to grow, and after a period of eighteen months or so, there are enough susceptibles to set off another epidemic. The process is somewhat different when measles is introduced into a relatively isolated, low-population community with a large number of susceptibles that will have arisen since the last epidemic. Here it moves like a fireball, consuming all susceptibles until there are no hosts left to maintain the disease, and there is complete fade-out as in the Icelandic example.

Within one country Bartlett (1957) considered the effects of population size by examining the mean period between epidemics for nineteen English towns. He classified the towns into three types:

type **A**: in which measles remains endemic; a population of over 250,000 is required to generate sufficient susceptibles to maintain the outbreaks;
type **B**: in which there is a complete fade-out, but there are regular epidemics; a population threshold of about 10,000 is required for such a pattern;
type **C**: in which there is again complete fade-out but the epidemics are now irregular; such towns have a population of below 10,000 and they do not generate sufficient susceptibles to partake in every epidemic that affects the country.

In a community or in a country with a sufficient population, one can anticipate regular and explainable epidemics. These patterns are very noticeable for measles (because of the infectivity of the disease and the lifelong immunity it provides) but similar, if not so remarkably regular, patterns are found for other diseases.

A simple mathematical model

Given the concepts of susceptibles, infectives and immunes and the rates of transition between them, it is possible to build a simple model that provides insight into periodicity of transmission and the herd immunity required to prevent outbreaks. It must be stressed at the outset that these models effectively treat the whole community as one, and do not consider geographical variations at all. The following accounting system can be recognised:

$$N = S + L + I + Z \tag{4.1}$$

where N is the total population of the community, S is the number of susceptibles, L is the number that are in the latent period, I is the number of infectives, and Z is the number that have been isolated, plus the number that have recovered and are now immune.

We also need to define the rates at which people move from one state to another, thus: μ is the per capita, per annum death-rate as people die and are removed from the system altogether; life-expectancy is therefore $1/\mu$;

μ is also the per capita, per annum birth-rate (we did call this model simple!); in advanced countries, where the average life-expectancy is about 70 years, approximately 1/70th of the people will die every year to be replaced by a similar number of births;

λ is the *per capita* recovery rate and represents the rate which people become immune ($I \rightarrow Z$); the average length of the infectious period is therefore $1/\lambda$;

σ is the per capita rate of transfer from the latent to the infectious state ($L \rightarrow I$); the average latent period is therefore $1/\sigma$;

β is the transmission coefficient, the rate at which susceptibles become infectious ($S \rightarrow I$).

A key ratio is the reproductive rate (R) and this gives the expected number of secondary cases produced by an individual given S susceptibles; if it is less than 1 the disease cannot reproduce itself and it will therefore die out in the host population. R is given by the following equation (Anderson, 1982, p. 5):

$$R = \frac{\sigma \beta S}{(\sigma + \mu)(\lambda + \mu)} \qquad (4.2)$$

Accordingly, secondary infection will be produced at a rate βS throughout the lifespan, $1/(\lambda + \mu)$, of an infectious individual, and of these a fraction, $\sigma/(\sigma + \mu)$, will survive the latent period to become the second generation of infected individuals. To take an example, the life-expectancy $(1/\mu)$ in advanced countries is about seventy years, and the latent and infectious period for measles is so comparatively short that equation 4.2 can be reduced to:

$$R = \frac{\beta S}{\lambda} \qquad (4.3)$$

Assuming that all the rate parameters are independent of age, Deitz (1976) has shown that R can be estimated from the relation:

$$R = 1 + L/A \qquad (4.4)$$

where L is years of life-expectancy $(1/\mu)$, and A is the average age, in years, at which individuals acquire the infection. Thus, for measles in England and Wales we can estimate R as follows:

$$R = 1 + 70/4.8 = 15.6 \qquad (4.5)$$

Measles is clearly very infectious with an infected individual expected to give rise to roughly 16 secondary infections. If we are to reduce this reproductivity rate from this high level to below 1 to achieve a fade-out, a high proportion of the population will have to be immunised. Following Anderson (1982, p. 19), P is the proportion of the population that needs to be immunised if such a complete fade-out is to be achieved and this is given by

$$P = (1 - 1/R) \qquad (4.6)$$

which is in the British case

$$P = (1 - 1/15.6) = 0.935 \qquad (4.7)$$

Therefore, about 94 per cent of the susceptibles in England and Wales must be immunised if the disease is not to remain endemic. Theory accords well with reality for there are examples where a measles outbreak has occurred when over 90 per cent of the community was immunised.

Anderson and May (1979) show the recurrent period for an epidemic can be approximated by

$$T = 2\pi(AD)^{0.5} \qquad (4.8)$$

where T is the length of the recurrent period, A is average age of infection, and D is the sum of the lengths of the latent and infectious periods, which for measles is about 10 days when allowance is made for isolation once the disease is recognised. In proportion of a year D is 0.027, and given an average age of measles attack in England and Wales of 4.8 years, the expected recurrence period is therefore:

$$T = 2 \times 3.14 \times (4.8 \times 0.027)^{0.5} = 2.2 \text{ years} \qquad (4.9)$$

which is supported by Figure 4.1. The identities of equations 4.4 to 4.9 are clearly interrelated and they suggest that if the immunisation rate is below the critical threshold for fade-out (4.6) then epidemics will still occur but the reproductive rate (R) will be lowered, the average age (A) of infection must increase (4.4) and so must the length (T) of the inter-epidemic period (4.6). Since 1968 England and Wales have operated a voluntary immunisation policy in which roughly 50 per cent of children have been immunised; epidemics have indeed continued with an increased recurrent interval and increased age of attack (Fine and Clarkson, 1982).

Regularities over space

There has been considerable geographical research in the last thirty years on diffusion of phenomena over space. Much of the earlier research concerned itself with the diffusion of new ideas or innovations (for example, Hagerstrand, 1953). Two contrasting spatial forms of the diffusion process have been identified at the aggregate level even when transmission at the individual level remains from person-to-person. In one type, the diffusion process is determined by distance and is known as contagious diffusion; the innovation spreads from one place to near neighbours, and there is a wave-like form that moves across the country. The second type is known as hierarchical diffusion and the innovation spreads not between nearest towns but by trickling down the central-place hierarchy. Thus, an innovation may start in a small provincial town but while spreading to neighbouring hamlets it also jumps spatially to the capital city from where it spreads rapidly down to the major provincial cities and from these to the major towns and so on until all the country is covered and there are no remaining potential adopters of the innovation.

The actual form of the diffusion depends on the settlement hierarchy of the country. If it is not well developed and it has low interconnectivity we can expect slow spread dominated by distance, that is contagious diffusion. In contrast, hierarchical diffusion will occur rapidly in countries with a developed, integrated network of communications. Indeed, we can examine how the spatial economy is developing by studying diffusion processes within a country over time. To do this we require recurrent phenomena and innovations are clearly unsuitable, for a specific innovation can only pass through a country once! But recurrent epidemics of communicable diseases do allow such a study and Pyle (1969) examined the changing nature of cholera diffusion in the USA in the nineteenth century, a time of rapid development of the country.

Cholera in the USA

Pyle examined three great epidemics: 1832, 1848–9, and 1886. The earliest epidemic was introduced into the continent by two routes, one originating in Quebec, the other in New York, in June of 1832. The epidemic was relatively slow moving and it took over 150 days to reach New Orleans. It is possible to test the nature of the spatial form of the epidemic by a standard regression model with the number of days it took to reach a particular town as the dependent variable (Y) and two independent variables, distance from the origin (X) in tens of miles, and the population (Z) in thousands of the town at the time. In a simple model of the New York route using only distance (Cliff *et al.*, 1981, p. 31) the following model was estimated, the figure in brackets being the calculated t statistics:

$$Y = 34.9 + 0.77X \qquad R^2 = 0.78 \qquad (4.10)$$
$$(10.6)$$

which shows that some 78 per cent of the variation in the timing of cholera reaching a town could be accounted for by the distance from New York and every hundred miles from New York meant 7.7 days delay in getting the disease. The t value of 10.6 means that the estimated distance coefficient is significantly different from the zero value of no effect. The introduction of population size into the model hardly improved the fit at all:

$$Y = 44.3 + 0.71X - 8.5Z \qquad R^2 = 0.79 \qquad (4.11)$$
$$(10.6) \quad (1.6)$$

and the population coefficient is not significantly different from zero at conventional significance levels. This is strong evidence that it was distance that was controlling the spread of the New York epidemic and that contagious diffusion was operating;

similar results were found for the wave that began in Quebec.

The 1848 epidemic also had two points of origin, New Orleans and New York. The findings for the major New Orleans route can be summarised as follows:

$$Y = 21.7 + 0.98X \qquad\qquad R^2 = 0.40 \qquad\qquad (4.12)$$
$$(3.0)$$

$$Y = 78.0 + 0.99X - 50.1Z \qquad R^2 = 0.60 \qquad\qquad (4.13)$$
$$(3.6) \qquad (2.5)$$

Both variables are required for an adequate model and the negative coefficient for population size shows that the smaller the town the longer the time that cholera takes to get there. The 1848 New Orleans epidemic therefore spread by both contagious and hierarchical diffusion.

The third epidemic, that of 1866, again began in New York, and it yields the following models:

$$Y = 149 - 29.5Z \qquad\qquad R^2 = 0.44 \qquad\qquad (4.14)$$
$$(4.7)$$

$$Y = 155 - 0.03 - 31.4Z \qquad R^2 = 0.44 \qquad\qquad (4.15)$$
$$(0.4) \qquad (3.8)$$

The spread is now dominated by the effects of population size and the distance term does not lead to a substantial improvement of the model. (But there are some problems of estimation as distance from New York is closely related to population size and it is difficult to disentangle the separate effects due to this collinearity.) Overall, these results confirm our speculation that as the spatial economy develops and becomes integrated, diffusion will not be based on distance alone for population size will play an increasing role. In the days of the stage-coach and river-boat in 1832, cholera would move as a slow wave, but by 1866 and the beginnings of the railways, the disease could jump from one town to another. Today with the continued integration of the world spatial economy, rapid diffusion can be anticipated unless communicable diseases are controlled in some way.

Measles in Cornwall

Our second example of spatial diffusion is the work of Haggett (1976) on testing alternative models of diffusion during a measles epidemic in Cornwall. Before discussing his results we need a measure of spatial patterning or spatial autocorrelation as it is technically known. Figure 4.2(a) and (b) represents the patterning of measles in one week in a peninsular area; the black circles are the reporting districts with measles and the white circles are those without; the area of the circle is in proportion to the population size. The question is whether (a) or (b) shows spatial patterning? You will probably answer that (a) has strong spatial patterning for all the measles areas are adjoining each other in the west of the peninsula, while in (b) the measles are found throughout the area in no particular spatial patterning. If this is your answer you have been examining measles in the 'neighbourhood' or 'contagious' space of Figure 4.2(c) and (d) in which places next to each other have been joined. In the spatially patterned distribution of (c) there are a large number of black-black joins, a large number of white-white joins, and no black-white joins. Actually, there are a lot more same-colour joins than would be expected from chance alone, so we say the pattern is not random but it has marked spatial autocorrelation. In contrast, (d), which has about the number of same-colour joins as you would expect by chance, is deemed spatially patternless or random.

These results only apply to the specific join structure we have used, in this case based on distance. If we now join up the districts in hierarchical space, with the biggest place joined to the next biggest and so on, our findings are completely changed, for (e) is now unpatterned while (f), with its disproportionate number of same-colour joins, is strongly patterned. Therefore, by altering the join structure we are able to detect different types of patterning: (a) has patterning in distance or contagious space but is unpatterned in hierarchical space, while the converse is true of (b). Indices of spatial autocorrelation have been developed for such two-colour maps and they essentially assess, for a given join-structure, whether there are more same-colour joins than would be expected from chance; an index of greater than 1.64 signifies more patterning than would be expected 95 times out of a 100 by chance alone. Ebdon (1985) gives a straightforward introduction and provides a suitable computer program.

Measles is not endemic in Cornwall but it is in the nearby city of Plymouth, and epidemics occasionally move through the area, that is we are dealing with a type B situation in the classification of Bartlett (1957). Figure 4.3(a) shows the number of cases in the whole of Cornwall over a forty-week period for an epidemic that began in 1966. As can be seen from the graph the number of cases increased rapidly to reach a peak in week 15, which was also the week that the largest number of districts were reporting cases, Figure 4.3(b). Haggett wanted to test the spatial form this

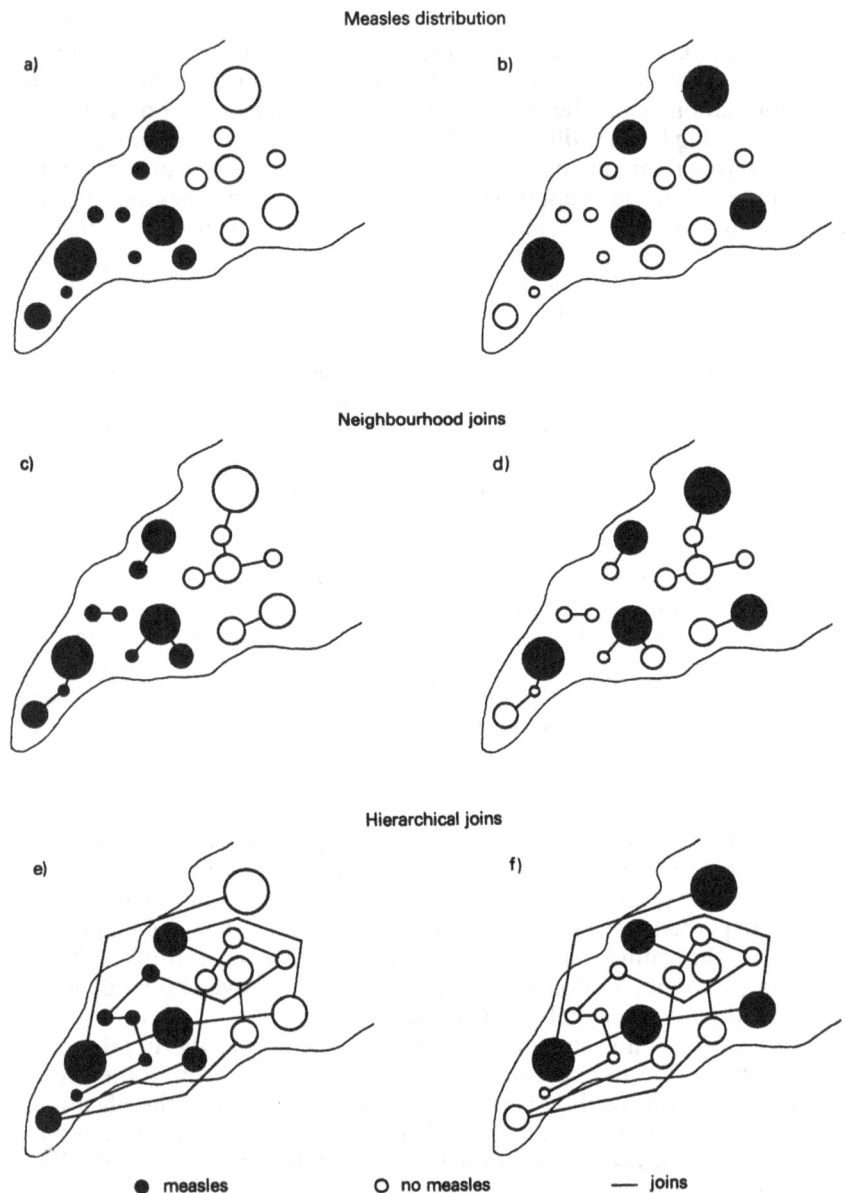

Figure 4.2 Patterns in space

epidemic took, and to discern whether there was a difference between epidemic build-up and fade-out. To examine the spatial form of the spread, Haggett constructed seven different join structures. They were:

G1: local contagion assuming a spread only between contiguous districts;

G2: wave contagion assuming spread by shortest path from Plymouth;

G3: regional contagion assuming spread occurs within two regional subsystems based on east and west Cornwall;

G4: urban and rural hierarchical spread within separate sets of urban and rural districts;

G5: hierarchical assuming spread down the population-size hierarchy;

G6: hierarchical assuming spread down the population-density hierarchy;

G7: journey-to-work contagion assuming spatial spread from urban centres to surrounding rural areas.

A district was coloured black if it had measles in that week or in the preceding three weeks to allow for the possibility of continued infection; white otherwise. The two-colour spatial autocorrelation statistic was evaluated for each type of join structure for each of the forty weeks, and they are plotted against time in Figure 4.3(c); the significance level of 1.64 and the epidemic peak in week 15 is also shown. The greater the degree of correspondence between the pattern in a week and the join-structure, the larger the value plotted. As the figure shows, the epidemic build-up is dominated by hierarchical spread with high values on graphs G4, G5 and G6 while the fade-out is dominated by contagion, G1, G3, G7. These results imply that diffusion is based on the central-place hierarchy in the early stages but localised spread takes over once the epidemic is established. An effective vaccination strategy would require widespread immunisation if an epidemic is to be checked.

Measles in Iceland

The team that had been examining the diffusion of measles in the south-west peninsula of Britain next turned their attention to Iceland (Cliff *et al.*, 1981). They had found that the diffusion pattern was complicated in the south-west because measles did not fade out completely, and it was therefore difficult to separate the end of one epidemic from the start of another. Iceland was

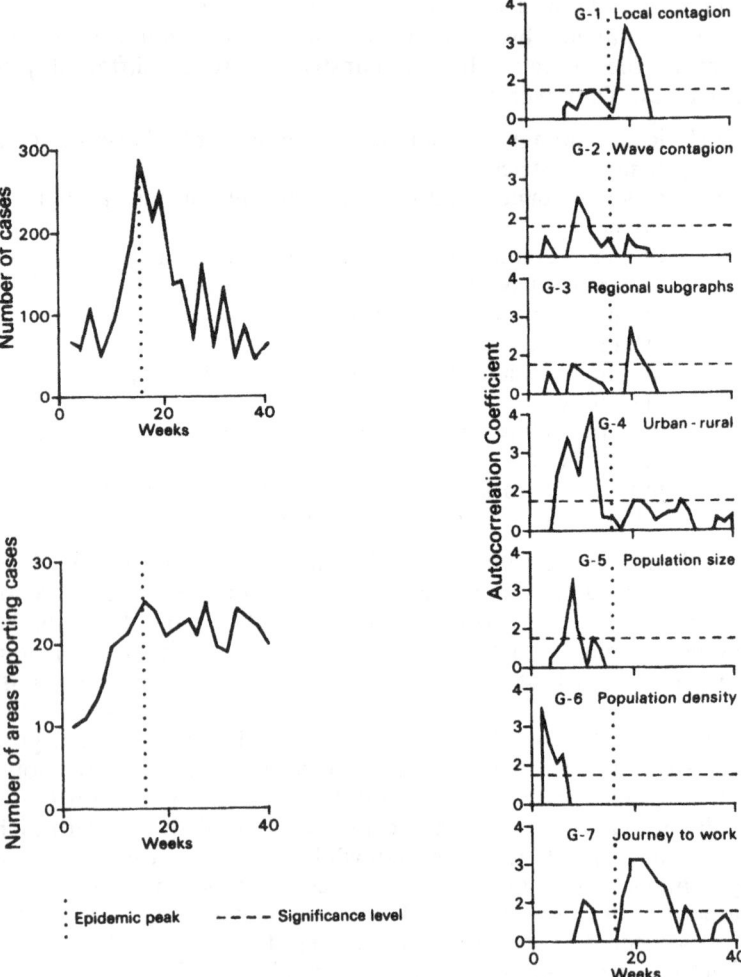

Figure 4.3   Measles in Cornwall: patterns in space and time
Source: Haggett (1976)

chosen as a more appropriate geographical laboratory due to two favourable attributes. First, the quality of the records was high and there were detailed data for fifty relatively fine geographical areas on sixteen distinct epidemics that had occurred in the first seventy-five years of this century. Second, as we have seen, Iceland has a complete fade-out between regular epidemics. The basic findings on the spatial form of the Icelandic diffusion process are summarised in Figure 4.4 for a typical outbreak and five stages can be recognised:

(1) introduction from abroad as a boat containing an infective lands at Reykjavik;
(2) the disease is spread to the smaller towns and
(3) gets into the schools from where it is spread to
(4) the homes where younger brothers and sisters become infected;
(5) the epidemic fades as the number of susceptibles runs out.

They also examined the changing nature of the sixteen epidemics in relation to the evolving Icelandic space-economy. For example, they found an increasing frequency of epidemics after 1945 (which reflects increasing international contact), a faster spread outside Reykjavik (a result of urbanisation, improved transport and the establishment of fixed schools) and a slower velocity within the capital (which was a result of improved medical care and decreasing population densities).

Forecasting

The research team did not stop at description but were concerned to develop forecasting procedures that would predict the size and time of onset of an epidemic in Iceland, and when and to what extent it would affect different communities. In their own words they employed an 'armada' (Cliff *et al.*, 1981, p. 181) of techniques, which can be grouped into two major types: space-time models and stochastic process models. The underlying concept of the space-time models is that the past is the key to the future. One can begin with the simplest of auto-regressive models:

$$E(Y_{i,t}) = b_1 + b_2 Y_{i,t-1} \qquad (4.16)$$

in which the expectation (E) of the level of measles (Y) in place (i) at time (t) is a function of number of measles in place (i) one time period earlier (t−1). Such models are highly reminiscent of regression models (Chapter 3) but cannot be estimated by the usual OLS procedures (Haggett *et al.*, 1977, Chapter 16). This model is termed a first-order auto-regressive model but it can be easily extended so that measles depends not only on the previous week but on the number of measles a fortnight before, that is a second-order model:

$$E(Y_{i,t}) = b_1 + b_2 Y_{i,t-1} + b_3 Y_{i,t-2} \qquad (4.17)$$

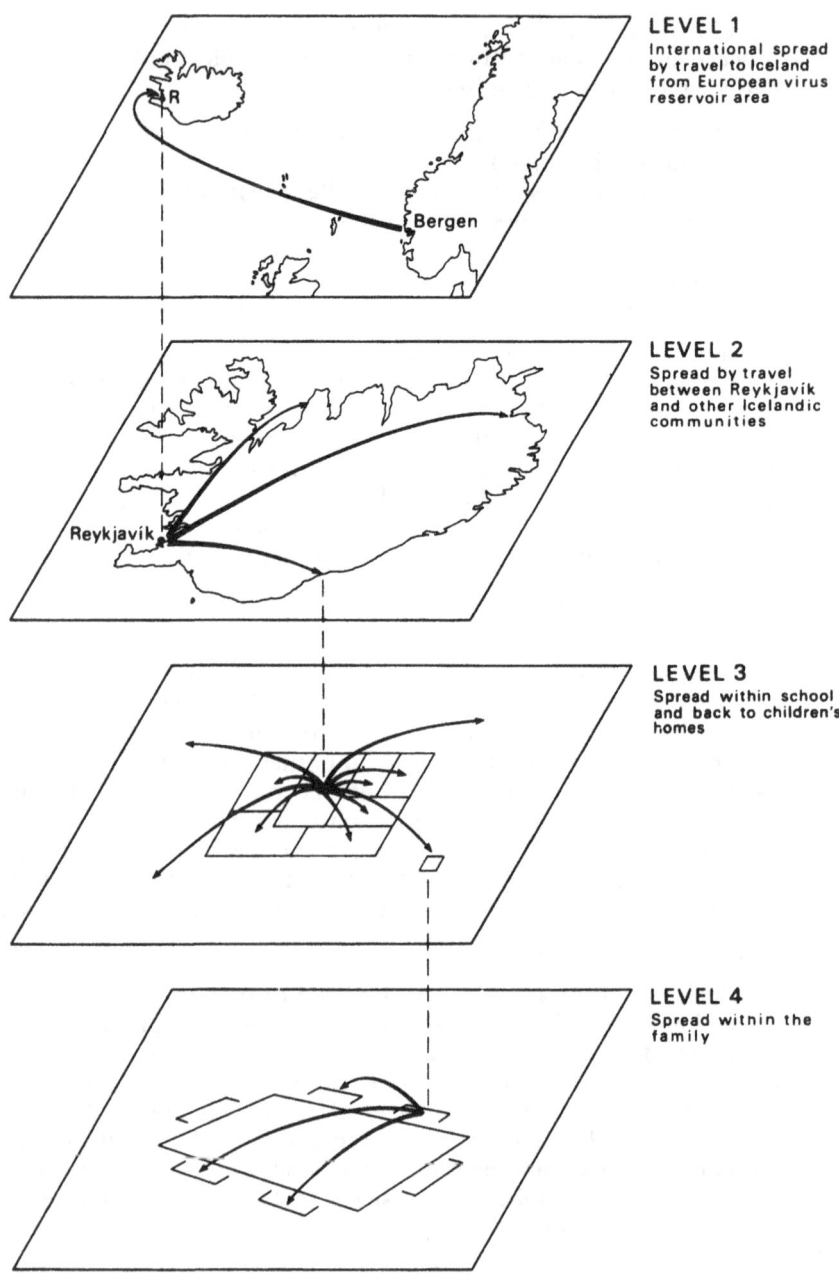

Figure 4.4   Measles diffusion in Iceland   *Source*: Cliff *et al.* (1981)

The model can also be extended to take account of space-time variations in a model of the form:

$$E(Y_{i,t}) = b_1 + b_2 Y_{i,t-1} + b_3 Y_{i,t-2} + b_4 \Sigma w_{ij} Y_{j,t-1} \qquad (4.18)$$

where the additional term means that the number of measles in place (i) is not just a function of measles in that place a week and a fortnight earlier but is also a function of the number of measles in 'neighbouring' places a week earlier; w_{ij} represents the degree of connection between places (i) and (j). The model can be extended by incorporating further space-time lags, by allowing the b parameters to vary over time and by including terms that allow random shocks to affect the system. Estimation techniques have been developed for these models and so have identification procedures that suggest the appropriate degree of space and time lagging that is required. Once a successful model has been calibrated it can be 'run on' to predict the future levels of measles.

Such space-time models are general procedures that can be used for the forecasting of any phenomena that vary over time and space. The stochastic process models that were used in the Icelandic study had to be specifically developed for the measles case. Although these models come in a very wide variety of forms, their basic underlying concept is the so-called mass action principle in which the speed of the spread is proportional to the product of the density of susceptibles multiplied by the density of infectives. Thus, it is possible to derive a regression-type model in which the expected number of new measles cases at time (t) at place (i) is dependent on the interaction between the susceptibles and infectives in place (i), plus the interaction between susceptibles and infectives in neighbouring areas. Once the model is calibrated, it can again be 'run ahead' to forecast the number of infectives which can then be used to update the number of susceptibles and thereby produce the next set of forecasts.

Despite the large number of models used and the technical sophistication of the work, the quality of the forecasts was disappointing when they were compared to reality. For example, if a model forecast the recurrence of an epidemic quite accurately then it tended to overestimate the size of the epidemic. Overall, they found that the best models were the most complicated; they required time-variant parameters to handle the major differences between epidemic build-up and fade-out, spatial lag information led to some improvements in forecasting recurrences, while estimates of the number of susceptibles led to better forecasts of epidemic size.

Although the measles forecasts have been deemed inadequate by the researchers themselves, Soviet work on influenza has been much more encouraging, as reported in Bailey (1975). Let us begin by defining the various symbols and parameters of the model. The number of individuals in a city (i) are subdivided so that:

$$N_i = S_i + I_i + U_i + Z_i \tag{4.19}$$

where N_i is the total population; S_i is the number of susceptibles; I_i is the number of circulating infectives; U_i is the number of isolated infectives; and Z_i is the number of immunes. β and λ are global epidemiological parameters for the specific influenza strain and represent respectively the general rate of infection and the recovery rate for all cities. The remaining parameter is w_{ij} and this symbolises the degree of connection or amount of traffic between city (i) and city (j). Given such a formulation it is possible to derive a set of differential equations that describe the operation of the system over time:

$$\frac{dS_i}{dt} = \frac{\beta S_i I_i}{N_i} + \overset{G}{\Sigma} \left(\frac{w_{ji}S_j}{N_j} - \frac{w_{ij}S_i}{N_i} \right)$$

$$\frac{dZ_i}{dt} = \lambda I_i + \overset{G}{\Sigma} \left(\frac{w_{ji}Z_j}{N_j} - \frac{w_{ij}Z_i}{N_i} \right) \tag{4.20}$$

$$\frac{dI_i}{dt} = \frac{\beta S_i I_i}{N_i} + \overset{G}{\Sigma} \left(\frac{w_{ji}I_j}{N_j} - \frac{w_{ij}I_i}{N_i} \right) - \lambda I_i$$

Although such equations look complicated the basic ideas are simple if not simplistic. For example, the third equation merely states that the rate of increase of circulating infectives in city (i) depends on

(1) the proportion of the population of that city that are converted from susceptibles to circulating infectives: the 'mass-action' principle;
(2) the proportion of infectives in all G other cities in the system and how 'connected' they are to the particular city (i);
(3) the decrease in circulating infectives as they recover and become immune.

This model as used by the Soviet team ignored the number of isolated infectives because of the lack of data, but available records for the early period of the epidemic in one city, Moscow, were used to estimate suitable values for β and λ for the 1965 influenza epidemic. Estimates of the connectivity between each pair of cities were based on a simple gravity model formulation in which the amount of traffic depended on the size of each of the cities and the distance between them (Foot, 1981). Put simply, the greatest connectivity would exist between the largest cities that were closest together.

On this basis, and given that the epidemic appeared to have begun in Leningrad, the model was used to forecast the course of the epidemic for all 128 cities in the system. For Moscow and twenty-three other cities the records were good enough to check the quality of the forecasts and extraordinarily good qualitative agreement was claimed in terms of size, time of onset and the maximum point of the outbreak. The model has been used to forecast more recent outbreaks and the available 'flu vaccines have been used for key personnel (firemen, police and teachers) in those cities that are predicted to suffer an epidemic. The model has been subsequently refined with, for example, the recovery rate now varying according to the age of the infected individuals, and by better estimates of the interconnectivity of the cities, but despite these refinements it remains an aggregate 'naïve', 'black-box' model.

The success of the Soviet model and the relative failure of the models developed for Iceland can be explained in two main ways. Firstly, the Icelandic forecasters were trying to predict when the epidemic would start, while the Soviet team knew that it had begun (and where) and were trying to predict when it would get to other cities. Secondly, the Soviet epidemic was based only on large cities and large numbers of cases; in contrast, the Icelandic models were based on relatively small numbers of cases which would inherently have a larger stochastic or 'chance' element (see the 'small number' problem in Chapter 2).

Alternative perspectives: the importance of the host

The Soviet influenza model is a clear illustration of the Cartesian and biomedical approach applied to epidemiology. The approach aims to find regularities in the natural world by analysis and disaggregation; the total population in the city being subdivided into susceptibles, infectives and immunes. Having found regularities, the model is then run into the future to predict occurrences that should occur if past patterns hold; control is achieved by biomedical intervention in those cities that are likely to experience sizeable outbreaks. The model assumes that if an individual is not already immune then the chances of becoming infected are based on the infection rate in the whole system, and on how many infectives live in their own and nearby cities.

The underlying theory of the model is captured in the β term; every susceptible individual has the same propensity to become infected by the disease. The model accepts the doctrine of specific aetiology in its original form of the germ theory in which contact with a specific virulent micro-organism will produce a specific disease. The model is socially blind in that the same rate of infectivity is presumed to apply if the individual is Algernon Smythe, living in a penthouse with a substantial bank balance or Fred Crudge, unemployed, former manual worker, living in poor housing on an inadequate diet (or indeed, their Soviet equivalents). From this perspective, the model must be deemed 'instrumentalist' (in the sense of Sayer, 1984, p. 115) in that there is no real social causal explanation of the mechanisms of how an individual becomes infected; the model is merely a calculating device, albeit a successful one in its own terms.

This social blindness is even more marked in the biomedical research where attention has been concentrated on micro-organisms and people are largely ignored. This is despite the recognition by the early biomedical investigators, notably Pasteur (1822–95), Virchow (1821–1902) and Pettenkofer (1818–1901), that something more than a micro-organism is involved. Pasteur is supposed to have said on his deathbed that 'the germ is nothing, the terrain is everything' in the belief that the germ could only flourish if the host was accommodating, otherwise it would die or at least not multiply to a level to induce disease. For example, *Meningococci* are frequently found in people's throats without any sign of illness and yet the same bacteria are said to produce meningitis (inflammation of the layers covering the brain and spinal cord) in other hosts. In fact, infection can be considered the norm while disease is rare; there are few infections that produce disease in a simple causal relationship (the organism always causing the disease; botulism appears to be one). For the majority of infections the key factor is the reaction of the host and this depends not only on genetics and physical constitution of the individual but also on psychological, emotional, environmental and social conditions. Indeed, Sencer and colleagues (1967) found that the threshold for herd immunity for measles differed between urban areas according to the ethnic and social composition of the areas.

Virchow, regarded by many as the founding father of pathology, was sent to Upper Silesia in 1848 to act as an 'outside expert' on the typhus epidemic that was affecting this Prussian province. His report (for a more detailed account and translation, see Taylor and Reigler, 1985), which was suppressed by the Prussian government, begins with a geographical, anthropological and sociological account of the province. Virchow considers the housing conditions, the diet and education of the people and shows how the severe material deprivation of certain social groups is linked to the role of the Catholic Church, and ethnic discrimination in terms of occupation, in producing a caste-like system of social stratification. After a clinical account of typhus, Virchow carries out an epidemiological analysis reporting on the accuracy of the disease notification and its morbidity and mortality by age, sex, social class and occupation. He attributes the marked differences he finds not to the individual susceptibility of the host but to the deeper structure and organisation of society. Accordingly, he does not recommend more medicine and health care but a whole series of sweeping societal reforms, believing that 'if you want to achieve anything you have to be radical'.

His long-term recommendations include unlimited democracy, agricultural improvement, industrial development, full employment, taxation reform, higher wages, the establishment of agricultural co-operatives, universal education, and the disestablishment of the Catholic Church. Underlying this classic report (Waitzkin, 1981) are specific views on infectious diseases. He refers to typhus and the other diseases that were endemic in Silesia (such as dysentry, measles, tuberculosis) as 'crowd' or 'artificial' diseases to emphasise that while they had their origin in specific bacteria, their geographical spread and individual susceptibility were determined by social factors. Even though he recognised the insights of bacteriological research, he was never to accept the simple causal relationship that a micro-organism causes disease (Galdston, 1954, pp. 49–54).

A few years after Koch had proclaimed a bacillus as the cause of cholera, Pettenkofer was trying to persuade the city of Munich to eradicate the disease by social reforms and changes. He wrote (translated in Sigerist, 1935, p. 572) that 'political and social conditions are also influential upon the health and mortality of a population. All over the world the rich generally enjoy better health and live longer than the poor.' He offered to demonstrate that the cholera vibrio alone could not possibly be the cause of cholera by swallowing the stuff. He obtained a culture that Koch had taken from a fatal case, he drank it and although huge numbers of the micro-organism were found in his faeces, he suffered no more than mild diarrhoea, despite being over 74 years old. The experiment was repeated by others with the same result. The stock answer to the question of 'how come?' is that he had acquired immunity in some way but two other explanations are possible. First, he was fit, healthy and well nourished and thereby he was constitutionally strong enough to see off the disease. Second, his unshakeable, presumptuous self-confidence did not permit the disease to develop. Such psychological and social themes will be further developed later in this chapter, for we will now consider the evidence that many apparently infectious diseases are spread by extra-terrestrial forces.

Diseases from space?

This is the title of a book (without the question mark) by two astronomers, Hoyle and Wickramasinghe (1979). Their argument is a kaleidoscope of astronomy, meteorology, the history of disease, warfare and much else including the anatomy of the human nose. The book contains imaginative speculation as well as empirical analysis of spatial patterns at scales varying from the whole world to the sleeping arrangements in a four-bedroom dormitory. Although it is difficult to do justice to the breadth of their ideas, their basic argument can be summarised as follows. There is a possibility that micro-meteorites that are raining down on the earth contain micro-organisms. Such pathogens may have a widespread distribution over the earth but would be locally patchy, with some areas receiving and others not, according to the prevailing meteorological conditions. People out of doors at the appropriate time may swallow the micro-organisms and become infected, those organisms that do not find a suitable host dying within a short time.

If this theory is correct, the observed spatial and temporal clusterings that are associated with an epidemic are not the outcome of person-to-person transmission and a large pool of infectives and susceptibles, but are the result of the pathogenic cloud raining down on a large area and infecting a large number of people simultaneously. They have found it difficult to provide direct evidence of micro-meteorites laden with pathogens, so they attempt to show that the observed patterns of cases cannot be produced by direct transmission. The marshalling of the arguments is done in detail for influenza and then is applied to other diseases.

One source of their arguments is the experimental studies that have tried to induce person-to-person transmission. For example, Andrewes (1965) found it very difficult to induce spread by exposing volunteers to those already with colds. He concluded that colds are 'by no means a very infectious kind of disease', that is colds do not easily spread from person to person. Their second argument is to examine detailed descriptions of diseases in the past that do not exist now or have been intermittently non-existent for centuries, thereby implying that there is no reservoir on the earth and that the particular disease has to be (re)introduced from space. Again they find support in that Thucydides gave a detailed description in 400 BC of a disease that no longer exists while there appears to be no record of cholera symptoms before the sixteenth century.

Their third set of arguments derives from the spatial distribution of cases and their timing. The incubation period of influenza is about 2–3 days and under normal patterns they expect it to take 10 days for about 50 cases to develop. But Creighton (1894) in his monumental survey of British epidemics dismissed the contagious theory because of the rapidity of the spread, with influenza being found simultaneously in isolated individuals over wide parts of the country, the epidemics of 1833, 1837 and 1847 covering Britain in one or two weeks. Similarly, Weinstein (1976) found that the 1918 influenza pandemic appeared on the same day in widely separated parts of the world and yet it took weeks to spread relatively short distances. It took four weeks to travel from Chicago to Joliet, a distance of 38 miles, three weeks to travel from Boston to New York (despite the considerable 'traffic' between these cities) and yet it appeared in Bombay and Boston on the same day.

The fourth set of arguments examines the spread of influenza to isolated communities. They noted that the 1947 outbreak in Sardinia was contemporaneous for isolated shepherds and town dwellers, while in Alaska the 1919 outbreak occurred at such a time that travel over the ground was virtually impossible. They also considered isolated communities such as Tristan da Cunha which had four outbreaks in the 1960s which could not be associated with boat landings. They also examined one group that was not isolated, married couples, and found that if one partner became infected there was no increased chance of the spouse becoming ill.

For the 1978 influenza epidemic in Britain, they collected data on the number of cases in public schools. According to their theory, the number of cases would peak when frontal systems passed over the country producing eddy activity that would draw micro-meteorites from the stratosphere. They correlated the peaks on a daily basis for the schools and they found remarkable agreement with an index of storm activity. They were also able to analyse these data on a detailed spatial basis. In St Donat's Castle, a school in South Wales, there were essentially 85 dormitories with four people sleeping in each. They reasoned that if there was person-to-person transmission there would be spatial clustering for the infected would be sleeping in the same room for many hours every night with the unaffected. If the virus was swallowed out of doors and not transmitted there would be no spatial clustering by dormitory and the observed distribution would be random. Using binomial probability theory the expected random distribution would be 31 dormitories with one, 7 with two, and 1 with three cases. In fact, the observed distribution was 35 dormitories with one, 5 with two, and 1 with three cases; that is even less clustering (but not significantly so) than random. There appeared to be no spatial patterning at the micro-scale as predicted by person-to-person transmission.

Turning finally to measles, they argue that there is no evidence that the disease existed before AD 160 and they can find no descriptions of the easily-characterised symptoms in Hippocratic writings of the fifth century BC nor in Indian medical writings of the sixth century BC. Further, they doubt that the early spring outbreaks that are found in temperate countries are adequately explained by current theory. Anderson (1982, p. 17) reports that seasonal variation in transmission is required for otherwise the mathematical epidemic models that have been developed do not 'lock' in on an integral number of years. He further reports that 'the mechanisms that generate such patterns are poorly understood at present. The main causes are probably climatic . . . and children returning to school.' The two astronomers prefer an explanation that is based on the earth's regularly experiencing micro-meteorites at a particular time of the year which triggers off an outbreak if there are sufficient susceptibles; at this stage person-to-person transmission may play its part in this disease.

It is difficult to assess the validity of their arguments for no micro-meteorites have been netted and found to have pathogens, but they remain convinced that influenza, at least, 'is not a transmitted disease' (1979, p. 78), and that 'it is driven directly from space' (p. 62). If their global findings for influenza are to be believed, then those arguing for person-to-person transmission have to rely on exceedingly high reproduction rates and very effective hierarchical diffusion. The general implications of the extra-terrestrial view are immense; if they are correct we can expect entirely new diseases to rain down (AIDS? Legionnaires' Disease?) while old ones will return (smallpox?).

The emotional host

Another body of literature is equally challenging to the accepted biomedical theory of communicable diseases. There is growing evidence that the psychological condition of the host has a powerful bearing on the effects of infection. Totman (1977) injected forty-eight volunteers with rhinoviruses and told them to expect a mild cold. Half of the group were then offered an anti-viral drug but told that this would require them to undergo an intubation to sample their gastric juices. In reality there was no anti-viral drug and those who accepted were given a placebo. The results were intriguing in that the highest rates of infection were found in those volunteers who had been offered the choice of the treatment; the researchers postulated that the increased anxiety of having to make a choice had induced the more severe symptoms. Further research by this group involved fifty-two volunteers who were interviewed and classified according to their stress levels and examined and classified according to their levels of antibodies. All the volunteers were then injected with rhinoviruses and kept in isolation. The resulting colds were monitored subjectively and objectively by nasal washes and it was found that the severity of infection was more closely related to stress levels than to the degree of immunity conferred by antibodies.

Kissen (1958) has examined the effects of emotional stress on another infectious disease, tuberculosis, in three case-comparison studies (see Chapter 2). In the first he interviewed 267 patients who were attending hospital complaining of chest disorders. Before they were examined and diagnosed, he identified a severe emotional stress group who had experienced such life events as recent bereavement, divorce or an unhappy love-affair. The results were impressive for of the third that were found to have TB, 65 per cent had been classified into the severe stress group, while the remaining two-thirds without TB contained only 26 per cent of those that had been classified as having severe emotional problems. In the second study he examined those that had relapsed from the disease. He interviewed and classified a group of patients that had previously been adjudged as cured but had not been examined for a year. Over three-quarters of those in which the infection had recurred had experienced preceding emotional stress, while in the group that had remained quiescent for over a year, only 12 per cent were categorised in the high-stress group. In the third study more detailed interviews were undertaken with a group of patients before they underwent examination and diagnosis. The interviews were designed to explore the personality of the patients and they were asked about their families, friends, and childhoods and on this basis they were classified into having 'normal' and 'inordinate' need for affection. The results were quite remarkable for one hundred per cent of those suffering from TB had been categorised as having 'inordinate' need while only 16 per cent of those without TB fell into this category.

If the emotional condition of the host is a key factor, how do we get the observed rapid spread of the disease associated with epidemics? Inglis (1981), in his thought-provoking discussion, considers that the answer lies in group pathology; some as yet unexplained force (he calls it 'mood convection' and suggests pheromones or extra-sensory transmission) which is capable of transmitting diseases or perhaps the signal at which a disease process starts up. In fact, the work of Kissen (1958) may give us a clue that the 'emotional' changes are due to social changes. He examined historical variations in the mortality of tuberculosis and he found that the death-rate rose rapidly in the Channel Isles at the time of the German invasion and yet there was no initial lack of resources and malnutrition. Moreover, the rate fell just as rapidly following the German surrender even though the material conditions did not improve for several years. Similarly, in mainland Britain he found that in the depression at a time of material deprivation there was no increase in mortality, but the outbreak of war in 1939 lead to an upturn in the death-rate. Could it be fear of the massive change and disruption of social life that led to these fluctuations in the death-rate?

The social context

In this final part of this chapter, the importance of the social context of communicable diseases will be illustrated by two examples. The role of social factors in the production of disease will be discussed in terms of the relationship between under-development and disease, while we shall examine the preventative strategy of vaccination to reveal the importance of social factors in determining the consumption and outcome of medical care.

Health and underdevelopment

The importance of social conditions and social change in determining the outcome and pattern of communicable diseases is most vividly and viciously seen today in the economically backward countries of the tropics. In such countries even the childhood disease measles takes on a different form: the rash becomes dark red-purple and exfoliates extensively, exposing large areas of 'skin' to bacterial infection, and there are much higher rates of complication, including blindness and persistent diarrhoea. The fatality rate for measles in the USA is currently about 0.01 per cent (that is 1 in 10,000 cases die), but in tropical countries the rate can be hundreds of times higher with, for example, a rate of 26 per cent being found in the late 1970s in Indonesia (Walsh, 1983). Crucially, the virulence of the virus does not change between temperate and tropical climates and a well-nourished child in the tropics undergoes a similar course of disease to that normally experienced in Britain and the USA. Indeed, high measles death-rates are taken as a sure sign of malnutrition. The course and outcome of the disease are thus largely determined by the health and nutritional status of the host, and this, in turn, is largely determined by social causes and societal organisation. While there is insufficient space for a detailed account, we need to show how the current distribution of the severe and killing communicable diseases can be directly related to the social and environmental changes that began in the colonial period and continue today.

The tropics are commonly perceived as an inherently unhealthy place. While it would be foolish to present the pre-colonial countries as a disease-free utopia, it would also be fundamentally wrong to believe, because of climatic or environmental factors, that the tropics are naturally rife with killing diseases. Indeed, there is evidence that the tropics were healthier before colonisation; early travellers in Africa, for example, commented favourably on the physical well-being of the indigenous people (Doyal, 1979, p. 101). It was Europe that was essentially disease-ridden and it was the Europeans that took the diseases to the New World where the Amerindian population died in their millions because they had not acquired any immunity to the introduced pathogens. The diseases that are now found in the tropics, such as TB, cholera, leprosy and plague, were until recently widespread in Europe, and their disappearance (as discussed in Chapter 1) was related to improved living and nutritional standards. Ironically, these improvements in the European standard of living were facilitated by the flow of materials from the colonies.

In the early period of colonialism, the European powers were fighting among themselves and against the indigenous population to gain control. In some areas and particularly when faced with stubborn resistance, the Europeans deliberately devastated large areas to remove the supplies that were sustaining the warring tribes. For example, in Tanganyika the German military campaign systematically destroyed large areas of farming land, and although such activities occurred over eighty years ago the results have been long-lasting, with South Province still sparsely populated until today. As Doyal (1979, p. 103) has concluded: 'much of the malnutrition and disease which came to characterise rural Tanganyika in the twentieth century was in a very real sense a product of early colonial repression'.

The changes associated with colonialism are even deeper than this, for the whole social, economic and spatial organisation was transformed to allow the colonies to play their major role of providing raw materials for the people and industries of the European countries. To take just one example, sleeping sickness is today widespread over large areas of tropical Africa. In May's terminology it is a four-factor disease with a protozoan agent (*Trypanasoma*), human and cattle hosts, an insect vector (tsetse flies) and a natural reservoir in wild game. The disease commonly ends in fatality for cattle and there is a high case mortality for humans. Many of the areas over which the disease is currently endemic, and where cattle cannot now be kept, were used for large-scale cattle-grazing in the pre-colonial times (Kjekshus, 1977). At such times it appears that indigenous farming and herding were well adapted to the ecology of the tsetse, for by the labour-intensive means of regular bush-clearing and game control, it was possible to keep the vector in check. However, changes associated with colonialism such as forced labour migration, war and other diseases led to depopulation and the breakdown of the intensive farming system. The result was the spread of the bush and tsetse breeding-grounds. Matters were made worse by the strategy adopted by the imperial powers who ordered the mass-evacuation of the tsetse belts, depopulating the areas and further expanding the habitat of the tsetse. The long-term effect is still being felt in the lack of animal protein and an impoverished diet which is characteristic of these areas.

The two key current health problems facing the under-developed countries are poor nutrition, and inadequate water-supply and sanitation; it is these problems that underly the poor condition of the host and hence the high level of infectious disease. The current widespread food shortages have not always been a continual problem (although there have always been natural disasters such as locusts and drought), but there is evidence to suggest that the inadequate food supply has been made qualitatively worse by colonialism. A large percentage of effort was diverted from subsistence farming to cash crops for export and increasingly the subsistence crops have been pushed into marginal land. For example, Foster-Carter (1985) gives the example of 'socialist' Tanzania in the early 1970s where people who had a protein-deficient diet were forbidden to pick cashews because they were earmarked for export to earn foreign exchange.

It has been estimated that as much as 80 per cent of disease in underdeveloped countries is associated with water. Few people have access to an adequate quantity or quality of water supply or to an effective sewage disposal. This applies to both crowded urban areas and rural areas, and the result is the high level of faecally-related diseases such as hookworm, cholera and chronic dysentry. This too can be related to the colonial legacy, for in the rural areas there was little investment by the colonial powers in infrastructure such as piped water and sewerage systems. In the urban areas there was enforced spatial and social segregation so that the colonists lived in the 'sanitary district' (with such amenities as water and electricity, drainage and sewerage) which was separated from the 'septic fringe', where the indigenous populated lived, by a *cordon sanitaire*. It is these septic fringes of the colonial period that have turned into the vast shanty-towns of today with their continued lack of basic services. As Southall (1971, p. 251) has commented, 'slum living has been conferred upon Africa by the colonial experience . . . developing in African cities most of the characteristic social evils of the poorer areas of western cities during the industrialisation period'.

To conclude, high levels of communicable diseases are not a natural part of being born in the tropics or for that matter of being black. The problem is not just one of biology and medicine but it is inextricably linked to material deprivation and the legacy of the colonial past and the current world order. A world order in which, under neo-colonialism, a large part of the economic surplus continues to be removed, thereby further limiting resources for development. As Navarro (1983) has argued, health and disease in the underdeveloped world can be seen to be rooted in imperialist and capitalist relations. Communicable diseases are social problems and not just biological ones.

Vaccinations

While later chapters deal in detail with medical care, we will now examine the role of vaccinations as a means of dealing with communicable diseases, from a social perspective. We will discuss both the social patterning of vaccination uptake and question the whole enterprise of preventing disease by this form of biological intervention. Pyle (1973) examined the social and geographical patterning of measles cases in Akron after the introduction of vaccination in 1965. He found that the highest rates of measles attack were in the city centre and the south-east of the city, the traditional black residential neighbourhoods. Acknowledging the problem of the ecological fallacy, his findings suggest that poor people make the least use of preventative immunisation. The uptake of measles vaccination has subsequently increased to very high levels in the USA because of legislation; children enrolling at school or kindergarten must have a vaccination certificate (Kirby, 1984).

In Britain, however, with an entirely voluntary procedure, the result is a partial vaccination campaign and the rates of both measles and whooping cough vaccination have remained relatively low (see Figure 4.5). The predictable result is that both diseases remain endemic, and the longer gap between epidemics ensures a higher average age of attack with a consequent increased risk of long-term complications in those who are affected by measles later in life. Moreover, results from longitudinal studies in Britain show that the lower social classes have consistently lower uptake rates. The results from the National Child Development Survey of children born in 1958 when they were seven is given in Table 4.1. Similar results were found in the later *Child Health and Education in the Seventies* (Butler, 1977); the variations in uptake were most marked when the children were classified not on the father's occupation but on a composite social index which was based on such factors as

Table 4.1 Use of immunisation by children under 7 by occupational class of father, Great Britain, 1965

Percentage not	*Occupational class*					
immunised	*I*	*II*	*IIIN*	*IIIM*	*IV*	*V*
Smallpox	6	14	16	25	29	33
Polio	1	3	3	4	6	10
Diphtheria	1	3	3	6	8	11

Source: Davie *et al.* (1972).

Table 4.2   Typical schedule of immunisation

Age	Immunisation
6 months	Triple immunisation (diphtheria, whooping cough and tetanus) plus oral poliomyelitis (1)
8 months	Triple immunisation (diphtheria, whooping cough and tetanus) plus oral poliomyelitis (2)
12 months	Triple immunisation (diphtheria, whooping cough and tetanus) plus oral poliomyelitis (3)
14 to 18 months	Measles
5 years	Booster: Diphtheria, tetanus and poliomyelitis (oral or inactivated vaccine)
11–13 years	(girls only) rubella (german measles)
10–13 years	BCG vaccine (to protect against tuberculosis)
15 years	Booster: Tetanus and poliomyelitis (oral or inactivated vaccine)

housing and parental education. The 'advantaged' had very nearly 100 per cent uptake of the triple vaccine (diphtheria, tetanus and whooping cough, Table 4.2) and over 80 per cent for the measles vaccine; the comparable rates for the 'disadvantaged' were 80 and 48 per cent respectively. An analysis by district revealed a particularly low rate of only 73 per cent for the triple vaccine by the 'urban' poor. In some parts of the country these differences may be even greater; Scott-Samuel (1977) found that the rates of uptake for the triple vaccine in Liverpool varied from nearly 50 per cent in high-status owner-occupied areas to only 15 per cent in inner-city local authority estates when the children were 14 months old.

Whether one adopts an individualistic or socio-structural explanation for these social patterns in uptake (see Chapter 7), there appears to be fairly general agreement that vaccination is a 'good thing'. Even the Black Report (DHSS, 1980) which explains health inequalities in terms of material and social deprivation argues for an increased uptake of immunisation as a valuable preventative measure. But from a social viewpoint, the appropriateness of this biological technical procedure as a way of dealing with a social problem, can be challenged. To illustrate such arguments we will consider the vaccine which has been subject to much controversy, that for whooping cough.

It has been estimated that over one-third of all those who caught whooping cough had already been immunised with the

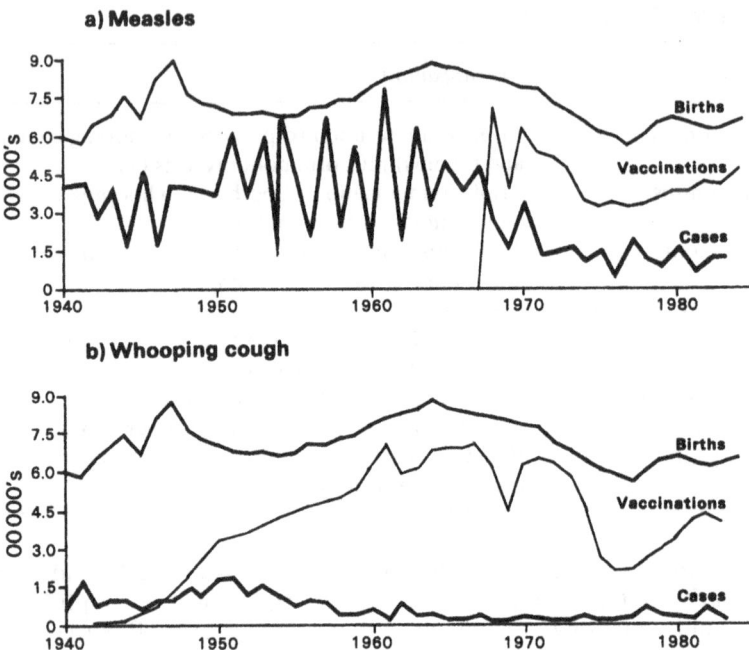

Figure 4.5 Births, vaccinations and cases: measles and whooping-cough – England and Wales

vaccine, while Stewart (1977, 1979) found that vaccinated children (though their symptoms were less severe) were as likely to catch and transmit as the unvaccinated with the result that mildly infected children may play a great role in spreading the disease. Stewart also found that immunisation did not appear to make any difference to the duration and complications of the illness. Moreover, the vaccination is unlikely to reach those that have the highest risk because those with 'contra-indications' or 'special conditions' (for example, a history of epilepsy or convulsions) will not be vaccinated, and while vaccination starts at three months, the median age of death from the disease is 30 days. Unlike measles, for which immunity can be easily passed across the placenta, the mother's immunity is locked into white blood-cells in the case of whooping cough and therefore the baby has very limited acquired immunity at birth.

It may be argued that whooping cough is a relatively trivial childhood illness unless there are other problems with the host, and this may relate directly to social conditions. Stewart in his study of Glasgow found that high social status was three times more important than vaccine uptake in determining the attack rate and almost all cases requiring hospital treatment were in social class IV and V. Indeed, in international comparison, Sweden had a similar rate of incidence to England and Wales during the 1977–9 epidemic, but while the latter countries had twenty-seven deaths, Sweden had none. Could it be that these differences can be explained by the greater social equality in Sweden? Sweden suffered no deaths at all from whooping cough in the 1970s and vaccination was ceased in 1979.

The greatest harm from vaccination may come from the resultant limited outlook, the reliance on medicine and (echoing Virchow) complacency over social conditions. Undoubtedly, the greatest achievement of the vaccination strategy is the global eradication of smallpox, announced by the World Health Organisation in 1979, following the last community case in Somalia in 1977. This success can be related to a number of rather special features in that the present-day vaccine is very effective, smallpox is a two-factor disease and therefore there are no vectors nor animal reservoirs, nor is there a carrier status (in which the infectious host does not show visible symptoms) so that the infecteds are easily discernible. Since 1948 the WHO has concentrated on specific diseases (especially smallpox and malaria) and specific preventative measures. Historically these measures have been applied without a great deal of concern for raising the general standard of living and improving environmental conditions. The result of this and the general importation of 'western' medicine, according to Doyal (1979, p. 290), is to

avert death without improving life . . . it has produced large number of mere survivors, handicapped by constant illness and incomplete recovery . . . curative medicine actually kills by diverting scarce resources away from other projects that would be more immediately beneficial to health.

Overall, a qualified success may be achieved from a vaccination campaign but if it ignores the social condition of the host, it cannot hope to achieve a long-lasting and meaningful victory over disease and illness.

Conclusions: the case of TB

The bacillus associated with tuberculosis has many advantages: it spreads by transmission through the air but unlike many other micro-organisms it has a waxy coat so that it does not dry out and can lie dormant in dust for many years until it finds a suitable environment for multiplication. Given such advantages we must ask why, in the industrialised countries, is there not more TB today, why has it declined to this low level when it was a common killer a hundred years ago, and what is the cause of tuberculosis? The answer to the last question can take several alternative forms as shown in Table 4.3.

The first explanation is the immediate, proximal biological one of the introduction of an external agent, the second is based on the state of the individual host, while the remaining explanations are increasingly based on the nature of society. We contend that it is the latter explanations that are more fundamental for it is distal causes that render humans susceptible to disease. Thus, the bacillus (whether from space or not) may not harm a well-nourished and otherwise healthy person and this would account for the low level of the disease in industrialised countries today and the high levels previously associated with lower standards of living a hundred years ago. As McKeown (1979) has argued (see Chapter 1) this improvement cannot be explained in terms of the declining virulence of the bacillus, nor by improvements in medical technology, but it reflects societal changes.

Moreover, a social explanation would not only help us to explain the endemic nature of TB in many underdeveloped countries, but it would also help us to understand why there are pockets of high incidence of TB in advanced countries today. For example, the tuberculosis rate among Asians in Leicester in 1975 was 632 per 100,000, while that for the non-Asian was only 8.5; indeed this relatively small ethnic group had over four times the

Table 4.3   The causes of tuberculosis

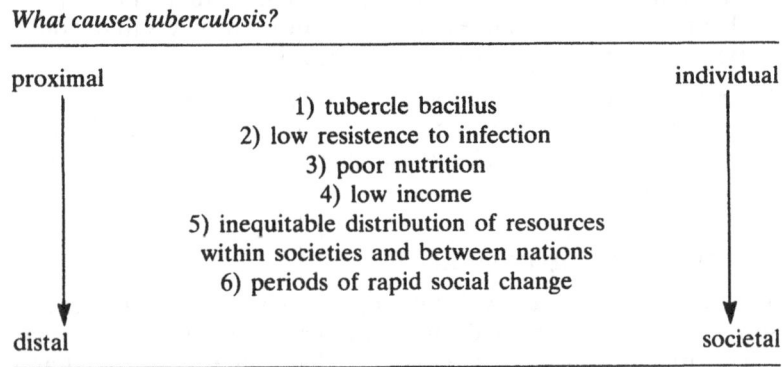

What causes tuberculosis?

proximal   individual

1) tubercle bacillus
2) low resistence to infection
3) poor nutrition
4) low income
5) inequitable distribution of resources
within societies and between nations
6) periods of rapid social change

distal   societal absolute number of notifications as the rest of the population (Lilienfeld and Lilienfeld, 1980, p. 184). It is not the biology of these immigrant groups that produces such high levels but the social conditions under which they live. Finally, this social perspective would lead us to question the reliance on the BCG vaccination as a routine technical procedure, particularly when Holland, the country with the lowest rate of TB in Europe, has never had a national immunisation programme and when large-scale controlled trials show that the effectiveness of this commonly-used vaccine is zero (Editorial, 1980, pp. 73–4).

As Stark (1977, p. 684) has argued:

> epidemics occur less and less because of natural events . . . and increasingly as the result of the private and unequal appropriation of an expanded surplus through war, slavery, trade, urban-industrial exploitation of colony and countryside and, of course, factory and slum organisation.

We fully realise that we have left many questions unanswered in this chapter. This is in marked contrast to the biomedical position which, in its most naïve form, assumes that all that is required is an understanding of the biology of infection, and control will follow. The continued high level of disease in the Third World is a testament to the fallaciousness of this view. The theme of the primacy of social conditions in the production of ill-health will be further developed in the final chapter.

Guided reading

The biology of communicable diseases is well covered in Open University (1985b) and in Burnet and White (1972). The importance of such diseases in the past is discussed in Howe (1972), Woods and Woodward (1984) and in the context of early capitalism by Doyal (1979, Chapter 2). The disease ecology tradition is considered in Learmonth (1975); Howe (1977) provides an encyclopaedic account of a large number of diseases on a world scale, and Pyle (1979) on Rocky Mountain spotted fever and Brownlea (1967, 1972) on infectious hepatitis show how quantitative methods can be used in such work. The most comprehensive account of the diffusion modelling approach is, from a geographical viewpoint, Cliff *et al.* (1981) and Bailey (1975) from the mathematical. Geographical diffusion studies on diseases other than measles include Morrill and Angulo (1979) on smallpox; Hunter and Young (1971) and Pyle (1979) on influenza; Kwofie (1976) and Stock (1976) on cholera. The relationships between health and underdevelopment is considered in Open University (1985a), Navarro (1983) and in Doyal (1979); the latter deals with the production of ill-health in Chapter 3 and the consumption of health care in Chapter 7. The importance of social factors in whooping cough is examined in Bassili and Stewart (1976) while a comprehensive account of the variations in uptake of vaccination is given by Blaxter (1981, Chapter 10) for the USA and Britain and by Walsh (1983) for the under-developed world.

References

Anderson, R. M. (1982), 'Directly transmitted viral and bacterial infections of man', in R. M. Anderson (ed.), *Population Dynamics of Infectious Diseases*, pp. 1–37, London, Chapman & Hall.

Anderson, R. M. and May, R. M. (1979), 'Population biology of infectious diseases', *Nature*, vol. 280, pp. 361–7.

Andrewes, Sir C. (1965), *The Common Cold*, London, Weidenfeld & Nicholson.

Bailey, N. T. J. (1975), *The Mathematical Theory of Infectious Diseases*, London, Griffin.

Bartlett, M. S. (1957), 'Measles periodicity and community size', *Journal of the Royal Statistical Society*, Series A, vol. 120, pp. 48–70.

Bassili, W. R. and Stewart, G. T. (1976), 'Epidemiological evaluation of immunisation and other factors in the control of whooping cough', *Lancet*, vol. 1, pp. 471–3.

Blaxter, M. (1981), *The Health of Children*, London, Heinemann.

Bradley, D. J. (1982), 'Epidemiological models – theory and reality', in R. M. Anderson (ed.), *Population Dynamics of Infectious Diseases*, pp. 320–33, London, Chapman & Hall.

Brownlea, A. A. (1967), 'An urban ecology of infectious disease: city of Greater Wollogong', *Australian Geographer*, vol. 10, pp. 169–87.

Brownlea, A. A. (1972), 'Modelling the geographic epidemiology of infectious hepatitis', in N. D. McGlashan (ed.), *Medical Geography: Techniques and Field Studies*, pp. 279–300, London, Methuen.

Burnet, Sir F. MacFarlane and White, D. O. (1972), *Natural History of Infectious Disease*, Cambridge, Cambridge University Press.

Butler, N. R. (1977), 'Community and family influences on 0 to fives: utilisation of pre-school day care and preventive health care', *Child Health and Education in the Seventies*, mimeograph, University of Bristol.

Cliff, A. D., Haggett, P., Ord, J. K., Bassett, K. and Davies, R. B. (1975), *Elements of Spatial Structure*, Cambridge, Cambridge University Press.

Cliff, A. D., Haggett, P., Ord, J. K. and Versey, G. R. (1981), *Spatial Diffusion: an Historical Geography of Epidemics in an Island Community*, Cambridge, Cambridge University Press.

Creighton, C. (1894), *History of Epidemics in Great Britain*, Cambridge, Cambridge University Press.
Davie, R., Butler, N. and Goldstein, H. (1972), *From Birth to Seven: The Second Report of the National Child Development Study*, London, Longman.
Deitz, K. (1976), 'The incidence of infectious diseases under the influence of seasonal fluctuations', in J. Berger (ed.), *Mathematical Models in Medicine*, Berlin, Springer-Verlag, pp. 1–15.
Department of Health and Social Security (1980), *Inequalities in Health*, Report of a Research Working Group chaired by Sir Douglas Black, London, DHSS.
Doyal, L. (1979), *The Political Economy of Health*, London, Pluto Press.
Ebdon, D. (1985), *Statistics in Geography: a Practical Approach*, 2nd edn, Oxford, Blackwell.
Editorial (1980), 'BCG: bad news from India', *Lancet*, vol. 1, pp. 73–4.
Fine, P. E. M. and Clarkson, J. A. (1982), 'Measles in England and Wales: an analysis of factors underlying seasonal patterns', *International Journal of Epidemiology*, vol. 11, pp. 5–14.
Foot, D. (1981), *Operational Urban Models*, London, Methuen.
Foster-Carter, A. (1985), *The Sociology of Development*, Ormskirk, Causeway.
Galdston, I. (1954), *The Meaning of Social Medicine*, Cambridge, Mass., Harvard University Press.
Girt, J. L. (1974), 'The geography of non-vectored infectious diseases', in J. M. Hunter (ed.), *The Geography of Health and Disease*, pp. 81–100, Chapel Hill, University of North Carolina.
Hagerstrand, T. (1953), *Innovation Diffusion as a Spatial Process*, trans. A. Pred, Chicago, University of Chicago Press.
Haggett, P. (1976), 'Hybridizing alternative models of an epidemic diffusion process', *Economic Geography*, vol. 52, pp. 136–46.
Haggett, P., Cliff, A. D. and Frey, A. (1977), *Locational Models*, London, Edward Arnold.
Howe, G. M. (1972), *Man, Environment and Disease in Britain*, Newton Abbott, David & Charles.
Howe, G. M. (1977), *A World Geography of Human Diseases*, London, Academic Press.
Hoyle, F. and Wickramasinghe, C. (1979), *Diseases from Space*, London, Dent.
Hunter, J. M. (1966), 'River blindness in Nangodi, Northern Ghana: a hypothesis of cyclical advance and retreat', *Geographical Review*, vol. 56, pp. 398–416.
Hunter, J. M. and Young, J. C. (1971), 'Diffusion of influenza in England and Wales', *Annals of the Association of American Geographers*, vol. 61, pp. 637–53.
Inglis, B. (1981), *The Diseases of Civilisation*, London, Hodder & Stoughton.
Jarcho, S. and Van Burkalow, A. (1952), 'A geographical study of

swimmer's itch in the United States and Canada', *Geographical Review*, vol. 42, pp. 212–6.

Kirby, A. M. (1984), 'The use of primary medical facilities', in M. Clarke (ed.), *Planning and Analytic Methods in Health-Care Systems*, pp. 78–89, London, Pion.

Kissen, D. M. (1958), *Emotional Factors in Pulmonary TB*, London, Tavistock.

Kjekshus, H. (1977), *Ecology Control and Economic Development in East African History*, London, Heinemann.

Knight, C. G. (1971), 'The ecology of African sleeping sickness', *Annals of the Association of American Geographers*, vol. 61, pp. 23–44.

Knight, C. G. (1974), 'The geography of vectored diseases', in J. M. Hunter (ed.), *The Geography of Health and Disease*, pp. 46–80, Chapel Hill, University of North Carolina.

Kwofie, K. M. (1976), 'A spatio-temporal analysis of cholera diffusion in Western Africa', *Economic Geography*, vol. 52, pp. 127–35.

Learmonth, A. (1975), 'Ecological medical geography', *Progress in Human Geography*, vol. 7, pp. 201–26.

Learmonth, A. (1978), *Patterns of Disease and Hunger*, Newton Abbott, David & Charles.

Lilienfeld, A. M. and Lilienfeld, D. E. (1980), *Foundations of Epidemiology*, New York, Oxford University Press.

McKeown, T. (1979), *The Role of Medicine*, Oxford, Blackwell.

May, J. M. (1950), 'Medical geography: its method and objectives', *Geographical Review*, vol. 40, pp. 9–41.

May, J. M. (1960), *Disease Ecology*, New York, Hafner.

Morrill, R. L. and Angulo, J. J. (1979), 'Spatial aspects of a smallpox epidemic in a small Brazilian town', *Geographical Review*, vol. 60, pp. 319–30.

Navarro, V. (1983), *Imperialism, Health and Medicine*, New York, Baywood.

Open University (1985a), *The Health of Nations*, Course U205, Book III, Milton Keynes, Open University Press.

Open University (1985b), *The Biology of Health and Disease*, Course U205, Book IV, Milton Keynes, Open University Press.

Phillips, D. R. (1981), *Contemporary Issues in the Geography of Health Care*, Norwich, Geo Books.

Pyle, G. (1969), 'The diffusion of cholera in the United States in the nineteenth century', *Geographical Analysis*, vol. 1, pp. 59–75.

Pyle, G. (1973), 'Measles as an urban health problem: the Akron example', *Economic Geography*, vol. 49, pp. 344–56.

Pyle, G. (1979), *Applied Medical Geography*, New York, Wiley.

Sayer, A. (1984), *Method in Social Science*, London, Hutchinson.

Scott-Samuel, A. (1977), 'Social area analysis in community medicine', *British Journal of Preventive and Social Medicine*, vol. 31, pp. 199–204.

Sencer, D. J. H., Dull, B. and Langmuir, A. D. (1967), 'Epidemiologic basis for eradication of measles in 1967', *Public Health Reports*, vol. 82, pp. 253–5.

Sigerist, H. E. (1935), 'Introduction to Pettenkofer's "The Value of Health to a City" ', *Bulletin of the History of Medicine*, vol. 10, pp. 570–609.

Southall, A. (1971), 'The impact of imperialism upon urban development in Africa', in V. Turner (ed.), *Colonialism in Africa 1870–1903*, vol. 3, pp. 230–68, Cambridge, Cambridge University Press.

Stark, E. (1977), 'The epidemic as a social event', *International Journal of Health Services*, vol. 7, pp. 681–705.

Stewart, G. T. (1977), 'Vaccination against whooping cough: efficacy versus risks', *Lancet*, vol. 1, pp. 234–7.

Stewart, G. T. (1979), 'Toxicity of pertussis vaccine: frequency and probability of risk', *Journal of Epidemiology and Community Health*, vol. 33, pp. 150–6.

Stock, R. (1976), *Cholera in Africa*, London, International African Institute.

Taylor, R. and Reigler, A. (1985), 'Medicine as social science: Rudolf Virchow on the typhus epidemic in Upper Silesia', *International Journal of Health Services*, vol. 15, pp. 547–59.

Totman, R. (1977), 'Cognitive dissonance, stress and virus-induced common colds', *Journal of Psychosomatic Research*, vol. 12, pp. 55–63.

Waitzkin, H. (1981), 'The social origins of illness: a neglected history', *International Journal of Health Services*, vol. 11, pp. 77–103.

Walsh, J. A. (1983), 'Strategies for control of disease in the developing world: measles', *Review of Infectious Diseases*, vol. 5, pp. 330–40.

Weinstein, L. (1976), 'Influenza – 1918, a revisit?', *New England Journal of Medicine*, vol. 294, pp. 1058–62.

Woods, R. and Woodward, J. (1984), *Urban Disease and Mortality in Nineteenth Century England*, London, Batsford.

CHAPTER 5

Concepts and issues in mental illness

Introduction

Mental illness in Britain, and many other parts of the world, today constitutes an issue of vast importance. Howells (1975) claims, for example, that it is the leading cause of disability, and ultimately death, in most western societies. The British Department of Health and Social Security (DHSS) describe it as 'a major social problem, perhaps the major health problem of our time' (DHSS, 1975). One woman in eight, and one man in twelve, will, at some time during their life, enter psychiatric care. The resource usage represented by these figures is considerable (Table 5.1). Kennedy (1983) endorses the importance of mental illness in contemporary society:

> One third of hospital beds are occupied by those characterised as mentally ill. . . . Five million people in England consult their family doctors each year about their mental health. Six hundred thousand are referred to specialist psychiatric services. There are 21,000 cases each year of compulsory detention in mental hospitals. Some have spoken of an epidemic of mental illness. (Kennedy, 1983)

By any standards then, mental illness is a phenomenon on such a scale that we should give it specific attention in its own right.

In this chapter, there are three study themes to which we will devote our attention. Firstly, we will introduce the general concept of mental illness and give some consideration to just what exactly does constitute psychiatric ill-health, Secondly, we will critically examine the claims of the various theories and models which purport to offer a means of understanding the cause of mental illness. We will develop this theme by focusing, in some depth, on an assessment of social and spatial aspects of the incidence of mental illness. Finally we will consider two recent themes in the study of mental illness: the need to focus on

Table 5.1   Resource usage on mental health services in England and Wales

	All hospitals	Mental illness hospitals	% accounted for by mental hospital care
Medical staff	33,416	2,836	8.4
Nursing staff	352,246	72,674	20.6
Hospitals	1,923	458	23.8
Beds	343,091	113,241	33.0

Source: DHSS (1985) *Health and Personal Social Services Statistics.*

the individual patient within her or his social context; and the importance of retaining an historical perspective on the way in which society has viewed mental illness.

Defining mental illness

Social scientists generally view mental illness as an essentially sociological construct. They suggest that it reflects an assessment of an individual's behaviour in relation to the accepted norms of the society in which the individual lives (Smith, 1977). When individuals are defined as mentally ill, they are transgressing the accepted tenets of normality within the particular society. This definition is, of course, not unchallenged; there are, for example, people who hold that mental illness is a precise clinical entity. This diversity of interpretation allows us to make links back to Chapter 1 and note that we can contrast a biomedical and sociological model of, in this case, mental illness.

To remain, however, for the moment within our initial sociological definition, one of the key questions must be what exactly constitutes normality? In absolute terms, of course, normality, in this context, is incapable of definition. Thus, in the last resort, the diagnosis of mental illness must inevitably be subjective. Behaviour which in one context may be highly abnormal, may be perfectly acceptable in another social situation (Rose *et al.*, 1984). For example, people who wander down a street shouting to no one in particular are generally thought to be probably mentally ill. Within a church, however, talking to one's self is a perfectly normal act.

The social model

One way in which we can shed some light on the social construction of normality is by considering the title 'mentally ill' as a form of label denoting abnormal behaviour. Lemert (1951) suggests that the process of labelling an individual as mentally ill comprises two stages. Primary deviancy is the first stage and is concerned with the acquisition of the label. This process mainly occurs either through self-diagnosis by the individual concerned, or via an assessment made by a doctor or a psychiatrist. In both cases societal conditioning is at work in a number of ways. The particular individual or professional involved in the case makes value judgements concerning normality; the social context in both time and space is important. Behaviour indicative of mental illness in one country may be acceptable elsewhere – contrast for example nudity in Papua New Guinea and Central London. Similarly current attitudes to homosexuality (Chapter 1) and masturbation are relatively liberal in British society, whereas in former times this was not so:

> Melancholia, stuporous insanity, katakonia and insanity of pubescence are the forms most frequently found in masturbators, and the essential characters are always recognisable in these circumstances . . . such lunatics are usually shy, retired, suspicious, hypochondriacal, mean and cowardly. (Spitzka, 1883, p. 379)

Secondary deviance is the term applied to the behaviour of a person who has previously been labelled as a deviant. Thus, in the context of mental illness, the initial act of labelling a person as mentally ill means that all subsequent behaviour by that person will be forced into a role conforming to societal preconditions of behaviour in madness. One of the longer-term consequences of this secondary deviance is institutionalisation, whereby long-term residents in mental hospitals become progressively more dependent upon an institutional environment. Incarceration in a mental hospital removes an individual from any contact with the norms of the society against which he or she was once deemed to have offended. The reference point of 'normality' is replaced by the internal hospital norms of regimentation. Secondary deviance therefore reinforces the previous label of insane with the result that the patient has even greater difficulty, on release, in adjusting once more to life in mainstream society.

Labelling, the concept of normality in society, and the

definition of mental illness are inextricably interlinked in the social model of mental illness. Szasz (1962) has produced one of the more extreme interpretations of this model. He claims that all mental illness can be defined as the imposition of a label, arguing that the alternative definition, making reference to a biochemical mechanism, has not been adequately conceptualised. For Szasz, mental illness is a convenient labelling device which society uses as a form of social control facilitating the incarceration of social misfits.

Empirical evidence exists to support the centrality of labelling theory in the definition of mental illness. In particular the imposition of secondary deviance onto previously normal people has been known to result in institutionalised behaviour. For example, pregnancy out of wedlock at the start of the twentieth century in Britain was viewed as immoral. In consequence it was an offence against social norms and a manifestation of mental illness. It was managed by incarceration in asylums. When, some fifty years later, social norms changed, the victims of this labelling were completely institutionalised and experienced considerable difficulty in coping with release.

Labelling allows us to define mental illness as originating within the functioning of society itself. A detailed consideration of the symptomology of mental illness provides evidence of the precise conditions which society currently finds abnormal. Mental illness may be crudely divided into two broad diagnostic categories: within the framework of the social model, neuroses refer to phobias, depression, obsessive compulsion and hysteria, while psychoses refer to alcohol- and drug-related disorders, organic decay, senility and schizophrenia.

The medical model

As we suggested earlier, the social model of mental illness is not unchallenged. The alternative medical model posits firm bio-chemical origins for both forms of mental illness. Under this framework neuroses are functional disorders of the nervous system affecting the ways in which individuals react to external stimuli; psychoses in contrast have their origin in the mind rather than the nervous system.

What distinguishes the social from the biomedical model of mental illness is that the latter places mental illness firmly within the realm of the internal biology of the body. Essentially mental illness is a malfunctioning of an individual's biochemistry. There is presumed to be something in the biology of the individual mental patient which predisposes him or her towards the disorder (Rose *et al.*, 1984). By reifying the problem in this way, treatment within the terms of traditional medical practice becomes possible. The psychiatrist can search for a *real* disorder internal to the particular individual under observation.

Causality and mental illness

In defining mental illness we emphasised the importance of the sociological construct of labelling. However, we also noted the existence of definitions placing mental illness firmly in a medical or biochemical context. A similar diversity of schools of thought exists if we turn to a consideration of the causes of mental illness; that is, if we shift our attention from what *is* mental illness to the study of how it comes about. For the most part these issues of causality and definition are related, so just as we found when trying to define mental illness, so too, when we seek to identify its cause, we may also distinguish broad general frameworks for our investigation. Thus, if we were to enquire *why* an individual was mentally ill, and leaving aside whether we feel that mental illness is a biological fact or merely a labelling device, we would need to adopt a particular theoretical standpoint.

As social scientists we again place greatest emphasis on a social model. We must not, however, disregard the potentialities afforded by the medical model. There is also a third alternative to consider: the psychological models. The medical model, to reiterate, emphasises a biologically rooted causation of mental illness, and may be equated with the medical model of disease causation (see Chapter 1). The psychological models are similarly predicated on siting the condition with the individual rather than the group, and with a general emphasis on the need to investigate the mind in the search for cause.

Medical models assume that mental illness is a direct or indirect result of a pathological condition. The extent and severity of this condition is reflected in the gravity of the illness. Biological or organic malfunctionings internal to the human body constitute the potential causes of mental illness according to this model, and their ultimate origins lie in an individual's genetic make-up and inheritance. It is only recently that attempts to verify this model have proved even partially successful. Three examples may be cited. Paralysis was once thought of as a mental illness in certain situations. Now, however, it is known that some forms of paralysis originate as an organic consequence of syphilis. A biochemical argument is also suggested in the linkage of stress to depression via mono-amine oxide release. Finally, the genetic argument finds support in data on the identification of schizophrenia. Identical twins stand a 50 per cent chance of both being schizophrenic if one twin is so diagnosed.

Psychological models of mental illness, of which there are a considerable number, draw on theories of individual human behaviour. All these models hinge around the twin factors of the extent to which people are able to adapt their behaviour to that of other individuals or particular situations, and the timing of the learning of this ability. The differences between the models reflect differing assessments of the importance of aspects of behaviour adaption and its timing. The moral model, for example, argues that mental illness is a relic of past learning patterns (Taylor, 1974). Any abnormal behaviour previously learned can, through therapy, also be unlearned. A variant on this approach is provided by Freudian psychology through the argument that mental illness is a response to the emotional stresses of a patient's past life. Other models place more stress on contemporary interaction and personal imagery. Laing's theories of the self provide one example, claiming that the mentally ill picture themselves as inadequate, often because of poor intercommunication within families, and, in consequence, withdraw into a fantasy world.

Social models are superficially similar to the psychological models. The difference, however, is that the psychological models are centred on the individual, exploring the individual's reaction to a group, while the social model operates in the opposite direction; society's conceptualisation of the individual is under consideration. Two distinct arguments are of key importance, both of which concern group social reactions, and can be linked back to labelling theory. Firstly, we can suggest that the definition or labelling of an individual as mentally ill can be equated with the cause of mental illness. Society can exclude an individual from the social construction of reality through the imposition of the label insane and by this act the insanity of the individual is created. Further frustration and resignation on the part of the labelled individual can cause increased loss of contact with social reality and the adoption of yet more socially offensive behaviour. Secondly, the place and acceptance of certain social groups within the context of society as a whole may be of relevance. Susser (1968), for example, suggests that more integrated groups may be better able to cope with stress.

The aetiology of mental illness, if indeed there is such a thing, is complex and the subject of considerable debate. We are faced with the competing claims of three main models and a number of sub-models. Each facilitates the study of particular causal influences in psychiatric ill-health, yet neglects other equally plausible factors. The most useful conclusion that we might make is that research in this area is dealing with an essentially multifactorial issue, where each model has something to contribute. We shall develop our discussion by focusing on social aspects.

The social context of mental illness

A useful point to begin our analysis is with the recognition that the incidence of mental illness varies between different social and demographic groups. The nature of this variation is, of course, a reflection of both the way in which psychiatric problems are defined, and also the way in which social and demographic status is defined. These definitions vary between countries (Table 5.2), although, as Haug and Sussman (1971) remark, trans-national comparisons are fraught with complications. Nevertheless some broad generalisations are possible.

First, there is a sexual division in susceptibility to mental illness (Table 5.3). Higher rates for schizophrenia occur, for example, among young males, and the suggestion that marriage may be associated with increased schizophrenia among women can be clearly related to the imposition or reinforcement of a traditional female sex role with its stressful juxtaposition of homebuilding

Table 5.2 International variations in mental illness definition

Diagnostic indicator for schizophrenia	*Selected significant spatial aspects to diagnosis*
Catatonia	not frequently cited by Euro-American psychiatry
Visual hallucinations	frequently cited by African and Near Eastern psychiatrists; not frequently reported by Euro-Americans
Social and emotional withdrawal	important for Japanese psychiatrists

Source: Adapted from H. Murphy *et al*. (1963), 'A cross-cultural survey of schizophrenic symptomology', *International Journal of Psychiatry*, vol. 9, pp. 237–49.

Table 5.3   Gender and mental illness

| Age | First admission age-specific rates per 100,000 | | | |
| | Schizophrenia | | Depressive disorders | |
	Male	Female	Male	Female
15–19	21	15	3	6
20–24	38	21	6	12
25–34	31	26	9	17
35–44	21	25	12	23
45–54	15	15	16	31
55–64	8	14	21	33
65–74	8	18	24	37
75+	11	16	22	28

Source: DHSS (1972), *Psychiatric Hospitals and Units in England and Wales: Inpatient Statistics from the Mental Health Enquiry*, London, HMSO.

and work demands (Gove and Tudor, 1972). For depressive disorders, there is evidence of an overall greater susceptibility among females. Again an explanation may be sought in the female sex role. Thus, depression can be explained as a psychological reaction to childbirth, menopause and pregnancy. Marriage itself appears to be less important as a correlate of depression.

Notwithstanding the inherent difficulties of defining social status, there is evidence for a link between low status and schizophrenia (Giggs, 1973). Other factors are also closely correlated with this association (Table 5.4), for example unattractive housing, private-renting tenure classes, single-person households and high levels of personal mobility. For depressive disorders there is some evidence that this relationship is reversed:

> Unlike the pattern for schizophrenia and other psychoses, the
> distribution of manic depressive psychosis in the community,
> as judged by first mental hospital admissions, is not materially
> affected by social status or its numerous correlates.
> (Silverman, 1968, p. 135)

In fact, when neuroses as a total category are considered, the evidence indicates an enhanced prevalence among upper status groups (Daiches, 1981).

Cultural factors also appear to have some bearing on the incidence of mental illness (Table 5.5). Bastide (1972) has indicated the varying susceptibility of different races and

Table 5.4   Social dimensions to mental illness

| | Relationship to mental illness† | |
	Positive	Negative
% single persons	*	
% married persons		*
% divorced	*	
% single mobile	*	
% social class V (male)	*	
% unemployed	*	
% owner occupied housing		*
% furnished rented housing	*	
% sharing household amenities	*	
% outside WC	*	

† Derived from Giggs (1973); selected variables loading highly on the first factor from the analysis; all the positive loadings also related to twelve indicators of schizophrenia.

religions. Durkheim (1951), in a classic study, interweaves those social and cultural themes. He argues that mental illness – and, in an interesting example of the changing social construction of mental illness, he includes suicide within this definition – develops in stressful environments, notably those where society is characterised by social and residential mobility and non-adherence to the ideal of a nuclear family. As a converse to this argument, low rates of mental illness would be expected in areas with stable, integrated family-based lifestyles. Empirical justification for this theory was forthcoming in the relatively low rates of mental illness found in socially integrated Catholic and Jewish communities.

It is, however, unwise to over-generalise about the social context of mental illness. In addition to definitional problems noted earlier, three further difficulties should be noted. Firstly, psychiatrists tend to apply more severe judgements to lower status persons (Labedun and Collins, 1976). The incidence of schizophrenia among low status groups may therefore be overestimated. Secondly, 'nosocomial' factors such as the availability of treatment and willingness to undergo treatment, may bias data concerning admission to mental hospital. In particular Taylor (1974) claims that women are more willing to enter hospital, and that this may perhaps inflate the differential sex rates for the incidence of mental illness. Finally, it must be

Table 5.5 Cultural variations in mental illness in the USA

| | % of mentally ill with disorder | | | |
Condition	Irish	Italian	Jewish	Negro
Character disorder	15.5	23.4	43.9	11.1
Addictions	10.9	0.8	0.0	8.9
Schizophrenia	47.5	48.4	36.3	55.6
Affective disorder	7.1	8.8	7.6	1.1
Psychosis (with mental				
deficiency)	3.7	5.3	5.4	3.3
Senility	13.0	9.1	5.4	10.0
Epilepsy	0.0	0.8	0.4	1.1
Other organic disorders	2.2	3.5	0.9	8.9

Source: Roberts, B. and Myers, J. (1954), 'Religion, national origins, immigration and mental illness', *American Journal of Psychiatry*, vol. 110, pp. 759–64.

The table was originally compiled in the 1950s and reflects the then contemporary views of what could be termed mental illness.

emphasised that the usual way of defining the mentally ill in statistical terms is by the use of hospital admission rates. As well as a sex-based variation in these rates, there is also a social status variation (Hollingshead and Redlich, 1958), a variation for in-patient and out-patient treatment rates, and a variation between public and private hospitals. Furthermore, much mental illness goes unreported unless detected in small-scale local community based surveys.

A spatial perspective on mental illness

When the incidence of mental illness is plotted on a map it frequently exhibits a strong pattern. Schizophrenia is more common in the urban areas where its social correlates are most clearly represented. The relationship between urbanisation and schizophrenia has long been recognised. Rosen (1968), for example, argues that, as the United States changed from a colonial to an industrial society during the nineteenth century, so a parallel rise in detected 'madness' occurred.

A major focus of current research attention among medical geographers is the identification of the zones within urban areas where mental illness is most prevalent. Such research links available knowledge of the social characteristics of psychiatric conditions to the known existence of clearly defined social areas

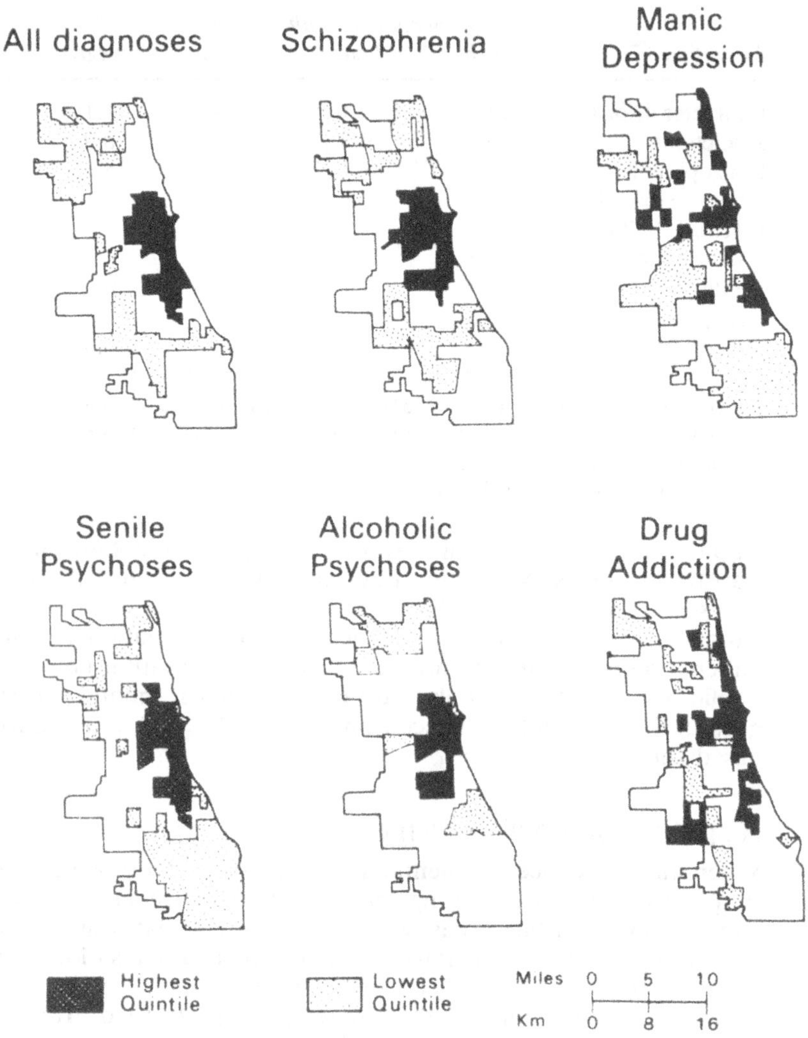

Figure 5.1   Mental illness in Chicago

within cities. Schizophrenia, with its clear social correlates, and thus a clear map pattern, has received most research attention; neuroses, with a more random association and patterning, have been less widely studied. Schizophrenia, it will be recalled, occurs mainly in highly mobile, low status areas. In cities conforming to the classic Chicago School model of residential geography (Burgess, 1967), these areas are to be found on the fringe of the central business district, in the 'Zone of Transition'. Empirical illustration of this spatial concentration has been provided for Chicago itself by Faris and Dunham (1939) (Figure 5.1). More recently Giggs (1973) and Dean and James (1981) have argued for the continued relevance of the Chicago School model in British cities (Figure 5.2). The latter authors for example state: 'Male schizophrenics tend to be single, separated or divorced. Their concentration in the inner parts of Plymouth where appropriate accommodation is to be found is thus understandable' (Dean and James, 1981, p. 49). In other cities, particularly in the USA, schizophrenia is also high in older neighbourhoods of municipalities which have recently been incorporated into adjacent cities (Jaco, 1954). Further factors complicating the spatial patterning of mental illness are urban renewal and, in Britain, public sector housing. Dunham (1965) suggests that the social dislocation occasioned by renewal may encourage schizophrenia, while Taylor (1974) rehearses a similar argument for newly constructed peripheral local authority housing estates.

While it would be folly to deny the existence of a spatial pattern for schizophrenia, we must, however, consider the importance of this pattern. Initially of course it is obvious that the spatial dimension, in purely static terms, is of little interest. Space itself cannot cause or explain concentrations of schizophrenia; rather it is the social and demographic structuring of that space. What is of interest though, is how the spatial concentration comes about. We know that at one level it reflects the social characteristics of the area. We are unsure, however, how social status, spatial location and schizophrenia come to interrelate in the observed way. Two contrasting theories have addressed this question (Figure 5.3).

The breeder theory argues that schizophrenics have always inhabited the parts of the city characterised by schizophrenia. By living in these locations they are more likely to become mentally ill because the characteristics of that environment breed mental illness. Two parallel hypotheses have been cited in support of the breeder theory. The social isolation hypothesis argues that social disorganisation and residential mobility causes breakdowns in

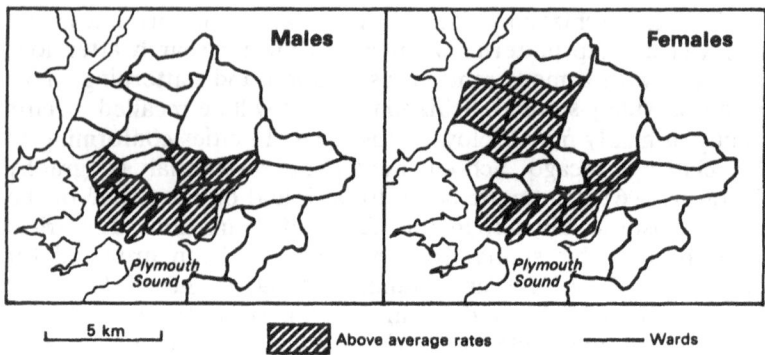

Figure 5.2   Mental illness in Plymouth
Source: Dean, K. and James, H. (1980) The spatial distribution of depressive illness in Plymouth, *British Journal of Psychiatry* Vol 136, pp. 167–180

supportive relationships, with consequent stresses and mental illness. The class hypothesis links the stress of living on the poverty line to psychiatric illness. Daiches (1981) suggests that the breeder theory can be placed firmly within the context of the work of the founding fathers of urban sociology. Thus, the concepts of *gesellschaft*, isolation and alienation all figure in the approach.

The drift hypothesis (Myerson, 1940) provides the alternative theory. In this case it is argued that, prior to becoming schizophrenic, the social status of an individual is immaterial, and the individual may inhabit any residential location. Upon becoming schizophrenic an individual will experience stigmatisation and loss of personal confidence. The individual's perceived and, perhaps, actual status within his or her initial residential community will decline, probably being accompanied by a move to a less salubrious area. In other words schizophrenia is accompanied by a downward drift in social status and, in consequence, a relocation to an appropriate social area. The residential location of schizophrenia is a result of acquiring the condition, rather than an inevitable consequence of initial location. The drift hypothesis may be refined to distinguish inter-generational status drift, where a family history of schizophrenia is accompanied by downward status drift measurable between generations, from intra-generational status drift occurring within an individual's own lifetime (Rushing, 1969).

Empirical research utilising breeder theory has a lengthy pedigree. Faris and Dunham (1939), as stated above, suggested

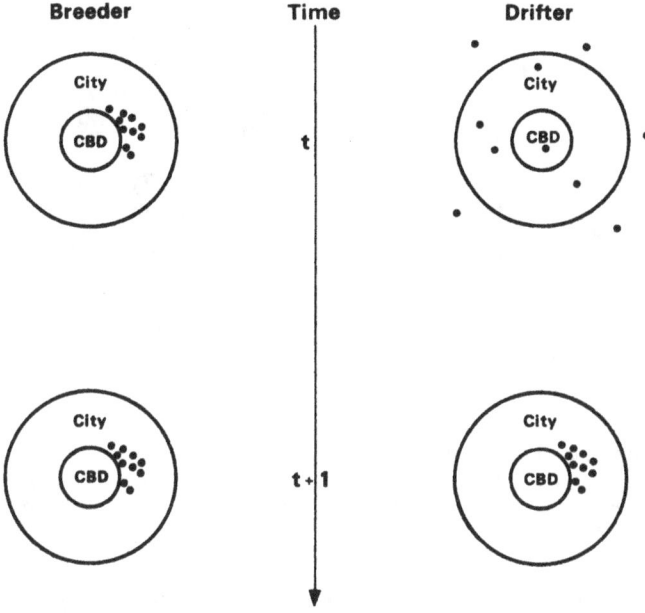

Figure 5.3   Breeder and drifter hypotheses of mental illness causation

that a broad concentration of mental illness could be identified in the fringes of the Chicago central business district. This assertion was further investigated by mapping the previous known addresses of persons admitted to hospital for a variety of mental disorders. They found that the inner city concentration and concentric zonation pattern remained the characteristic residential location for schizophrenics. Depressives exhibited a more random spatial distribution. In attempting to explain this spatial patterning, the authors drew upon the social isolation aspect of the breeder theory. The social conditions of inner city Chicago constituted an inherently unstable environment where 'the result is a breakdown of standards of personal behaviour and a drifting into unconventionality and into dissipations and excesses of all sorts' (Faris and Dunham, 1939, p. 7). Almost thirty years later Levy and Rowitz (1973) found that the distribution of psychiatric disorders in Chicago was still concentrated in a roughly similar position.

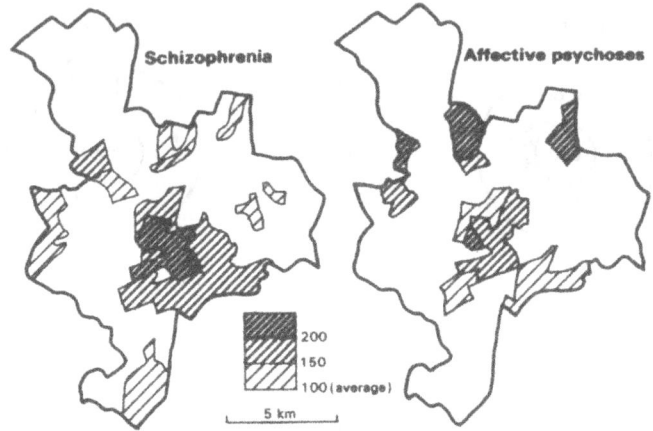

Figure 5.4   Mental illness in Nottingham
Source: Giggs, J. (1986) Mental disorders and ecological structure in Nottingham. Conference proceedings, joint symposium in medical geography, Nottingham, 15–19 July 1985

In Britain, research employing breeder theory has been pursued in Brighton (Bagley *et al.*, 1973; Bagley and Jacobsen, 1976) and in Nottingham (Giggs, 1973). The Brighton studies investigated a wide range of social pathologies and found strong statistical associations between measures of social disorganisation and mental illness. For example, in the four socially disorganised wards of central Brighton with their highly mobile populations, a clear link between suicide and mental illness was evident. In contrast, elsewhere, suicide correlated more strongly with physical illness. The Nottingham research, much quoted as a classic example of what has been called 'ecological' work, sought to explore the linkages between social process and spatial pattern. This study utilised data on the first hospitalisations of schizophrenics during a seven-year period extending from 1963 to 1969. The addresses of 444 patients were mapped and categorised according to the census enumeration district in which they were located. The distribution of the patients exhibited, as expected, a strong concentration in the inner residential areas of the city (Figure 5.4). The number of patients in each enumeration district was then related to twenty-nine socio-economic indicators for each enumeration district in an effort to effect an explanation for the observed distribution. A factor-analytic technique was used, which indicated that the distribution of schizophrenics could be broadly related to a set of unfavourable life circumstances evidenced by the enumeration district characteristics. It was then argued that the areas exhibiting concentrations of schizophrenics could be considered as breeding grounds for the disease with social disorganisation forming a possible cause.

In contrast to breeder theory, and despite a pedigree extending back to the 1940s, empirical evidence for the drift theory has only recently been collected. Perhaps the earliest study was by Gerard and Houston (1953). Working on a sample of 305 male schizophrenics in Worcester, Massachusetts, they found that most patients had only lived in poor inner city areas for a short time. When these recent arrivals in the inner city were removed from the sample, the spatial distribution of schizophrenics became unpatterned. Dunham (1965) and Turner and Wagenfield (1967) examined intra- and inter-generational status drift. The former claimed that the social status of schizophrenics in Detroit upon first admission to hospital was significantly below that of their parents, while the latter, in a similar study, found that 34.4 per cent of their sample showed downward inter-generational status mobility. On this evidence it would seem that the residential location of schizophrenics may be due to a drift process stemming from stresses associated with a failure to compete with parents, rather than a failure to meet self-set goals.

Breeder and drift theories both provide plausible explanations for the residential concentration of the mentally ill. Debate has, not surprisingly, focused on their relative merits and drawbacks. Breeder theory, for example, depends perhaps overmuch on the assumption that mental illness can be derived from the stresses associated with inner city life. Though Durkheim's *anomie* and Tönnies's *gesellschaft* (Durkheim, 1951; Tönnies, 1963) provide perfectly respectable theoretical antecedents for this assumption, it is all too easy to make the mistake of categorising the breeder hypothesis as a naïve form of environmental determinism.

This argument, that is that breeder theory is, in fact, claiming that location equals cause, was the essential subject matter of Gudgin's critique of Giggs's work in Nottingham (Giggs, 1973; Gudgin, 1975):

> [Giggs] . . . appears to ignore an intuitively more plausible reason why schizophrenics should be more prevalent in the central areas of cities. It is possible that sufferers from schizophrenia move into areas of poor quality housing, and may thus *not* be a product of that environment in any meaningful sense. (Gudgin, 1975, p. 148)

Gudgin's criticism arose as a result of Giggs's statement that 'there are pathogenic areas which seem to destroy mental health' (Giggs, 1973). Informed reading of the Giggs work would, however, theoretically situate the work within breeder theory as Giggs himself argued in a response to Gudgin (Giggs, 1975). Gudgin also criticised Giggs for relying upon ecological association and thus committing the ecological fallacy. This criticism is more justified, and indeed can be levelled at both breeder and drift theories. As we pointed out in Chapter 3, it is unwise to extend conclusions based on aggregate areal statistics to the individual. Just because an area contains a lot of poor housing or other unfavourable life circumstances, in conjunction with a high schizophrenic population, it does not mean that the schizophrenic population and the people with unfavourable life chances are identical. Similar conclusions could be made for an area with a high proportion of recent in-migrants; there would be no reason to assume individual schizophrenics were also individual in-migrants drifting to inner city areas.

An additional criticism which can be levelled at both breeder and drifter theories is their heavy reliance on available data. This is dangerous for two reasons. Firstly, confirmation of the theories requires the prior identification of schizophrenics. This is usually done via hospital admission statistics, yet we know much mental illness goes undetected. Secondly, and with particular reference to the drift theory, data generally provide information on a patient's location when mental illness was detected, not when its onset occurred. Though a schizophrenic might be a recent in-migrant to the inner city, drift theory has little relevance if the onset of schizophrenia did not occur prior to migration.

The individual mental patient in context

Much of the material which we have outlined above has been derived from quantitative analysis of readily available or specially collected statistical data. The analysis of this information has involved sampling theory and statistical tests, and the derivation and substantiation of results, theories and propositions from aggregates of the collected data. This is not surprising given the current construction of science and (much) social science, with their adherence to Popperian method and its requirement to falsify theories through rigorous large-scale empirical testing. Even within the context of the medical model, and, paradoxically, notwithstanding the emphasis placed by that model on illness causation internal to the *individual*, conclusions have been derived primarily from *aggregate* data. For example, the research on schizophrenia in twins cited earlier demanded large numbers of participants before a link could be acceptably established. Similar situations occur within the context of the social model: the mentally ill are regarded as a collectivity.

This group or aggregate approach to mental illness provides us with the generalisations which we expect from models and theories. Naturally deviations from these generalisations occur for individual cases. Furthermore the variety present in the experience of mental illness can easily be lost amidst the general trends. By shifting our focus to individual cases, we can therefore often add significantly to our knowledge of mental illness while at the same time retaining a recognition of the human dimension involved. This task involves more than simply collecting more variables on a large number of individual cases, although this can provide a worthwhile strategy. Rather, what can prove most effective is to develop research strategies employing such techniques as the case-history. Approaches like this involve developing knowledge about an individual and may be either retrospective – in that an account is obtained of the individual's recollections and history, or prospective – in which case an individual is monitored as events actually occur. Such studies should not be embarked upon without a word of caution. While the ecological fallacy indicates the danger of making assumptions about individuals on the basis of aggregate relationships, so the atomistic fallacy must remind us that it is unwise to found any general theoretical propositions on individual cases.

On the merit side, detailed case histories allow us to elucidate factors which may have contributed to the onset of mental illness in a particular individual. Furthermore, particularly with prospective studies, the time of onset of mental illness may be determined with some (albeit small) degree of accuracy, and the roles played by stress, disorganisation and downward mobility evaluated accordingly. In short, individual-based studies offer a vehicle by which we can consider, through time, the full social context of the onset of a disorder, and the social status and changing residential location of the individual sufferer from mental illness.

Studies founded on the individual and working within the context of the social model of mental illness are a comparatively recent innovation. Taylor (1974) provides an interesting comparison of aggregate and individual-based approaches. Schizophrenia in Southampton was found to concentrate in local authority housing estates in both the inner city and the urban periphery. On the aggregate level this was held to indicate the applicability of a social isolation hypothesis and breeder theory. A checking process using case histories, however, suggested a rather different interpretation. Examination of the case histories of eighty-seven male schizophrenics revealed that fewer than half had lived in their present address for longer than five years. The schizophrenics tended to be highly mobile with few even being born in their present county of residence. Additional conclusions from detailed case histories indicated a uniform experience of schizophrenia prior to moving to a current address, and also both inter- and intra-generational status drift.

Further support for the drift theory comes from Daiches (1981) using a rather different research methodology in a study of the Chicago standard metropolitan statistical area. He interviewed 2,300 residents comprising mentally ill and non-mentally ill sub-samples. This enabled him to identify non-reported mental illness (that is mental illness which was not officially known to the authorities, but which individuals themselves were aware of). He was able to confirm the presence of intra-regional status drift – although he also suggested that social isolation had a role to play. Dean and James (1981), in their study of schizophrenia in Plymouth, provide additional comparative information on eco-logical and individual approaches. They noted an evident sex differentiation in the incidence of schizophrenia, with male schizophrenics being associated with areas dominated by poor housing and single-person in-migration, and females correlating with low social status. Taken together these findings could be equated with unfavourable life circumstances and the breeder theory. Individual data however again indicated a drift hypo-thesis, related, in the case of males, to the availability of accommodation in inner city lodging houses.

Research employing individual data faces considerable prob-lems, not least in terms of ethics and confidentiality. Data will only be released after lengthy consultation, assurances that individuals will not be identified, and guarantees that any publications will be available for vetting by local medical authorities. Given these caveats it is hardly surprising that many studies have been conducted by research teams which include practising psychiatrists. The main methodological problem raised is that these studies tend to reveal mental illness in its full multifactorial complexity. Seldom will the symptomology or aetiology of mental illness be reduced to the broad generalisa-tions evident in aggregate studies. Despite their difficulties, however, this research approach does allow us to gain a clear insight into mental illness within the individual context. In the case of the spatial location of the mentally ill, this insight provides mounting evidence for the explanatory power of drift theory.

Society and treatment

Another direction, which has recently become important in geographical work on the mentally ill, involves the consideration of the way in which societal attitudes to the mentally ill have evolved over time, and the linkage between this evolution and the treatment of mental illness. Parallels can be drawn here with our discussion, in Chapter 1, of the overall evolution of attitudes to illness. The presence of the mentally ill within society involves diverse reactions; some positive but many negative. The assumptions and viewpoints which structure these attitudes have not been constant through time and, as we shall show, have had important consequences for the way in which the mentally ill have been treated.

In perhaps the most succinct exposition of the history of social attitudes to the mentally ill, Conrad and Schneider (1980) suggest a useful starting point in Classical Rome. At that time madness 'was viewed largely as a family problem to be dealt with by kin. People who could not function in society and were not dangerous to others were allowed to wander about and were cared for by family' (Conrad and Schneider, 1980, p. 40). Those for whom the family would not or could not care were categorised as the pauper insane and were ridiculed or stigmatised. Attitudes to the mentally ill were thus subsumed within a societal emphasis on the family; negative attitudes confronted only the mentally ill whose families failed them.

The classical tolerance of mental illness was rooted in the fact that, for the dominant classes at least, madness was a disease entity caused by an imbalance of the four Hippocratic 'humours' (yellow bile, black bile, phlegm and blood) thought to influence health. As a consequence of this, mental illness could be cured by balancing the humours. In medieval times this conception was largely abandoned. Social and political dominance in this period was exerted by the church through the medium of the feudal mode of production. Contemporary interpretation of the (often more lurid) parts of the Christian Bible came to influence attitudes to mental illness. Madness was seen as the result of sin and the work of the Devil. Szasz (1970) argues that this viewpoint became emphasised when the power of the church was bought into question during the Renaissance. At this time the definition of mentally ill was expanded to include the opponents of the church. This draconian construction of mental illness as the offending of the viewpoint of the dominant within society meant that those so defined could be classified as witches and generally repressed.

Gradually the progressive elements of Renaissance thought prevailed and a more tolerant attitude to mental illness returned. This toleration was however within the social context of the rise of capitalism (Foucault, 1965). The development of the work ethic meant that the mentally ill were viewed as potentially productive labour. The way in which this labour power could be exploited was through confinement in asylums. These facilities served several purposes. Firstly, the mentally ill could be removed from the offended gaze of normal society. Secondly, concentration of the mentally ill facilitated the efficient exploitation of the more able through coercion, regimentation and the imposition of mundane yet productive tasks. Thirdly, in terms of economic growth the asylums provided a reserve army of labour performing low level productive work. During economic slumps this labour force could easily be discarded. Rosen (1968) suggests that the rise of the asylum was accompanied by a jurisdictional crusade against the mentally ill who remained in society. The insane were often forcibly excluded from urban areas, sent back to their communities of origin and even transported to overseas colonies.

The asylum initially developed as a place of confinement for a number of antisocial or unproductive social groups. The mentally ill formed but a single sub-category of inmate. This situation had two results. Firstly, in the context of a growing interest in scientific medicine, it became evident that there were difficulties in treating mental illness in the stressful environment of the asylum. Within the infant science of psychiatric medicine a 'cult of curability' prevailed which held that, given the right circumstances, lunacy could be cured. Secondly, in the context of labour power theory, the mentally ill tended to disrupt any productive efforts by other inmates.

Foucault (1965) argues that separate asylums for the mentally ill were eventually developed primarily in response to the needs of capitalism but also, in some cases, as a moral, human crusade with rehabilitation as a goal. These initial humanitarian aims were however progressively discarded through the nineteenth century, and the asylums assumed a merely custodial role. In the main this changing attitude was occasioned by the impossibility of treating the growing numbers of mentally ill who were naturally to be found in the increasing population of the time with its accompanying social stresses and growing number of elderly with accompanying problems of senility.

The attitudes to the mentally ill which found an outlet in the asylum model of mental health care were essentially those of the nineteenth century. Yet this attitude and approach, involving the exclusion of the mentally ill from society, retains an important place in British and North American society. Elsewhere too the confinement and exclusion of the mentally ill is usual. Recently however there have been signs of some liberalisation in attitudes. In terms of care, the role of the asylum has decreased and a trend has developed towards placing the mentally ill within normal community settings where they can learn to cope with the stresses of living in modern society. This trend has, in a number of countries including the USA (Table 5.6) and the UK but most notably Italy, involved progressively closing the large mental asylums and opening smaller community-based facilities offering day care or hostel accommodation. The logical aim of this policy is ultimately to provide for the mentally ill in a way which, except on an out-patient basis, is completely non-institutional. The liberal good intent of this policy has however been usually thwarted by the fiscal stress of the present economic slump.

Rabkin (1975) suggests that present attitudes to mental illness are behavioural responses to three key factors. Firstly, the symptomology of mental illness is important in so far as negative attitudes are more likely to be shown towards patients who exhibit violent, strange or uncontrolled behaviour. Secondly, attitudes are influenced by the nature of treatment. Mild therapy and infrequent recourse to psychiatrists is viewed more favourably than in-patient treatment or radical therapies such as ECT or lobotomy. Thirdly, the socio-demographic status of the

Table 5.6 The rise of community care in the USA

	Patient care episodes in:	
	State mental hospital	*Community mental health centre*
1971	748	753
1975	599	1,832
1977	574	2,011
1979	529	2,249

Source: Statistical Abstract of the United States, 1985.

general public exerts an inevitable influence. In addition, the locus of mental health care seems to colour attitudes. Placing the mentally ill in community settings rather than isolated asylums encourages acceptance; tolerance develops when the general public are able to observe the mentally ill in normal settings. A complicating additional factor, which could be added, would stress the existence of a geographical dimension to attitudes (Smith and Hanham, 1981; Dear and Taylor, 1982). People may profess tolerance towards the mentally ill, and support for community-based care, but few are willing to accept concentrations of psychiatric patients within the vicinity of their own home.

To some extent the current attitudes held by the public towards the mentally ill have been led by governments and medical science. As early as the Second World War it became clear that attention would have to be given to the extent of mental illness. In the USA, for example, some two million men failed to reach the psychiatric standards for armed service. Legislation to address this problem followed in the form of the 1963 Community Mental Health Centres Act in the USA and, in Britain, the 1959 Mental Health Act. Both these acts aimed to remove, as far as possible, the stigma of mental illness and supported community-based services as an ideal way by which this aim might be realised. Asylums and incarceration were no longer in official favour. This change in official attitudes was largely facilitated by developments in chemotherapy. Psychotropic and tranquilliser drugs allowed a patient to be treated in the community and mental illness to be managed without the need to keep the patient in an institution.

The argument for the benevolence of government attitudes towards the mentally ill is however but one side of the coin. Scull (1978) propounds the alternative view claiming a key role for governmental self-interest. He suggests that by the mid-twentieth century, the asylum system was becoming prohibitively expensive. The costs of providing treatment and residential facilities had escalated beyond the cost-effectiveness necessary in capitalist economies. Community care was far cheaper as it removed, for the most part, costs for residential facilities and non-medical support staff. Obsolete asylum plant could be sold, non-productive personnel made redundant and the families of the mentally ill made to shoulder a greater fiscal burden.

The theme which runs through the study of attitudes to and treatment of the mentally ill is that it is a function of the basic tenets of contemporary society. As ideologies have changed, so have attitudes to the mentally ill. Our interpretation of this variation has laid stress on the macro politico-economic environment. For example, over the past four centuries, the rise of mercantile capitalism, and later industrial capitalism with its recurrent crises, can be clearly linked not only to attitudes to the mentally ill but also to modes of mental health care. We will return to this theme in Chapter 8, where we will consider the difficulties which are posed by planning a community-based mental health care strategy.

Conclusion

Throughout this chapter we have been concerned to demonstrate the extent of debate in research on mental illness. Controversy is widespread in this research with the proponents of different theoretical perspectives all being keen to air their claims. At the outset we face the question of whether or not mental illness is a disease entity. If it is not, alternative definitions must be found. We must then proceed to consider whether the origins of mental illness are biomedical, psychological or social. Then there is the problem of whether to follow breeder or drifter theories for the explanation of the spatial concentration of the mentally ill.

We have, ourselves, placed great emphasis on the social context of mental illness. We do not dismiss other perspectives, but would like to stress the insights offered by a social perspective. Labelling theory, for example, is important in both the definition and aetiology of mental illness. Stresses stemming from modern society also play a part in aetiology, in addition to their role in promoting a residential segregation of the mentally ill. On the larger scale, social constructs also have considerable importance in the determination, in conjunction with politico-economic factors, of attitudes to the mentally ill and modes of treatment.

In the final section of the chapter we argued that attitudes to the mentally ill could not be divorced from a consideration of the ways in which society cares for the mentally ill. This contention also applies to health and health care beyond the limited case of purely mental health. In earlier chapters we demonstrated the relationships between society and ill-health; in subsequent chapters we will shift our analysis to consider health *care* within a social context. In particular we will demonstrate that, although a social norm for health care may exist, as for example in readily available mental health care, the practical application of those norms, as they affect individual members of society, is subject to much variation.

Guided reading

Perhaps the most readable introduction to the material contained in this chapter is Kennedy (1983, Chapter 5). Those interested in a less popularist approach might try Smith (1977), who provides a comprehensive review from a geographical standpoint.

Faris and Dunham (1939) is, without doubt, the classic aggregate analysis of mental illness. The Giggs-Gudgin controversy (Giggs, 1973, 1975; Gudgin, 1975) together with the work of Daiches (1981) exemplify current work.

Attitudes to mental illness, the mentally ill and mental health care are all excellently covered by Conrad and Schneider (1980, Chapter 3). Those in a sterner frame of mind should consider consulting Foucault (1965), Scull (1978) or, a reference not cited in the text, Ingleby (1982).

References

Bagley, C., Jacobsen, S. and Palmer, C. (1973), 'Social structure and the ecological distribution of mental illness, suicide and delinquency', *Psychological Medicine*, vol. 3, pp. 177–87.

Bagley, C. and Jacobsen, S. (1976), 'Ecological variation of three types of suicide', *Psychological Medicine*, vol. 6, pp. 423–7.

Bastide, R. (1972), *The Sociology of Mental Disorder*, London, Routledge & Kegan Paul.

Burgess, E. (1967), 'The growth of the city', in R. Park (ed.), *The City*, pp. 47–62, Chicago, University of Chicago Press.

Conrad, P. and Schneider, J. (1980), *Deviance and Medicalisation: From Badness to Sickness*, London, Mosby.

Daiches, S. (1981), *People in Distress: A Geographical Perspective on Psychological Well-being*, Research Paper 197, Department of Geography, University of Chicago.

Dean, K. and James, H. (1981), 'Social factors and admission to psychiatric hospital: schizophrenia in Plymouth', *Transactions of the Institute of British Geographers*, vol. 6, pp. 39–52.

Dear, M. and Taylor, S. (1982), *Not on Our Street*, London, Pion.

Department of Health and Social Security (1975), *Better Services for the Mentally Ill*, London, HMSO.

Dunham, H. (1965), *Community and Schizophrenia: An Epidemiological Analysis*, Detroit, Wayne State University.

Durkheim, E. (1951), *Suicide*, New York, Free Press.

Faris, R. and Dunham, H. (1939), *Mental Disorders in Urban Areas*, Chicago, University of Chicago Press.

Foucault, M. (1965), *Madness and Civilization: A History of Madness in the Age of Reason*, New York, Random House.

Gerard, D. and Houston, L. (1953), 'Family setting and the social ecology of schizophrenia', *Psychiatric Quarterly*, vol. 27, pp. 90–101.

Giggs, J. (1973), 'The distribution of schizophrenics in Nottingham', *Transactions of the Institute of British Geographers*, vol. 59, pp. 55–76.

Giggs, J. (1975), 'The distribution of schizophrenics in Nottingham: a reply', *Transactions of the Institute of British Geographers*, vol. 64, pp. 150–6.

Gove, W. and Tudor, J. (1972), 'Adult sex roles and mental illness', *American Journal of Sociology*, vol. 78, pp. 812–35.

Gudgin, G. (1975), 'The distribution of schizophrenics in Nottingham: a comment', *Transactions of the Institute of British Geographers*, vol. 64, pp. 148–9.

Haug, M. and Sussman, M. (1971), 'The indiscriminate state of social class measurement', *Social Forces*, vol. 49, pp. 549–63.

Hollingshead, A. and Redlich, F. (1958), *Social Class and Mental Illness*, New York, McGraw-Hill.

Howells, J. (ed.) (1975), *World History of Psychiatry*, London, Baillière Tindall.

Ingleby, D. (1982), 'The social construction of mental illness', in A. Treacher and P. Wright (eds), *The Problems of Medical Knowledge: Towards a Social Construction of Medicine*, pp. 123–43, Edinburgh, Edinburgh University Press.

Jaco, E. (1954), 'The social isolation hypothesis and schizophrenia', *American Sociological Review*, vol. 19, pp. 567–77.

Kennedy, I. (1983), *The Unmasking of Medicine*, London, Granada.

Labedun, M. and Collins, J. (1976), 'Effects of status indicators on psychiatrist's judgements of psychiatric impairment', *Sociology and Social Research*, vol. 60, pp. 199–210.

Lemert, E. (1951), *Social Pathology*, New York, McGraw-Hill.

Levy, L. and Rowitz, L. (1973), *The Ecology of Mental Disorder*, New York, Behavioural Publications.

Myerson, A. (1940), 'Review of "Mental disorders in urban areas" ', vol. 96, pp. 945–97.

Rabkin, J. (1975), 'The role of attitudes towards mental illness in evaluation of mental health progress', in M. Guttentag and E. Struening (eds), *Handbook of Evaluation Research*, vol. 2, Beverly Hills, Calif., Sage.

Rose, S., Kamin, L. and Lewontin, R. (1984), *Not in Our Genes*, Harmondsworth, Penguin.

Rosen, G. (1968), *Madness in Society*, London, Harper & Row.

Rushing, W. (1969), 'Two patterns in the relationship between social class and mental hospitalisation', *American Sociological Review*, vol. 34, pp. 523–41.

Scull, A. (1978), *Decarceration: Community Treatment of the Deviant – A Radical View*, Englewood Cliffs, NJ, Prentice-Hall.

Silverman, C. (1968), *The Epidemiology of Depression*, Baltimore, Johns Hopkins University Press.

Smith, C. (1977), *The Geography of Mental Health*, Commission of

College Geography Resource Paper 76–4, Washington DC. Association of American Geographers.

Smith, C. and Hanham, R. (1981), 'Proximity and the formation of public attitudes to mental illness', *Environment and Planning A*, vol. 13, pp. 147–65.

Spitzka, E. (1883), *Insanity: Its Classification, Diagnosis and Treatment*, New York, Bermingham.

Susser, M. (1968), *Community Psychiatry: Epidemiological and Social Themes*, New York, Random House.

Szasz, T. (1962), *The Myth of Mental Illness*, London, Paladin.

Szasz, T. (1970), *The Manufacture of Madness*, London, Harper & Row.

Taylor, S. (1974), 'The geography and epidemiology of psychiatric illness in Southampton, with particular reference to schizophrenia', unpublished PhD thesis, University of Southampton.

Tönnies, F. (1963), *Community and Society*, New York, Harper & Row.

Turner, R. and Wagenfield, M. (1967), 'Occupational mobility and schizophrenia – an assessment of the social causation and selection hypothesis', *American Sociological Review*, vol. 32, pp. 104–13.

CHAPTER 6

Inequalities in health care

Introduction

There has been considerable interest in recent years in what has been termed inequality in health care. We take this as our theme in broadening our discussion of health care away from the relatively narrow issues which are covered in the previous chapter.

As a context for the chapter the words of a Government White Paper published prior to the founding of the British National Health Service (NHS) are instructive:

> The Government wants to insure that in future every man, woman and child can rely on getting all the advice, treatment and care which they may need in matters of personal health; that their getting these shall not depend on whether they can pay for them, or any other factor, irrespective of need. (HM Govt, 1944, quoted in Le Grand, 1982, p. 23)

Today, some thirty-seven years after the creation of this 'model' health service, the evidence from the Black Report indicates continued inequalities in health care in Britain (DHSS, 1980). The situation is similar elsewhere.

The chapter will start by examining the social context of equality and inequality. We will then proceed to show how these social constructs can be linked to a spatial consideration of health care. Empirical evidence for both social and spatial inequalities will then be identified at the international, regional and local scales.

Equality and inequality

The term 'equality' may mean different things to different researchers. It is essential for us to recognise, at the outset, at least five distinctive conceptualisations (Le Grand, 1982):

  (i) Equality of public expenditure;
  (ii) Equality of final real income;
  (iii) Equality of use;
  (iv) Equality of cost;
  (v) Equality of outcome.

Equality of public expenditure argues that expenditure from the public purse should be allocated to each individual on a *per capita* basis. Each and every individual should therefore receive the same proportion of available resources. The surface appearance of this form of equality is that it is just; however, different individuals will have different needs and so equality of public expenditure may, in reality, be inequality. Table 6.1 illustrates this point with an example, taken from Le Grand (1978), using data from the 1976 British General Household Survey (see Chapter 2). Clearly there is evidence of rough equality in terms of expenditure per person on health care; there is at least no consistent pattern of disadvantage against any particular class. When, however, need is taken into account, and expenditure is weighted according to the extent to which different social classes report illness, a very different pattern emerges. Much more is spent on the ill upper classes.

If we adopt a criterion of equality of final real income we assume that public services such as health care constitute

Table 6.1 Equality of public expenditure?

Socio-economic group	Health Service expenditure per person (% of mean)	Expenditure per person reporting illness (% of mean)
Professionals, employers, managers	94	120
Intermediate and junior non-manual	104	114
Skilled manual	92	97
Semi- and unskilled manual	114	85

Source: Adapted from Le Grand (1978).

Mean = 100

undeniable benefits. When they are added to an individual's wage or salary the resultant sum can be termed real income. Arguably the quest for equality should be concerned with equalising this real income; health service resources should therefore be allocated disproportionately to the poor in order to iron out wage differentials. Again, however, there are problems. Fiscal poverty need not imply a greater need for health care – although it usually does.

Moving on to the concept of equality of use, we make the supposition that the goal of health services should be to insure that individuals have an equal opportunity to use the system according to their needs. This definition differs from that above in that it shifts the emphasis from income to need. In making this shift there is also incidentally a contribution to efficient usage of resources through the presumption that those in need will make full use of resources. Allocation by income, in contrast, would be likely to lead to under-used facilities in the event of people being poor but unhealthy. Equality of use is the definition which approximates most closely to the founding ideals of the British NHS. Even in this case, where, as we have seen, equality was written into the 'constitution' of the service, inequality of use is evident (Table 6.2). The children of fathers in social class I are twice as likely to have visited a dentist, five times as likely to have been immunised against smallpox and yet more certain to have received polio or diphtheria immunisation.

Our fourth definition, equality of cost, assumes minimal state intervention in health care. Under this definition individuals would be expected to contribute to medical insurance to top-up any state expenditure. The level of these contributions would be the prerogative of the individual and thus, given that people place differential importance on health and have differential

Table 6.2 Inequality of use: children and preventative services

Social class	% never visiting dentist	% not immunised Smallpox	Polio	Diphtheria
I	16	6	1	1
II	20	14	3	3
IIIN	19	16	3	3
IIIM	24	25	4	6
IV	27	29	6	8
V	31	33	10	11

Source: Adapted from Townsend and Davidson (1982).

ability to pay, they would also be the probable subject of inequality. State expenditure would provide an equalising measure to iron out this inequality. The approach bears a superficial resemblance to the system of health care provision in the USA, and there are indications, as we shall see in the final chapter, that it also has support among certain British political factions.

Finally we should consider equality of outcome. All our definitions so far have revolved around the outputs of health care systems. They have been concerned to equalise, in some way, the resources concerned with service provision. This final definition takes the consideration of equality a step further by suggesting that our real interest should lie with the results, or outcomes of provision. As we saw in Chapter 2, there is indeed considerable evidence that despite the promotion of equality in health care provision in Britain, there remains a marked class gradient in health itself:

> Indeed that gradient seems to be more marked than in some comparable countries (though it must be said that the data for the United Kingdom almost invariably are fuller) and in certain respects has become more marked. During the twenty years up to the early 1970s . . . the mortality rates for both men and women aged thirty-five and over in occupational classes I and II had steadily diminished while those in IV and V changed very little or had even deteriorated. (Townsend and Davidson, 1982, p. 15)

Resources should thus be allocated to provide equality of health, not merely equality in the treatment of ill-health. We may term this definition the idealist model.

These conceptualisations of equality by no means constitute an exhaustive categorisation. Their existence, however, illustrates the difficulties of organising a service which aims to foster 'equality'. We have suggested above that it is equality of use which is relevant to equality as envisaged in the British NHS. This definition, often termed equity, places great emphasis on the role of need. The identification and measurement of need is, in fact, central to the promotion of equality of use. Unfortunately, however, need is just as hard to define as equality. Bradshaw (1972) provides perhaps the clearest exposition of the problem, arguing that need is neither an absolute concept, nor necessarily quantifiable. Normative need, for example, is defined by reference to professionally set or politically determined standards. It may make little reference to the real service needs of individuals. Comparative need, in contrast, finds its basis in comparisons with the more well provided. Expressed needs rely on an individual to articulate his or her needs and are, in many ways, analogous to demands rather than pure needs. Finally, felt needs which are, by their nature, of little use since they are individually perceived yet unexpressed wishes, can also be identified. In addition to these internal inconsistencies, and despite close interrelationship with equality, further problems can also be raised. Demand, preference or willingness to pay; each provide perfectly possible alternative criteria for service provision (Lucy, 1981), although they would, within the terms we have set out so far, lead to inequality.

It should now be clear that neither equality nor need can be viewed as precisely definable concepts. Even if they are accepted as goals within health service delivery, organisational problems must be overcome (Lee, 1979). Firstly, whilst equality, or more particularly equity, may benefit the consumer of a service, it may not be an efficient method of service delivery with regard to available financial investment. Secondly, the demands of the more articulate consumers must be balanced against the needs of less vocal service groups. Finally, the relative importance of horizontal and vertical equity must be considered. The former concerns equity within single services such as accident and emergency, while the latter involves us in the equalisation of all the constituent parts of a health service. Notwithstanding the theoretical debates and inconsistencies however, equality is a key concept in social scientific studies of health care, and more particularly the persistence of inequality is the subject of considerable research effort (DHSS, 1980; Open University, 1976).

Developing the spatial perspective

One thing which has been lacking from our analysis so far has been any specific consideration of the role of space in inequality. We can remedy this deficiency by recognising that the social groups for whom inequality is a problem occupy particular discrete locations in geographical space.

Spatial equality of use provides a useful vehicle to demonstrate the identification of patterns of equality. Equality of use, it will be remembered, demands that need be equated with expenditure. The spatial aspect may be incorporated by arguing that need should be equated with expenditure within pre-defined spatial units of analysis. Davies (1968) has termed this concept territorial justice to distinguish it from purely social justice (Harvey, 1975).

A graphical presentation of territorial justice is illuminating (Pinch, 1979). Figure 6.1 illustrates this point. Firstly, both need and expenditure must be quantified. The two indicators are expressed as comparable scales of measurement, such as standard scores or ranks. Thus in a ten-region system the need and expenditure values for each region could both be ranked. The use of comparable scales of measurement means that need and expenditure are perfectly matched, that is to say equality of use occurs along a line lying at 45° to the graph axes and passing through its origin. Deviations from this line represent departures from territorial justice and thus inequality in terms of over- or under-expenditure. Incidentally, expenditure could easily be replaced in the analysis by any other measure of resource output, for example facilities or manpower provided. The result of the exercise would still be the identification of territorial justice.

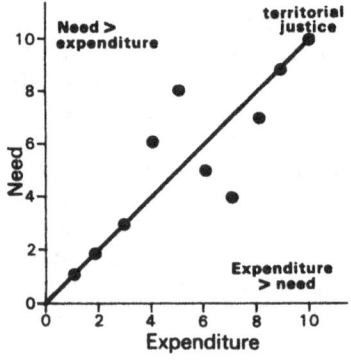

Figure 6.1   Territorial justice

The logical counterpart to territorial justice is inverse care. Again using the graphical approach (Figure 6.2) we can show that in this case areas with higher need receive proportionately less of the available resources. Hart (1971) provides the definitive statement on inverse care, arguing that:

In areas with most sickness and death, general practitioners have more work, larger lists, less hospital support and inherit more traditions of clinically ineffective consultation than in the healthiest areas; and hospital doctors shoulder heavier case-loads with less staff and equipment, more obsolete buildings

and suffer recurrent crises in the availability of beds and
replacement staff. These trends can be summed up as the
inverse care law: that the availability of good medical care
tends to vary inversely with the need of the population served.
(Hart, 1971, p. 412)

Hart continued his thesis by claiming that the prevalence of
inverse care in Britain is a consequence of the continued
unwillingness of successive governments to shield the National
Health Service from the operation of a market economy. To
combat inverse care, the needs of consumers need to be placed
before the ideological and financial ambitions of those medical
professionals for whom health care is merely an avenue to
personal rewards.

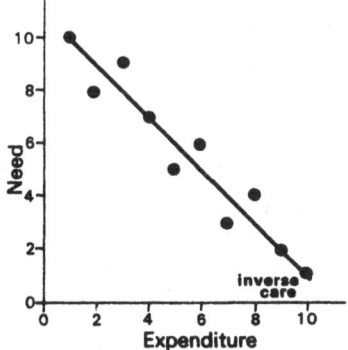

Figure 6.2   Inverse care

A further way in which we can show a spatial aspect to
inequality is by considering how it is evidenced at different spatial
scales of analysis. Conclusions concerning inequalities in the
provision and consumption of health care exhibit wide variation
according to the geographic scale at which the analysis is
conducted. At the international scale (Smith, 1979) inequalities
are, as we shall see in the next chapter, a reflection of broader
issues of national development. Comparative work at this scale
also allows us to make interesting contrasts in the way national
social and political-economic organisation affects health care. At
the regional level, that is within a country, the way the national
health care system is administered may affect equality. An
emphasis may be placed, for example, on certain health care
functions at the expense of others. At the local level we can focus

our concern on inequality in single jurisdictions. Here it may be the case that provision is inadequate, it may be of the wrong type, it may not be used by those for whom it is intended, or it may be in the wrong place. This last case is of particularly important significance geographically as it introduces the concept of physical accessibility to health care facilities. Access has been widely used by medical geographers to illustrate micro-scale inequalities as it is inevitable that medical facilities, discretely located at a single point in space, will be nearer some consumers than others.

In the following sections we will make some preliminary comments on inequality at the various spatial scales. We will concern ourselves primarily with setting out the evidence for inequality and providing some examples of the nature of the problem. This scene-setting will provide the grounding necessary to proceed to Chapter 7 where we will consider the ways in which researchers can explain inequality.

Perspectives on international inequality

The way in which a country organises its health care system can have a considerable effect on the dimensions of health care inequality. There are three basic issues of interest in discussing this effect. Firstly, the social and politico-economic ideology of the country will, with the occasional overlay of cultural factors, determine the general attitude to health care (Maxwell, 1974). In particular the role of the state in the provision of care and the financial responsibilities of the consumer of health care will be determined at this level. Secondly, a country may adopt a specific organisational partitioning of its national area to facilitate resource planning. The jurisdictions so created form the general basis for the geographical consideration of regional inequality. Finally, it is the national health care policy–makers who determine or decide the relative importance of the particular sectors of health care. Eyles and Woods (1983) exemplify this last issue in their discussion of the 'cinderella' services within the British National Health Service. Preventative care and services for such client groups as the mentally ill, the handicapped and the elderly are continually underfunded and accorded a low priority despite advocacy to the contrary. Of the three issues above, however, it is the first which is the most important and to which we will return in the next chapter when we seek an explanation for health care inequality. At this juncture it provides a useful

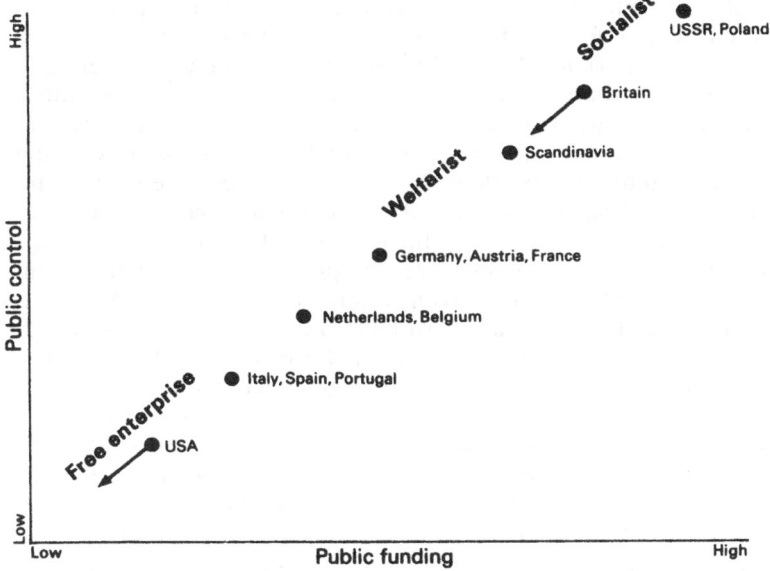

Figure 6.3 Ideology and health care organisation

framework for considering international variations in health care organisation (Figure 6.3).

The USA provides perhaps the best-known example of the situation where the ideology of health care has been one of minimal state intervention. Health service delivery, with a few exceptions, is on the basis of the unfettered operation of a free market economy. Consumers of health care are expected to pay, either personally or through an insurance scheme, for the services which they receive. The general public have little involvement in the regulation of the locational decisions of medical personnel. Such decisions tend to be made with regard to individual financial reward or prestige by the providers of care. A 'safety net' system of medical support for the very old and very poor is operated in the form of a limited welfare system and a network of public hospitals. The latter, situated close to their client population, in run-down inner city areas, invariably do not generally secure particularly high quality resources (De Vise, 1971). The insurance schemes reimburse hospitals for the treatment which patients receive. This reimbursement is a flat fee based on diagnosis; all treatments fall into a particular Diagnostically Related Group (DRG). The system is designed to inhibit over-elaborate treatment, but at the same time it places a clear onus on the patient to pay for anything costing above the threshold.

In the Netherlands the health care system exhibits greater state intervention than in the USA (Doeleman, 1980). A more extensive welfare system also exists. Nevertheless, the system remains essentially private with medical professionals being formally autonomous though, of course, restricted by state regulations designed to provide a regionally based system of planning norms. The state, through social security legislation, aims to ensure that people are not prevented by poverty from receiving health care, or driven to poverty as a consequence of a medical condition. On a macro-level therefore the Dutch health care system pays some attention to the problems of inequality. As a side-note, the Dutch system also indicates the impact of cultural factors on health care. In a society divided on religious lines, this cultural input has, not surprisingly, been evident in health care: 'the great majority of hospitals, psychiatric institutions local public health organisations . . . and community mental health services were privately organised on a Protestant, Roman Catholic or non-denominational basis' (Doeleman, 1980, p. 48). The indications are that this cultural division is now lessening.

The British National Health Service provides the clearest example of a service funded by general taxation and controlled by national policy (Drury, 1983). It is perhaps the most state-controlled of health services in welfare capitalist societies and, in this context, merits particular attention.

On a broad level we can view health care in Britain as a collective consumption service similar to those concerned with social welfare, education, housing and health (Pinch, 1985). A significant part of public expenditure in Britain is devoted to these services, and this within the context of a change in the public expenditure proportion of national Gross Domestic Product from 10 per cent in 1890 to 45 per cent in 1980. Health care itself is the third largest item of public expenditure. All the collective consumption services in Britain are predicated, to a greater or lesser extent, on a strategy of promoting equality and freeing individuals from dependence on market mechanisms. In particular the aim has been to ensure that no one should be excluded from basic comforts or necessities on the basis of income.

As we have stated at the start of the chapter, and as we shall see in the following chapters, these aims have proved more problematic in practice, and many measures have been enacted in their pursuit. One characteristic of the British health care system which may have contributed to these problems may be introduced at this stage. The system has an intensely hierarchical internal structure which partitions health care into primary and secondary sectors. The former concerns services where the consumer makes an initial contact with 'generalist' medical personnel, for example, family doctors or community services. The latter comprises those 'specialist' services to which the consumer is referred by the primary sector – basically the hospital service. Though officially the interdependence of these sectors is recognised (Office of Health Economics, 1977) there is disturbing evidence that prestige and resources have at times been disproportionately invested in the secondary sector (Table 6.3). Further inequality within the system arises from the continued persistence and even growth of a private fee-for-service service option (Townsend and Davidson, 1982).

The health care system in socialist countries provides the logical counterpart to the USA system in our spectrum of health care system types. In socialist countries health care is under centralised state control and financial barriers to access are removed (Ryan, 1978). Health services form important and well integrated parts of national planning, corresponding closely to other governmental frameworks in terms of their spatial organisation. In Poland, for example, despite economic and social unrest, the system is structured to provide continuity of care with the family doctor as the central figure (Millard, 1982). Area Health Complexes serving 30–150,000 people form the basic level of spatial organisation and correspond either to local government

Table 6.3 Sectoral inequality within the British National Health Service

Sector	% Expenditure (England)	% Expenditure (Wales)
Headquarters administration	4.2	3.9
Hospital services	62.3	61.0
Community health service	6.4	6.7
Family practitioners services	26.6	25.5
Ambulance service	2.2	2.6
Blood transfusion service	0.4	0.3

Source: *NHS Health Service Costing Returns, 1979* (HMSO, 1981).

areas or large factory plants. Hospitals, organised on a provincial level, provide specialist care and polyclinics, similar to those in the USSR, provide an intermediate point of access to the system, broadly similar to the role undertaken by health centres in Britain (Smith, 1979).

Third World health care systems stand somewhat outside the spectrum of health care systems which we have outlined here. Indeed the variation in organisational approaches to health care in the Third World is perhaps best approached by distinguishing an internally variable category of 'underdeveloped system' (Pyle, 1979). Three factors would seem important. Firstly, there is the degree to which indigenous or peasant medicines persist (Elling, 1981). This may have consequences for inequality if 'modern' or indigenous health care modes are functionally restricted to particular social groups. Secondly, the health care system may be slotted into the free-market/socialist health care spectrum through the dominating influence of either a former colonial power or a contemporary major giver of foreign aid (Doyal, 1979). This produces inequality by preparing doctors for situations in western urban medicine rather than the rural Third World, and also through the exploitation of medical manpower by developed countries. The latter situation is evident both in Australia (Connel and Engels, 1983) and Britain:

> the fact that former colonies have been actively encouraged to maintain British 'standards' of medical education means that they can supply doctors to meet the shortfall (in Britain) effectively subsidising high cost health care in Britain. Vast numbers of nurses, midwives and unskilled hospital workers are also exported from third world countries. During the 1960s, for example, only three out of every seven newly trained West Indian midwives were actually working in the Caribbean, while four were in Britain. (Doyal, 1979, p. 264)

Finally, political instability in the Third World may cause rapid changes in the ideology of health care. Post-revolutionary Cuba, for example, has moved rapidly towards a developed world health care system, on the socialist model, which has placed great emphasis on preventative care (*Radical Community Medicine*, 1982). In Nicaragua the overthrow of the Samoza regime occasioned the demise of a free-market system, under which the poor had been unable to afford health care, and the institution of a system leading to the eradication of polio and the setting up of an extensive network of preventative care. Chile, of course, provides an example of the process working in the opposite direction. Most of the advances and initiatives planned by the Allende regime have since been rejected as the ruling junta have led the country back towards an avowedly free-market system (Navarro, 1976).

Regional inequality

Spatial inequality at the regional scale has been widely researched, particularly in Britain. A pioneering study by Bosanquet (1971) has shown for example, that northern England is, in general terms, less well off in the case of health care resources. This finding was also noted by Coates and Rawstron (1971) who considered health care as well as a wider selection of other services. Particularly in terms of expenditure, large variations in *per capita* financing were noted between the richer and poorer regions. Regional patterns of health care inequality were a key focus of an Open University social science course (Open University, 1976) and, of course, have been identified in many countries other than Britain (Figure 6.4).

Recent examples of work in this tradition have been provided by Knox (1979, 1981). In the earlier work he used location quotients. These take the form:

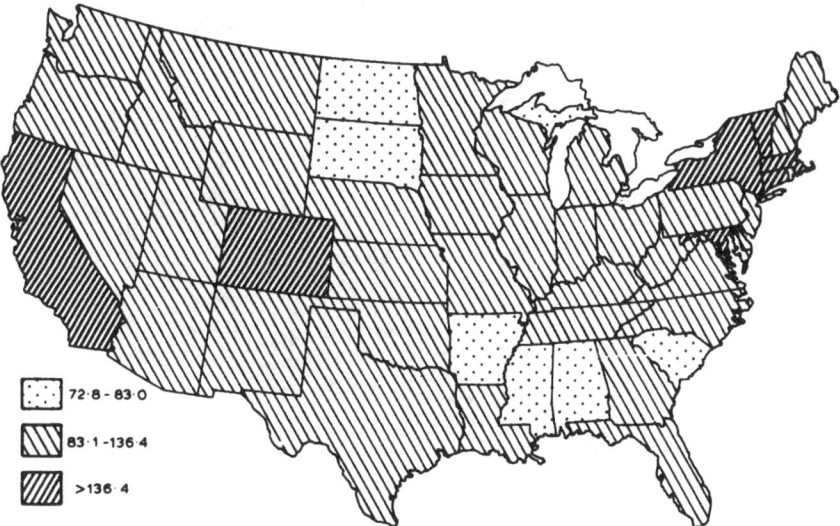

72·8 - 83·0

83·1 - 136·4

>136·4

Figure 6.4 Regional inequality in health care in the USA, general practitioners per 100,000 inhabitants 1970
Source: Shannon, G. W. and Dever, G. E. A. (1972) *Health Care Delivery*, McGraw Hill, New York

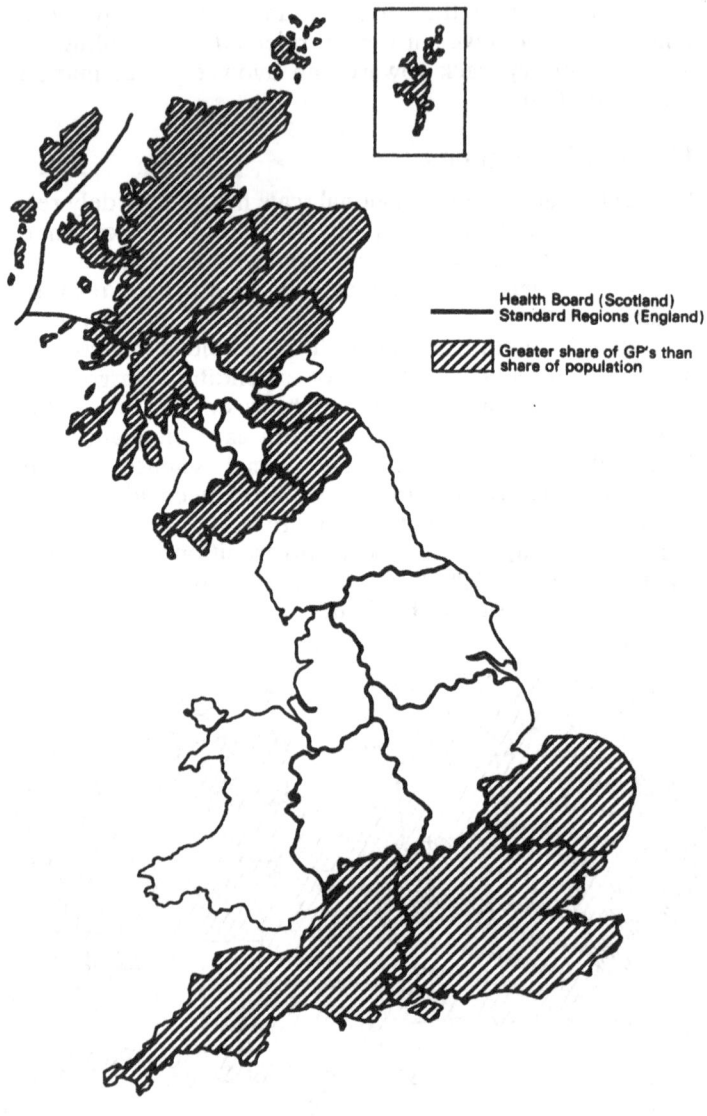

Figure 6.5 Regional inequality in health care in U.K., G.P. provision, 1975
Source: Original data from Knox (1979)

$$I_i = \frac{(Gi/Pi)}{(\Sigma Gi/\Sigma Pi)}$$

I_i = location quotient for region i
Gi = number of general practitioners for region i
Pi = population for region i

His indices enabled him to identify regions in England, Scotland and Wales where the distribution of general practitioners was approximately in balance with the population ($I = 1$) or where positive ($I > 1$) or negative ($I < 1$) inequality was evident (Figure 6.5). In the 1981 study a regression-based approach was used to investigate the relationship between population change and the changing provision of pharmacies between 1950 and 1980 for the old counties of Scotland (Figure 6.6).

Regional variation in mental health care in Britain provides a convenient way in which we can demonstrate the various ways in which regional inequalities can be viewed. Relevant data for such a task is readily available from the *In-patient Statistics from Mental Health Inquiries* and *Facilities and Services in Mental Illness and Mental Handicap Hospitals*, both published by HMSO. Table 6.4 sets out a selection of the available statistics

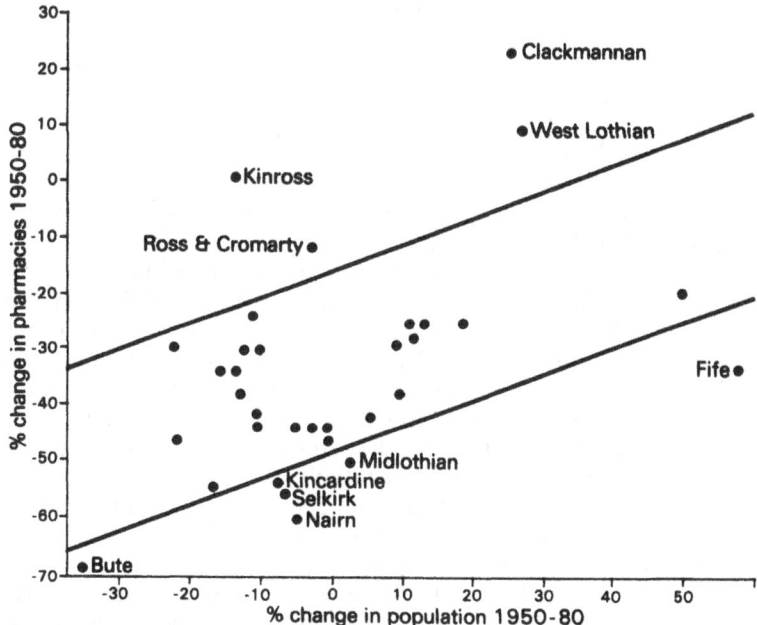

Figure 6.6 The changing distribution of pharmacies in Scotland
Source: Knox (1981)

Table 6.4   Regional inequality in mental health care: England

Region	Attenders at hospital per 100,000 catchment population[a]	Nurses per 100,000 catchment population[a]	Cost £/inpatient day[b]	Cost £/day-patient[b]
Northern	409	102.2	25.0	14.9
Yorkshire	435	102.7	26.0	11.5
Trent	336	81.1	27.6	14.6
East Anglia	353	97.8	27.3	11.2
N.W. Thames	405	121.3	28.5	15.3
N.E. Thames	384	104.3	29.7	13.2
S.E. Thames	392	113.6	30.4	20.5
S.W. Thames	450	136.6	27.6	18.4
Wessex	385	103.2	31.5	17.4
Oxford	287	71.6	32.9	13.2
South West	377	100.6	29.7	11.4
W. Midlands	341	87.6	28.7	12.7
Mersey	428	134.9	25.7	8.0
North West	381	94.6	27.7	9.5

Sources: a. DHSS (1980), *Facilities and Services in Mental Illness and Mental Handicap Hospitals in England 1876.*
b. DHSS (1982), *Hospital Service Costing Returns.*

concerning mental health care for the fourteen English Regional Health Authorities (RHAs).

Column one provides information on the number of people using these facilities. Expressed as a ratio of each 100,000 persons in the RHA, this measure allows us to make some crude generalisations about regional variation in the demand for mental health services. Attendance ratios are highest in Merseyside, Yorkshire and SW Thames. In column two we turn our focus to the staffing resources of mental health care. In this case regional variation is, to some extent, inevitable (MIND, 1977) as some areas of the country will always have a residential unpopularity which will cause recruitment difficulties in the health service. Nevertheless, we might well question the discrepancy in values between the Oxford Region and SW Thames. It might be, for instance, that the lower level of staffing might be perceived as adequate, with the consequence that well-provided areas would then be cut back. The last two columns utilise expenditure indicators. If we consider the cost expended in each region per in-patient day – that is basically the cost of keeping a person in a hospital for the mentally ill – we find a considerable degree of equality between the regions. This is only to be expected if it is recognised that detailed resource planning has formed a central part of recent National Health Service policy making. Nevertheless, there is some inter-regional variation in costs, ranging from £25.0 per day in the Northern Region to £32.9 in the Oxford Region. In passing we should also reflect that little of the cost of this care is actually connected with treatment; most is simply the result of providing accommodation. Furthermore, among all other hospital services, only the care of the mentally handicapped is cheaper (MIND, 1977). Day patient costs provide an interesting comparison, being, of course, markedly cheaper as bed costs are not included. Though, with our figures, such comparisons are not strictly valid, they may indicate a tentative confirmation of Scull's argument (set out in Chapter 5) that a crucial reason for the shift from in-patient modes of care has been the cheapness of alternatives (Scull, 1978).

Following any consideration of inequality it is always wise to reflect upon the wider context. In the case of the regional pattern of mental health care in Britain this might take two forms. Firstly, we can note that we identify only comparative inequality. Nowhere did we ask whether any region catered *adequately* for need. It may well be that those regions which appear to be well provided for actually *need* high levels of provision to deliver the necessary levels of care indicated by local demand or need. Secondly, we should note that the attitudes of the various authorities charged with providing facilities may vary, with consequences for the provision of care. A 1975 Government White Paper, for example, revealed that 24 out of 108 authorities surveyed provided no community-based mental health facilities (DHSS, 1975) and more recently a number of health authorities have shown an increasing propensity to devolve their responsibilities for mental health provision to the private sector (Moon, 1985).

A more general point concerns what the Black Report (DHSS, 1980) has termed 'the interaction of geographic and social disparities'. Earlier in this chapter we noted that a spatial aspect can be incorporated relatively easily in the analysis of the essentially sociological construct of inequality. Our argument was broadly that deprived people, in terms of health care as well as many other areas, tend to occupy specific locations in geographical space. Two empirical studies have demonstrated the validity of this linkage at the regional scale and provide added substantiation to the conclusions of Coates and Rawstron (1971) and the Open University (Open University, 1976). Noyce *et al.* (1974) examined expenditure on hospital and community services between regions in England (Figure 6.7). They found that, in both cases, there was greater expenditure in higher status areas. Their case study of hospital expenditure was, furthermore, focused on capital budgets, so it indicated that future prospects in higher status areas were likely to be enhanced. West and Lowe (1976) utilised the fifteen hospital board regions which existed in England and Wales prior to the 1974 reform of the health service, as the basis of their research and tried to relate provision to need. Only midwives appeared to be distributed equitably; the distributions of doctors and health visitors were particularly dissimilar to that of need.

A similar regional context to inequality is identifiable in other countries. In the USA, Joroff and Navarro (1971) note that 50 per cent of US cities are underprovided with doctors and a comparison of rural and urban areas reveals that, overall, the rural areas are yet more poorly provided for (Navarro, 1976). The 26 per cent of the USA population who live in rural areas are being served by just 12 per cent of the available doctors. Many rural areas have no medical provision at all. It appears that the rural areas do not offer sufficient opportunities for medical practitioners, for whom the health care system prevailing in the

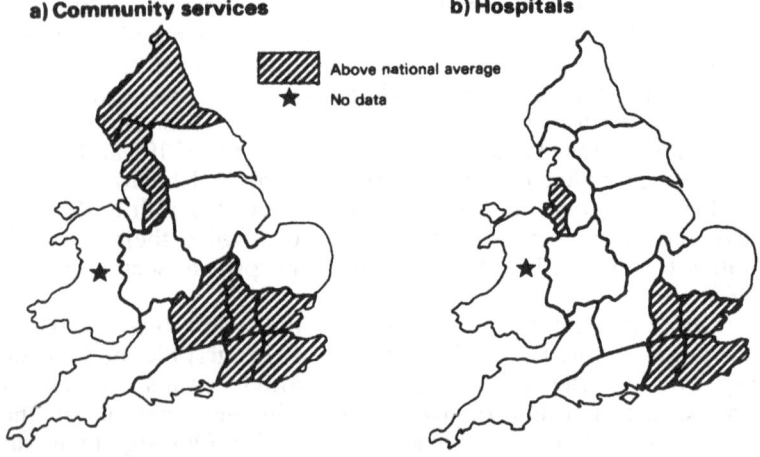

Figure 6.7 Spatial variation in hospital and community services expenditure
Source: Original data from Noyce *et al.* (1974)

Table 6.5 Regional inequality in socialist health care systems: doctors in the USSR

Republic	Number of doctors per 10,000 population			
	Year			
	1940	1965	1970	1975
RSFSR	8.2	24.8	29.0	34.8
Ukraine	8.4	24.3	27.6	32.0
Belorussian SSR	5.7	21.8	25.8	30.2
Uzbeck SSR	4.7	17.0	20.1	25.1
Kazakh SSR	4.3	18.7	21.8	27.3
Georgia	13.3	35.0	36.2	41.1
Azerbaijan	10.0	23.8	25.0	28.9
Lithuania	6.7	21.5	27.4	34.2
Moldavia	4.2	17.9	20.5	26.2
Latvia	13.2	31.2	35.6	39.2
Kirghiz SSR	3.8	19.1	20.7	24.4
Tadjik SSR	4.1	15.0	15.9	20.6
Armenian SSR	7.5	26.7	28.8	34.8
Turkmen SSR	7.6	21.2	21.4	25.7
Estonian SSR	10.0	29.5	33.1	36.8

Source: Lewytskyj, B. (1979), *Soviet Union, Facts and Figures*, Saur, London.

USA demands financial success in practice, as well as the satisfaction of providing care.

In socialist countries it might be expected that centralised health care planning would have led to less marked regional inequality. Unfortunately evidence from empirical studies (Table 6.5) suggests that even in the Soviet Union this is not the case (Ryan, 1978). It does seem, however, that a gradual process of equalisation is taking place – paramedics, for example, are being used to extend care into rural areas. Although arguably not strictly a socialist country, Jugoslavia provides an example of this convergence in regional patterns of provision (Kunitz, 1980). Jugoslavia comprises a federation of six provinces which, with the addition of the two autonomous areas of the Vojvodina and Kosmet, vary strikingly in their levels of economic development. Doctors have traditionally been attracted to the more developed provinces of Slovenija and Serbia, while the underdeveloped areas of Kosmet and Macedonija have been unpopular (Table 6.6). Since 1950, however, the inequality gap has been narrowing in most regions as a response to direction of physician

Table 6.6 Regional inequality in Jugoslavia

Republic/ province	1950 Population/ physician ratio	Index (Av = 100)	1973 Population/ physician ratio	Index (Av = 100)
Bosnia-Hercegovina	6,504	214	1,407	160
Montenegro	4,437	144	1,270	145
Croatia	2,324	75	738	84
Macedonija	5,217	170	1,000	114
Slovenija	2,049	67	697	79
Serbia	2,663	87	702	80
Vojvodina	3,132	102	783	89
Kosmet	11,372	370	2,502	285

Source: Kunitz (1980).

recruitment and a system of locally defined need assessment. By 1973, for example, the population:physician ratio in Kosmet had fallen from 3.7 times the national average to 2.8 times the national average – a startling but still inadequate improvement.

Local inequality

Smith (1982) has stressed the importance of a consideration of inequality at the local level by commending the study of 'The spatial organisation of service delivery, including differential physical accessibility of the local population to sources of health care such as hospitals and doctors' (Smith, 1982, p. 3). By studying regional inequality we may paint a broad brush picture, but the localised, humanised problems of having poor care, care situated in less accessible locations, or no care may be neglected. Indeed it has been widely argued that the most trenchant manifestations of inequality are to be found at the local or sub-regional level (Buxton and Klein, 1976; Jones and Masterman, 1976). A convenient, but in many ways arbitrary, way in which we can divide up the vast body of research which has considered local inequality is to consider separately secondary and primary care. As we have already stated, secondary care is concerned broadly with the hospital sector, centralised large facilities to which the health care consumer is referred from a primary sector comprising general practitioner and community services delivered in a generally more decentralised fashion.

Secondary care

Rationalisation of secondary care facilities has been occurring in many countries in the last twenty years. In Britain hospital care has developed from a historic legacy of provision by both the state and charitable bodies. By the end of the nineteenth century a situation had developed such that most larger towns possessed one or more hospitals located in city centre sites (Cowan, 1965). Often these hospitals had developed particular specialities – for example, surgery or maternity care. Provision in smaller towns was through cottage hospitals with a generalist function and, usually, some formal links to a larger hospital. In recent years a number of changes have occurred to this system. Changes in the distribution of population and growing urban traffic congestion have meant that hospital provision is now re-concentrating in city fringe locations. A new hierarchy of provision similar to a central place system has been developed. Single District General Hospitals provide a full range of specialisations while small numbers of more locally based community hospitals provide hospitalisation facilities for chronic illness (DHSS, 1969, 1974, 1977). Beyond this hierarchy, some hospitals continue to provide particular specialisations for consumers drawn from very wide areas – for example, the orthopaedic hospitals at Oswestry and Stoke Mandeville which draw on national catchments. As a consequence of these changes many small and very localised cottage hospitals have closed. The result is that secondary care is now more centralised and streamlined but for the consumer the journey to care is often far longer.

A seeming anomaly is posed, however, by secondary care for the mentally ill. As we saw in the last chapter, responsibility for mental health care was, for many years, discharged through large asylums. Many of these facilities were situated in greenfield locations away from major centres of population, or peripheral to those locations. The reasons for this locational choice, which was rather different from that pertaining for other hospital facilities, can be traced to the prevailing social construction of mental illness. Unlike other forms of ill-health, mental illness was not viewed as an inherent problem of society; rather it was abnormal and demanded the removal of the mentally ill from the rest of society. Nowadays day hospitals and halfway houses are coming to possess a very different residential geography, being predominantly in urban locations occupying, respectively, premises in general hospitals or houses subdivided for multi-occupation. In contrast then to secondary care in most cases, secondary care for the mentally ill is therefore now becoming more decentralised and physically more accessible to the consumer.

Let us return to the case of secondary care for those other than the mentally ill, and examine the major consequence of rationalisation – inequality in physical access. Strangely, the general indication is that, for patients booked into hospital by their doctor, the problem of travel to hospital is not an overwhelming issue (Rigby, 1978). This view fails, however, to consider the full picture. Distance is vitally important (often literally) in the case of accident and emergency patients (Ingram *et al.*, 1978) and is also an issue of concern to out-patients. Wheeler (1972), in a study of the patients attending clinics at Bury St Edmunds Hospital, found that 15 per cent lived between 24 and 45 km from the hospital, over 50 per cent of out-patient journeys to hospital lasted over one hour and 33 per cent of the out-patients had no access to public transport. Physical accessibility is also an important issue for hospital staff and visitors to patients, particularly in more rural areas. Cross and Turner (1974), for example, have noted that visits to long-stay geriatrics double if the visitors live within 24 km of the hospital.

The major British study of physical accessibility to hospital has been conducted in East Anglia (Haynes and Bentham, 1979). Hospital provision in the Kings Lynn Health District has been centralised in Kings Lynn and Wisbech – two towns thought to be sufficiently far apart to ensure reasonable physical access for all. It was found that most residents lived within 40 minutes of a hospital and, not surprisingly in an area of poor public transport, most had access to a private car. The problem of inaccessibility to hospital resources was thus specific to those without a car living in the more remote rural areas. Even in those cases patients generally were afforded the option of hospital transport. Visitors, on the other hand, were disadvantaged such that fewer came from the remote villages. This meant that some patients, particularly those facing long-term hospitalisation, received fewer visitors. Out-patients were also clearly affected by distance with fewer than expected travelling distances exceeding 16 kilometres.

Whitelegg (1982, 1983), working in north-west England, has emphasised the political nature of the conflicts which centralisation arouses in both the medical profession and the communities which are affected. When Rossendale General Hospital, in a small Pennine Valley, was scheduled for closure, resistance was forthcoming from both the local community and local doctors. Both groups foresaw the loss of a valuable community resource. Powerful support for closure came, however, from consultants at Burnley District General Hospital who had the advantage of greater importance in the relevant decision-making machinery and who emphasised the need for rationalisation to promote economy and enhance the importance of specialist care.

Mohan (1980) has rightly argued that, in the British context, the centralisation of secondary care facilities has had profound implications for inequality. Similar conclusions are evident in the USA where the situation is compounded by admissions policies which render hospitals socially as well as physically inaccessible to many low status groups. De Vise (1971, 1973), for example, reports that many of the eighty hospitals in Chicago discriminate on the grounds of race, religion and income. We may summarise thus: localised needs for secondary care are neglected.

Primary care

In moving to a consideration of primary care we confront a vast range of literature in medical sociology, medical geography and many other disciplines. Inevitably we must be selective and draw out only the main themes of relevance to our discussion of inequality. To this end we will consider inverse care, physical accessibility and social access. This approach will allow us to move from a consideration of inequality as a result of a lack of provision, through the role of distance from a facility in promoting inequality, to a realisation that inequality as exemplified in primary health care has a profound social context.

Inverse care is perhaps most strikingly manifested in the declining industrial cities of the USA. Particularly in the inner cities the supply of doctors has often actually fallen, for example by 37 per cent in Boston (Robertson, 1970). In these areas, which have anyway historically been under-provided, population decline has exacerbated inverse care (Guptill, 1975). A residual, marginalised population of ethnic, low income and elderly social groups, often with high health care needs, remains, to be serviced by similarly elderly doctors, often practising in ill-equipped single-handed practices. Elements of inner city inverse care may be found in Britain, but the evidence also indicates that rural areas and peripheral local authority housing estates are disadvantaged (Knox, 1978; Moon, 1983). For some areas national provision averages are seldom reached, for example only

Newcastle in the Northern Region exceeds the national average of persons per dentist (Carmichael, 1983). Dentists, who are governed by few sanctions concerning where they can practise, appear to favour two types of location – higher status, high income suburbs (Bradley *et al.*, 1978) or inner city zones close to potential customers in the city centre workforce (Jones and Kirby, 1982). In both cases, as with doctors, these spatially discrete distributions mean disadvantage to the working class.

In addition to inverse care, spatially discrete concentrations of provision also inevitably make physical access to care an important issue (Fielder, 1981). The lack of physical proximity to care has largely been considered in terms of distance (Knox, 1978; Jones and Kirby, 1982; Joseph and Phillips, 1984). Those who have to travel farther to care are disadvantaged such that patients who live close to a surgery consult up to a third more often than those living at distances of over 2½ miles away, and those with transport problems are particularly disadvantaged (Whitehouse, 1985). This approach is obviously limited and corrections to the crude distance factor are made to allow for facility opening hours, transport availability and journey times. Some workers have focused on these latter correction factors. Wachs and Kumagai (1973), for example, have compared the number of facilities which could be reached in 15 and 30 minutes by car and public transport from selected nodes in Los Angeles and Shannon, Skinner and Bashshur (1973) note the vastly increased time taken to reach health services by inner city residents in Cleveland, Ohio.

Knox (1978) provides us with one of the best-known examples of work concentrating on spatial accessibility. He first calculates a simple index of potential accessibility:

$$A_i = \sum_{j=1}^{n} \frac{S_j}{D_{ij}{}^k}$$

A_i = accessibility index for zone i
S_j = size of surgery (hours open)
D_{ij} = linear distance from zone to surgery
k = constant reflecting distance decay

This index was then used to investigate the relationship between spatial access and inverse care. Factors such as car ownership and population base were taken into account, and it was eventually possible to show that in Glasgow, Edinburgh, Dundee and Aberdeen, spatial access was inversely related to socio-economic status. Those who were socially disadvantaged were also disadvantaged in terms of physical access to health care. Furthermore, although some needy inner city areas had relatively high provision, quality was often poor. Seventeen per cent of general practitioners in the deprived East End of Glasgow were over retiring age and almost half were in substandard premises. Particular problems were evident in the peripheral local authority estates of all the cities.

We must, at this juncture, note that there are problems with the notion of physical accessibility. For example, work has often assumed that, given choice, people will prefer to use the nearest facility; evidence suggests that they do not (Phillips, 1979). Analyses of physical accessibility also suffer from boundary problems, with important and popular facilities often being excluded from analysis simply because they lie outside a study area (Moon, 1983). Critiques of quantitative models of physical accessibility abound; some condemn the lack of realism in the calibration of the models, others caution against over-elaboration:

> In essence, however, Knox's measure takes [the accessibility potential index] and divides this by population potential, and this will not, we suspect, yield results that are very different from dividing size of facility by local population; distance effects may fail to be manifested. (Cole and Gattrell, 1983, p. 9)

The centralisation and rationalisation processes which we identified as important in our discussion of inequalities in the physical accessibility of secondary care are also relevant in primary care. Status-based inequalities in physical access to doctors have been compounded by health care policy changes. For example, there has been a trend from single-handed practices towards health centres housing several doctors (Table 6.7). Health centres could only be established by combining a number

Table 6.7 Trends in medical practice organisation in England

Type of practice	1961	(%)	1971	(%)	1981	(%)
Solo practices	5,337	(28.3)	3,954	(20.4)	2,990	(13.4)
2 doctor partnership	6,384	(33.8)	4,552	(23.5)	4,004	(18.0)
3 doctor group	4,008	(21.2)	4,911	(25.3)	5,132	(23.0)
4 doctor group	1,984	(10.5)	3,232	(16.7)	4,255	(19.1)
5 doctor group	715	(3.8)	1,490	(7.7)	2,940	(13.1)
6+ doctor group	450	(2.4)	1,235	(6.4)	2,983	(13.4)

Source: Adapted from Joseph and Phillips (1984).

of previously separate practices. The result was that a once decentralised system offering accessible care was abandoned in favour of the financial economies of scale given by centralised, yet poorly accessible, facilities. For the poorer consumer, particularly with the onset of recession, the cost of reaching health care became very high (Sumner, 1971). Figure 6.8 shows the disadvantage imposed by centralisation for one such group.

An important factor related to physical inaccessibility and inverse care is that particular social groups consume less medical care than they might. Some argue that the use of health care resources is a direct consequence of proximity to those resources. This phenomenon has been termed Jarvis's Law, and was first noted because those living near mental hospitals were found to provide more patients than those living further away. Vaughan (1967) suggests that there is a distance decay in the number of patients registered with a particular doctor with distance from the doctor's surgery, and Phillips (1979), in a comprehensive study of medical care utilisation in West Glamorgan, claims that consultation rates may be significantly affected by distance from provision. A social context to these conclusions may be added by reference back to our suggestion that lack of proximity to medical care is closely related to social status.

We should also be concerned with more intrinsically non-spatial aspects of health service utilisation. While we might expect the lower status groups, who we know suffer from inverse care and poor physical accessibility, to make less use of medical care, in fact the reverse occurs. This is because there are class-based differences in attitude to particular services which are reflected in their consumption. Social classes IV and V, for example, are known to make quantitatively greater use of doctors. This usage however, is not as great as we would expect from the evidence of such need indicators as mortality and morbidity data (Alderson, 1970; Forster, 1976; Le Grand, 1978). So in reality, class-based inequality is still in evidence. In contrast, the middle classes use dentistry services more frequently (O'Mullane and Robinson, 1977; Taylor and Carmichael, 1980) and furthermore are also better attenders of antenatal, birth control and immunisation clinics (Bone, 1973; Brotherston, 1976). Figure 6.9 illustrates the pattern for the case of immunisation completions in one small British town. There is a clear difference between rates in middle-class and working-class catchments, despite a probably similar need for such preventative care.

In summary, inequality in primary medical care means that low status groups, often in particular locations, receive less care and

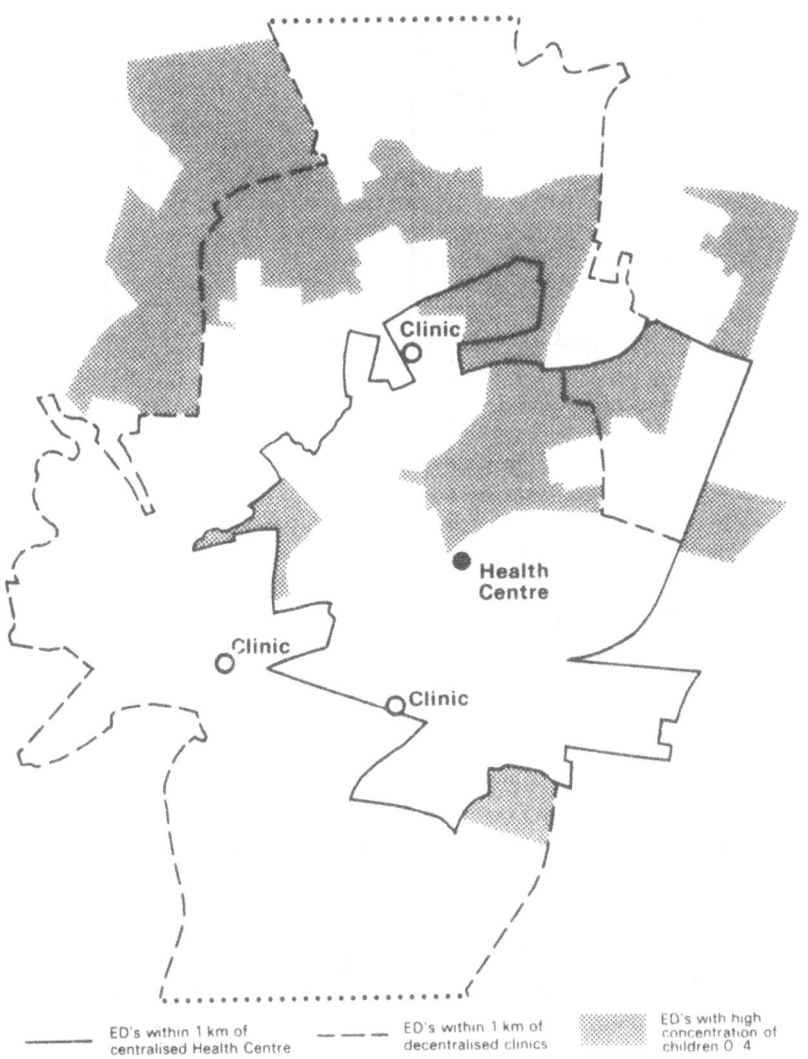

Figure 6.8 Physical access and health centre development, Havant, England

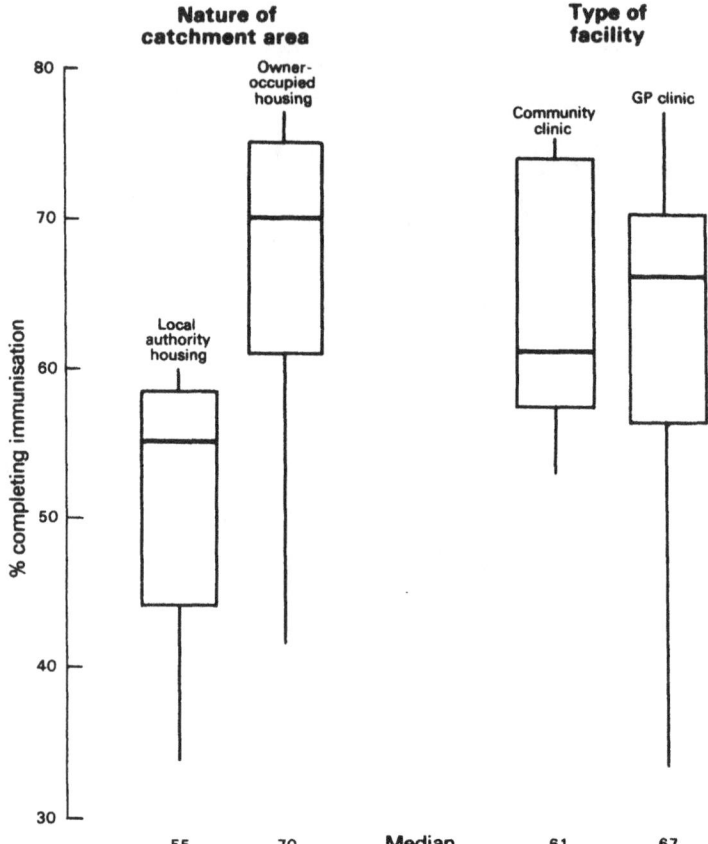

Figure 6.9   Local inequalities in immunisation uptake, Havant, England

make less use of the care they receive than might be expected. Distance, and thus the issue of geographical space, plays a role in this inequality, which is underpinned by wider social and organisational factors.

Conclusion

Our discussion of inequality in health care points to one seemingly self-evident conclusion. Inequality is widespread and pervasive at all levels and in all systems of health care.

In this chapter we have concentrated solely on setting out the evidence for inequality. Geographical manifestations of this inequality have provided an instructive basis to our discussion; however, we have tried to indicate that much of this geographical inequality which we have uncovered has, in fact, been linked more fundamentally to social, and indeed political or economic factors – most notably class and income. There is a grave danger that medical geographers, in particular, neglect this fact and attach over much importance to the superficial issue of spatial variation.

Guided reading

By far the most detailed study of inequalities in health care is the Black Report (DHSS, 1980), reprinted with an additional introduction as *Inequalities in Health* (Townsend and Davidson, 1982). Virtually suppressed by the Conservative government of Britain, it is essential reading.

For the original statement on the inverse care law, Hart (1971) is also essential reading. Maxwell (1974) provides one of the best-known typologies of international health care systems. Inequalities in care in a number of countries are covered by Navarro (1976) although particular concern is given to the USA. The classic studies of what we have termed local inequalities are Haynes and Bentham (1979), Knox (1978) and Phillips (1979). All are avowedly geographical. Whitelegg (1982) provides a disjointed but vaguely alternative viewpoint, while Le Grand (1982), Stacey (1976) and Carter and Peel (1976) give the viewpoint from mainstream medical sociology. Overviews are provided by Phillips (1981), Joseph and Phillips (1984) and Eyles and Woods (1983).

References

Alderson, M. (1970), 'Social class and the health service', *The Medical Officer*, vol. 124, pp. 50–2.

Bone, M. (1973), *Family Planning Services in England and Wales*, London, HMSO.

Bosanquet, N. (1971), 'Inequalities in the health service', *New Society*, vol. 450, pp. 809–12.

Bradley, J., Kirby, A. and Taylor, P. (1978), 'Distance decay and dental decay: a study of dental health among primary school children in Newcastle-upon-Tyne', *Regional Studies*, vol. 12, pp. 529–40.

Bradshaw, J. (1972), 'The concept of social need', *New Society*, vol. 496, pp. 640–3.

Brotherston, J. (1976), 'Inequality: is it inevitable?', in C. Carter and

J. Peel (eds), *Equalities and Inequalities in Health*, pp. 73–104, London, Academic Press.

Buxton, M. and Klein, R. (1976), 'The distribution of hospital provision: policy themes and resource variations', *British Medical Journal*, vol. 1, pp. 345–9.

Carmichael, C. (1983), 'General dental service care in the northern region', *British Dental Journal*, vol. 154, pp. 337–9.

Carter, C. and Peel, J. (eds) (1976), *Equalities and Inequalities in Health*, London, Academic Press.

Coates, B. and Rawstron, E. (1971), *Regional Variations in Britain: Selected Essays in Economic and Social Geography*, London, Batsford.

Cole, K. and Gattrell, A. (1983), 'Access to libraries in Salford: explorations using census data', paper given at Institute of British Geographers Annual Conference, University of Durham.

Connel, J. and Engels, B. (1983), 'Indian doctors in Australia: costs and benefits of the brain drain', *Australian Geographer*, vol. 15, pp. 308–18.

Cowan, P. (1965), 'Hospitals in towns: location and siting', *Architectural Review*, vol. 137, pp. 417–21.

Cross, K. and Turner, R. (1974), 'Patient visiting and the siting of hospitals in rural areas', *British Journal of Preventative and Social Medicine*, vol. 28, pp. 276–80.

Davies, B. (1968), *Social Needs and Resources in Local Services*, London, Michael Joseph.

Dear, M. and Taylor, S. (1982), *Not in Our Street*, London, Pion.

Department of Health and Social Security (1969), *The Functions of the District General Hospital* (The Bonham-Carter Report), London, HMSO.

Department of Health and Social Security (1974), *Community Hospitals: Their Role and Development in the National Health Service*, London, HSC(IS).

Department of Health and Social Security (1975), *Better Services for the Mentally Ill*, London, HMSO.

Department of Health and Social Security (1977), *The Way Forward*, London, HMSO.

Department of Health and Social Security (1980), *Inequalities in Health*, Report of a Research Working Group chaired by Sir Douglas Black, London, DHSS.

De Vise, P. (1971), 'Cook County Hospital: bulwark of Chicago's apartheid health care system', *Antipode*, vol. 3, pp. 9–20.

De Vise, P. (1973), 'Misused and misplaced hospitals and doctors', Commission on College Geography Resource Paper 22, Washington, DC, Association of American Geographers.

Doeleman, F. (1980), 'The health care system in The Netherlands', *Community Medicine*, vol. 2, pp. 46–56.

Doyal, L. (1979), *The Political Economy of Health*, London, Pluto Press.

Drury, P. (1983), 'Some spatial aspects of health service developments: the British experience', *Progress in Human Geography*, vol. 7, pp. 60–77.

Elling, R. (1981), 'Political economy, cultural hegemony and mixes of traditional and modern medicine', *Social Science and Medicine*, vol. 15A, pp. 89–99.

Eyles, J. and Woods, K. (1983), *The Social Geography of Medicine and Health*, London, Croom Helm.

Fiedler, J. (1981), 'A review of the literature on access and utilisation of medical care with special emphasis on rural primary care', *Social Science and Medicine*, vol. 15C, pp. 129–42.

Forster, D. (1976), 'Social class differences in sickness and in general practitioner consultations', *Health Trends*, vol. 8, pp. 29–32.

Guptill, S. (1975), 'The spatial availability of physicians', *Proceedings of the Association of American Geographers*, vol. 7, pp. 80–4.

Hart, J. T. (1971), 'The inverse care law', *Lancet*, vol. 1, pp. 405–12.

Harvey, D. (1975), *Social Justice and the City*, London, Edward Arnold.

Haynes, R. and Bentham, C. (1979), *Community Hospitals and Rural Accessibility*, Farnborough, Saxon House.

HM Govt. (1944), *A National Health Service*, Cmd 6502, London, HMSO.

Ingram, D., Clarke, D. and Murdie, R. (1978), 'Distance and the decision to visit an emergency department', *Social Science and Medicine*, vol. 12, pp. 55–62.

Jones, D. and Masterman, S. (1976), 'NHS resources: scales of variation', *British Journal of Preventative and Social Medicine*, vol. 30, pp. 244–50.

Jones, K. and Kirby, A. (1982), 'Provision and well-being: an agenda for public resources research', *Environment and Planning A*, vol. 14, pp. 297–310.

Joroff, S. and Navarro, V. (1971), 'Medical manpower: a multivariate analysis of the distribution of physicians in the urban USA', *Medical Care*, vol. 9, pp. 428–38.

Joseph, A. and Phillips, D. (1984), *Accessibility and Utilization: Geographical Perspectives on Health Care Delivery*, London, Harper & Row.

Knox, P. (1978), 'The intraurban ecology of primary medical care: patterns of accessibility and their policy implications', *Environment and Planning A*, vol. 10, pp. 415–35.

Knox, P. (1979), 'Medical deprivation, area deprivation and public policy', *Social Science and Medicine*, vol. 13D, pp. 111–21.

Knox, P. (1981), 'Retail geography and social well-being: a note on the changing distribution of pharmacies in Scotland', *Geoforum*, vol. 12, pp. 255–64.

Kunitz, S. (1980), 'The recruitment, training and distribution of physicians in Jugoslavia', *International Journal of Health Services*, vol. 10, pp. 587–609.

Lee, K. (1979), 'Need versus demand: the planners dilemma', in K. Lee (ed.), *Economics and Health Planning*, London, Croom Helm.

Le Grand, J. (1978), 'The distribution of public expenditure: the case of health care', *Economica*, vol. 45, pp. 125–42.

Le Grand, J. (1982), *The Strategy of Equality*, London, Allen & Unwin.

Lucy, W. (1981), 'Equity and planning for local services', *Journal of the American Planning Association*, vol. 47, pp. 447–57.

Maxwell, R. (1974), *International Comparisons of Health Needs and Services*, London, King Edward's Hospital Fund.

Millard, F. (1982), 'Health care in Poland: from crisis to crisis', *International Journal of Health Services*, vol. 12, pp. 497–515.

MIND (1977), *Mental Health Statistics*, London, MIND.

Mohan, J. (1980), 'Hospital location in Durham Health District', paper given at annual conference of Regional Science Association, University College, London.

Moon, G. (1983), 'Perspectives on the spatial analysis of community health care services', paper given at the Institute of British Geographers Annual Conference, University of Edinburgh.

Moon, G. (1985), 'Community care and privatisation', in *Into the Community?*, Social Services Research and Intelligence Unit Occasional Paper 11, Portsmouth Polytechnic.

Navarro, V. (1976), *Medicine under Capitalism*, London, Croom Helm.

Noyce, J., Snaith, A. and Trickey, A. (1974), 'Regional variations in the allocation of financial resources to community health services', *Lancet*, vol. 7857, pp. 554–7.

Office of Health Economics (1977), *The Reorganised NHS*, London, Office of Health Economics.

O'Mullane, D. and Robinson, M. (1977), 'The distribution of dentists and the uptake of dental treatment by schoolchildren in England', *Community Dentistry and Oral Epidemiology*, vol. 5, pp. 156–9.

Open University (1976), 'Patterns of inequality: health and inequality', in *Inequality within Nations, D302, Unit 13*, Milton Keynes, Open University Press.

Phillips, D. (1979), 'Spatial variations in attendance at general practitioner services', *Social Science and Medicine*, vol. 13D, pp. 169–81.

Phillips, D. (1981), *Contemporary Issues in the Geography of Health Care*, Norwich, Geo Books.

Pinch, S. (1979), 'Territorial justice and the city: a case study of the social services for the elderly in London', in D. Herbert and D. Smith (eds), *Social Problems and the City*, pp. 201–23, London, Oxford University Press.

Pinch, S. (1985), *Cities and Services: the Geography of Collective Consumption*, London, Routledge & Kegan Paul.

Pyle, G. (1979), *Applied Medical Geography*, New York, Wiley.

Radical Community Medicine (1982), 'Health in Cuba', *Radical Community Medicine*, vol. 9, pp. 1–6.

Rigby, J. (1978), 'Access to hospitals: a literature review', Transport and Road Research Laboratory, Report 853.

Robertson, L. (1970), 'On the intra-urban ecology of primary care physicians', *Social Science and Medicine*, vol. 4, pp. 227–38.

Ryan, M. (1978), *The Organisation of Soviet Medical Care*, Oxford, Blackwell.

Scull, A. (1978), *Decarceration: Community Treatment of the Deviant – a Radical View*, Englewood Cliffs, NJ, Prentice-Hall.

Shannon, G., Skinner, J. and Bashshur, R. (1973), 'Time and distance: the journey for medical care', *International Journal of Health Services*, vol. 3, pp. 237–44.

Smith, D. (1979), *Where the Grass is Greener: Living in an Unequal World*, Harmondsworth, Penguin.

Smith, D. (1982), 'Geographical perspectives on health and health care', in Health Research Group (eds), *Contemporary Perspectives on Health and Health Care*, Occasional Paper 20, pp. 1–11, Department of Geography, Queen Mary College, University of London.

Stacey, M. (ed.) (1976), *The Sociology of the NHS*, Sociological Review Monograph 22, University of Keele.

Sumner, G. (1971), 'Trends in the location of primary medical care in Britain: some social implications', *Antipode*, vol. 3, pp. 46–53.

Taylor, P. and Carmichael, C. (1980), 'Dental health and the application of geographical methodology', *Community Dentistry and Oral Epidemiology*, vol. 8, pp. 117–22.

Townsend, P. and Davidson, N. (1982), *Inequalities in Health: The Black Report*, Harmondsworth, Penguin.

Vaughan, D. (1967), 'The dispersion of patients in urban general practice', *The Medical Officer*, vol. 23, pp. 337–40.

Wachs, M. and Kumagai, T. (1973), 'Physical accessibility as a social indicator', *Socio-Economic Planning Sciences*, vol. 7, pp. 437–56.

West, R. and Lowe, C. (1976), 'Regional variations in need for and provision and use of child health services in England and Wales', *British Medical Journal*, vol. 2, p. 843.

Wheeler, M. (1972), 'Hospitals, accessibility and public policy', *Hospitals and Health Services Review*, vol. 68, pp. 82–5.

Whitehouse, C. (1985), 'Effect of distance from surgery on consultation rates in an urban practice', *British Medical Journal*, vol. 290, pp. 359–62.

Whitelegg, J. (1982), *Inequalities in Health Care: Problems of Access and Provision*, Retford, Straw Barnes.

Whitelegg, J. (1983), 'Health care planning and the politics of reorganisation in the Rossendale Valley', paper given at Institute of British Geographers Urban Geography Study Group conference on public service provision, University of Reading.

CHAPTER 7

Explaining health care inequality

Introduction

The material which we set out in the last chapter was essentially concerned with detailing the evidence for health care inequality in contemporary society. We now turn our attention to the deeper question: what are the causes of this inequality? Our task in this chapter is therefore one of explanation. In particular, we will be considering the way that societal structure closely and intimately affects both the provision and the consumption of health care. The still broader issue of health outcome will be considered in the final chapter.

In involving ourselves with explanation, we are embarking on a complex search. A satisfactory explanation for health care inequality must not only relate to the observable factors associated with inequality; it must also, crucially, link inequality and those factors in a causally significant way. A valid explanation must involve the recognition of a causal mechanism linking a particular manifestation of inequality and the issue(s) which are presumed to constitute the cause. Explanation, then, is not merely the cataloguing of relationships or associations; if this were the case we would only have to present the material in the previous chapter. We could say that lower status persons live further from medical facilities, therefore low social status causes poor geographical access to health care provision; but this would be insufficient. We need a causal mechanism linking status to health care provision.

This chapter draws on current research in the social sciences to develop an explanation for health care inequality. In our study of mental illness and mental health care in Chapter 5 we hinted at elements of the perspective which we will employ, and we will return, in some detail, to our central themes in Chapter 9. At this juncture we would like to offer a brief apologia for the selectiveness of the approach which we will set out. Others might adopt very different explanatory strategies, and, although we will briefly note these, it is for the reader to evaluate their merits. In the next section we will outline these competing approaches, drawing attention to their shortcomings. We will then turn our attention to an analysis of the important structuring role that is played in determining health care inequality by the macro politico-economic structure of society. We will exemplify this role through a selection of case studies of health care under capitalism. A final section will examine the case for more locally based explanations.

Non-explanation and partial explanation

The material which we developed in Chapter 6 enables us to make, with some certainty, the assumption that inequality in health care provision and consumption exists. Having accepted this 'fact', we must now proceed to question why this should be so. In doing this we must confront several alternative theories and approaches. With a number of these there is an intrinsic problem: they do not offer a full explanation within the context, cited above, of identifying a causal mechanism.

Quantitative methodology provides a useful starting point from which to consider this issue. Such an approach gives us an effective indication of an association or relationship but, unless accompanied by the necessary input of theory, this indication is not causal. Any observed association is merely an artifact of statistical manipulation, and we may call this situation 'non-explanation'. To return to the example which we were using in the introduction to this chapter, the existence of a significant correlation between measures of low social status and geographical access to health care provision does not mean that the two are causally linked. A theoretical reason for an identified statistical link must be found and specified for non-explanation to be translated to explanation (Keat, 1979). Similarly, a high correlation between lower social class and low uptake of immunisation does not indicate an explanation. It is merely a finding which needs to be explained. Indeed, on its own, the finding is susceptible to differing explanations reflecting different theories about why and how the high correlation comes to exist.

For other researchers, a behavioural study of the activities of individuals involved in the consumption of health care has provided a potential road to explanation (Phillips, 1981; Joseph and Phillips, 1984). The behaviouralists are concerned primarily with the way individuals make decisions concerning their health care needs and actions. Thus, an individual may choose to visit the surgery following upon a self-assessment of her health; the GP then makes her own decision about whether or not the patient is treated. Although it is recognised that both decisions take place within a wider environment, for example it may be a long way to the surgery and this may constrain the patient in her willingness to attend, there are severe problems with this approach. First, the patient and the GP in our example are both viewed as capable of autonomous action; the environment is only a ground on which action occurs and its constraining influence is always secondary to the paramountcy of choice. Manifestly, however, autonomy is not always possible. British general practitioners, for example, are legally constrained in the choice of drug treatments which they can prescribe, and the consumers of care can face very real constraints in attending for surgery appointments. These constraints need to be recognised and understood. Second, by focusing on the individual, the behaviouralists draw attention away from this constraining aspect, with its roots in socio-economic structures, and attaches a crucial importance to the actions of the individual. The individual's behaviour becomes a determining factor in health care inequality. The danger here is that victim-blaming may be facilitated (Chapter 1). Vaccination uptake again provides a useful illustration of this point. A behavioural approach would stress the importance of the decision of individual parents to have or not have their baby vaccinated. Those who decide on the latter might be categorised as either forgetful, and therefore feckless, or actually opposed, and therefore antisocial. Where the behavioural approach is used in conjunction with a statistical approach further problems can arise. People who do not have their children vaccinated tend to be lower social status, so we can blame the feckless poor for non-vaccination. All this would, of course, miss the crucial role played by socio-economically induced constraints.

For a third alternative mode of explanation we can turn to welfarism. It is this approach which provided the background to the Black Report (DHSS, 1980); it is well represented in the research traditions of geography (Smith, 1977) and sociology (Townsend, 1979). Essentially welfarism argues that social groups have unequal access to the benefits of society and this inequality is associated with class and income. Government intervention is seen as the solution to the problem acting through an enhanced welfare system. The exact mechanisms underlying the approach were seldom specified in early studies, although researchers in this tradition have now usually developed away from what was, at times, rather naïve description.

An important aspect of more recent welfarist works has been a recognition that health care inequality is a reflection of a complex of many factors. A tendency has remained, however, to see things in terms of real income over which autonomy is possible, although, in contrast to behaviouralism, the approach does ascribe a central role to the social group rather than the individual:

> unequal use is explicable not in terms of non-rational response to sickness by working class people, but of rational weighting of the costs and benefits to them of compliance with the prescribed regime. These costs and benefits differ between social classes both on account of differences in way of life, constraints and resources, and of the fact that costs to the working class are actually increased by the lower levels and perhaps poorer quality of provision to which many have access. (Townsend and Davidson, 1982, p. 89)

The central role played in welfarism by class allows some understanding to be developed of the way in which class may compound inequality, but this understanding is often rather superficial; it seldom penetrates to the root cause of inequality and it does not say why class *should* compound inequality. For example, low use of dentists and negative attitudes to dentistry are both associated with low social status. Class is therefore observed to compound inequality of use, but we never learn why. Lower status groups also make less effective consultations. For example, Cartwright and O'Brien (1976) claim that the middle classes are better informed on medical matters and know more accurately how to describe their symptoms and when to refer themselves to a doctor, but, again, welfarism does not adequately question why this situation persists.

Our final example of an approach to the explanation of health care inequalities concerns those studies which have seen inequality as a consequence of administrative and organisational aspects of the operation of the service-delivering body. This work draws on organisational theory and, to a limited extent, Weberian sociology. Some of the research in London on the problems and results of National Health Service resource allocation policies (LHPC, 1979, 1980) illustrates this approach. Inequalities are argued to flow from the decision to transfer resources from inner city areas to non-metropolitan health districts. The advantage of this standpoint is that it places the blame for inequality squarely on the shoulders of the administrators of health. It does not, however, recognise the constraints under which they themselves operate.

The theme which we have been developing in this cursory review has been that explanation needs to be firmly founded on cause and it needs to identify underlying determining constraints and structures. The approaches which we have reviewed have provided partial or non-explanation and are, on their own, just not enough. We will now outline the basic framework of an approach which comes to terms with these needs and sites the production of inequality firmly within the context of social structure and social relations.

Society and health care inequality

The way in which society is structured plays a crucial role in explaining health care inequality. We saw, in Chapter 1, how the claims of a biomedical model of ill-health, which perceives illness and disease internally to the body, could be challenged by a social model placing ill-health within the context of social conflict (Doyal, 1979). In this section we want to draw on the many parallels between the understanding of ill-health and the understanding of health care inequalities and address a similar task: to challenge the shortcomings of the non-explanations and partial explanations set out above by developing a political economy of health care. This structurally determined political economy will provide the causal framework for an understanding of the nature and extent of health care inequality in its various manifestations.

The analysis of contemporary society which we used in Chapter 1 in our discussion of Lesley Doyal's work, and to which we will return in some detail in a consideration of medical futures in Chapter 9, provides a convenient starting point for our approach. Society, we suggested, revolves in very crude terms around the interplay of the interests of class and capital. In many cases these two interests will coincide, and we can basically assume that society contains people, indeed institutions, which dominate others by virtue of their power. This power reflects control over capital; it is economically based and the aim, once you have it, is to keep it. Obviously only the few benefit from this and it is in their interests to try to justify and maintain the structure of a society from which they have benefited.

The relevance of all this to health care inequality is that the provision of good care and the ability to consume that care are social benefits which exist within a social context. It is inevitable, if we follow the marxist analysis of the social system which we briefly summarised above, that those who dominate within the system will be able to secure more, and more effective, care. They will also strive to maintain this position and deflect any reforms which might radically alter the status quo. This, then, is the structural root cause of inequality. To elaborate this explanatory thesis further we must introduce the marxist concepts of legitimation and reproduction. They are essential to understanding the reasons for health care inequality. In a very simple sense, these terms can be equated with what we earlier termed justifying and maintaining the structure of society. Legitimation entails ensuring that the existing order of things can be shown to be for the best. Reproduction involves the facilitation of continued capital accumulation by the dominant class. As far as health care is concerned, this means that both the provision and the consumption of care must be seen to be functional for capital. Thus, health care by the state is a legitimation device demonstrating the humanity of the capitalist system. It is also a reproduction mechanism producing (hopefully) healthy workers for the future. Inequality comes into the picture because, as we suggested above, some people are dominant within the political economic system, and correspondingly benefit more than others from the health care system.

If we look at the way that certain aspects of health care inequality in the past have been maintained, the role of legitimation and reproduction becomes clearly evident. Despite governmental commitment, for example in the guise of national health services, some inequalities have continued, suggesting that there is an underlying force surmounting welfare initiatives. For the marxist, this force is undoubtedly the capitalist economy, and its power is an indication of the practical inability and political unwillingness of governments to combat inequality. Past inequalities existed because of fee-for-service systems where utilisation reflected income and provision would only be located where income could be guaranteed. The dominant class benefited from this situation and, to an extent, the spatial patterns of health care which were entailed, as in the Harley Street area of London, have been maintained until today.

Contemporary health care also reflects the processes of legitimation and reproduction. This result comes about in a number of ways, and we will set out just four to illustrate the point. First, there is a considerable discrepancy between hospital-

and community-based services in terms of the commitment and investment which they respectively receive. Following Navarro (1976, 1978), this is a reflection of a medical, as opposed to health care, ideology which continues to be strongly rooted in the cult of the individual (see Chapter 1). Progress in medicine is often viewed as the result of the achievements of a limited number of free-wheeling individuals; transplant surgeons are a case in point. These achievements reflect individual enterprise and contrast with the monolithic bureaucracy which critics claim characterises the British National Health Service. Support for a contemporary view of progress as the outcome of interventionary medicine and the biomedical model is thus facilitated, but little attention is given to broadly defined health care for cases requiring less dramatic treatment. The success of the few legitimises a system in which capital structures its rewards to the benefit of those working where investment potential and possibilities are greatest. As far as health care inequality is concerned, developments need not be sought with any great urgency in the less lucrative fields of community health care; inequality in investment and intra-professional inequality are the outcomes.

Inequalities in consumption and the continuation of unequally available private medicine in predominantly state-run health care systems provide a second case example. The ethic of freedom of choice promotes inequality within this context, working to legitimate the capitalist system. If we examine the rhetoric of right-wing governments concerning the consumption of care, the centrality of freedom of choice is clear. Essentially this means that people are free to make their own decisions concerning their consumption of care. Inequality is acceptable as it corresponds to the decision not to consume care. Societal barriers to care consumption are not recognised or are redefined to suggest that, if the need is great enough, the individual should be able to overcome them. Freedom of choice links to the position of private care within national health services because the ability to sustain a private sector, manifestly indicative of inequality and of benefit only to the rich or the employees of certain companies, is justifiable in terms of the maintenance of freedom of choice.

Provision inequalities can also be explained in this way. Titmuss (1971), writing in a rather different context, provides the basis for this third case example. He unwittingly indicates how the provision of health care, through the legitimating role, can become subject to inequality. Health care, as it is currently constructed in the British National Health Service, is promoted as a benevolent activity in which an altruistic state indulges with the goal of helping the sick. The reality is that it is a gift of the state; the act of provision is emphasised but not the effectiveness of the means of distribution. Furthermore, because of the lack of attention to the means of distribution, it is unlikely to be universally or uniformly provided.

This last viewpoint is essentially a welfarist argument, putting forward only a partial explanation of health care inequality; it pays little attention to the mechanisms at work. Our final contribution to explanation takes the thesis a bit further, incorporates the reproduction of the labour force argument and offers an altogether more cogent approach. Doyal (1979) argues that the altruism involved in the healing and care of the sick is a smokescreen obscuring a hidden agenda involved with the reproduction of a labour force which is fit and available for the production of profit. Paradoxically, of course, the workforce only needs to be fit enough to produce effective returns; there is no need to maintain health levels above the minimum required, in fact it would be costly and against the needs of capital. At certain times, for example during economic crisis, adequate labour power may exist without the need to maintain the health of the sick. Given this, it is possible to conceive of a health care provision system in which facilities might be located specifically to benefit the functional needs of the economy. Inequalities in health care provision therefore reflect the underlying dynamics of capitalism.

Two short examples will help to show the way in which all the cases which we have just considered indicate themes that can interact together to compound inequality. We will look at the relationship between health care inequality and the socio-economic status of women and ethnic minorities in capitalist society. Taking the case of women first, Doyal (1979) has suggested that their role is conceptualised extensively in terms of their place in ensuring the reproduction of labour power. Thus, women are seen as the producers and rearers of future workers, and their health care needs are predefined as gynaecological, obstetric and otherwise related to their efficiency as mothers. It is only in these areas that there is any concentration on women's health care needs; in other situations the reproduction of the labour force is not at stake and so health care for women can be neglected. In this way there has come about a large industry of clinics associated with the birth process, but a relative lack of provision for the well woman. Health care for ethnic minorities (Rathwell and Phillips, 1986) exhibits a similar history of domination and exploitation in the interests of capital. Health care in general is structured to cater for illnesses and attitudes to ill-health which characterise the dominant group in a society. The consequences of dietary variation and misunderstandings about differing cultural practices inhibit the effective consumption of health care by ethnic minorities. Furthermore, following the argument that health care exists to facilitate the existence of a healthy workforce, the level of provision necessary to ensure the fitness for labour of an ethnic minority usually confined to a fairly lowly position in the economic hierarchy is often minimal, and even reducible when that labour is not needed. The health services themselves exemplify this latter point (Connel and Engels, 1983). Immigrant doctors, welcomed in times of manpower shortages, are often exploited by being constrained to practise in areas unpopular with the dominant ethnic group. They are also dispensed with more easily when demand for manpower falls.

Capitalist health care

Thus far it is evident that our approach to explanation ascribes a crucial role to capitalism in structuring the underlying determination of health care inequality. We will now set out some further examples to demonstrate the utility of this approach.

In Chapter 1 we gave a brief outline of the social history of medicine. At that stage we were demonstrating the rise of the dominant biomedical model of ill-health. We can also use that social history to show how, particularly in the USA, a spatial pattern of health care provision developed which continues strongly to reflect the prerequisites of the biomedical model. This situation came about because the rise of the biomedical model indicated the fruitful association of medicine with the influence of large-scale capital through the investments of late nineteenth-century tycoons in drug-based laboratory medicine. Present-day health care is reflective of the parallel adoption of capitalist ethics by the medical professions fuelled by a concurrent subscription to individualism and professionalisation.

Before the advent of laboratory medicine, health care in the USA was a confused, crowded and uncontrolled free market of qualified doctors, paramedics and outright charlatans (Knox and Bohlund, 1983). After the Flexner report this situation changed radically and clinical medicine attained ascendancy. This restructuring laid the roots of present inequalities in provision because it professionalised the practice of medicine and turned it into a highly marketable skill and commodity. Dispensaries providing cheap and relatively effective care were phased out, and qualified doctors, being relatively few in number in comparison to the situation in the years prior to licensing, assumed a near-monopoly position. The defining of good care as clinically based care led directly to the setting up of inequalities in care because the American Medical Association saw that there were considerable financial benefits to be reaped and medical personnel, being free to sell their prized services for the highest reward, not surprisingly usually chose those high income areas where rewards promised to be greatest. The present locational inequalities in health care in the USA are therefore an outcome of the past impress of capitalist ethics coupled with a calculated professionalisation; they are not merely the outcome of sets of individual behavioural decisions.

Navarro (1976) provides us with an equally clear example in his excellent study of the way that class conflict underlies health care inequalities in Chile. Chilean society is, in essence, comprised of an extensive proletariat, a 'middle class' and a small bourgeoisie. Each occupies a specific position regarding health care:

> the governmental health service or National Health Service (NHS) covers the working class, peasantry, the unemployed, and the poor – groups which, together with a small fraction of the lowest-paid white collar workers also covered by the NHS, represent 70% of the Chilean population; voluntary health insurance (SERMENA) covers the middle class, who represent approximately 22% of Chilean people; and the fee-for-service, out-of-pocket market medicine covers the lumpenbourgeoisie, approximately 8% of Chileans. (Navarro, 1976, p. 36)

At first glance this situation seems equitable: there is a health service for the poorest members of society. Yet, while we saw in the last chapter that such people would be expected to have greatest need for health care, 60 per cent of Chile's governmental health care expenditure goes on the wealthiest 30 per cent of the population. Obviously state medicine in Chile is seriously underfunded. Navarro adopts the reproduction of labour power argument, and claims that the Chilean NHS was founded only to ensure the productive efficiency of the workers. This he couples with a legitimation thesis suggesting that a minimal system of state-provided health care was instituted in order to subdue labour unrest during the industrialisation programmes of the 1950s. Hall and Diaz (1971) provide details of the inevitably regressive geography of health care which follows from this politico-economic derivation.

As we stated earlier, similar arguments can be advanced to account for the establishment and subsequent shortcomings of the British National Health Service. Orthodox writings see the NHS as a move against inequality in health care, a victory for the working class and an example of the benign nature of welfare capitalism. Alternative explanations are possible, however. These see the NHS as functional for continued capital accumulation and characterised by continued inequality. Navarro (1978), for example, sees the creation of the NHS as a specific response to the demands being made on the ruling class by a working class radicalised by a war which had demonstrated the fruitful consequences of organised solidarity against oppression and by post-war election successes. Superficially the Labour government could demonstrate its concern with the health of its supporters by sponsoring the creation of the NHS. In reality what was created was however structured in capital's favour, fragmented, individualised and intensely hierarchical. It ensured the availability of a healthy workforce for post-war reconstruction, and did not threaten the process of capital accumulation. There was to be little or no democratic control through the popular vote, general practitioners maintained their cherished status as independent contractors, and, particularly importantly for the study of inequality, a significant private health care sector was maintained. The wealthy were thus able to utilise their higher incomes to secure preferential treatment and enhanced health care.

While there is now a considerable body of literature on gender divisions in illness (Clarke, 1983), there is a relative lack of work offering anything more than description of the provision and consumption of health care for and by women. As we suggested in the previous section, this is one particular area where structural explanations of inequalities are clearly evident. Coupland (1982) is one writer who has tried to address this issue. She has carefully analysed women's health care consumption in terms of the interplay of such factors as lack of time, poor access to transport and the imposition of sole responsibility for child care, with the availability of health care facilities, their opening hours and their management in terms of waiting times. It is evident that, while women in general suffer inequality from these interactions, working-class women suffer more than most. Lupton *et al.* (1985) have provided specific detail on one particular service where this inequality is marked. Women's opinions of antenatal care were found to be significantly affected by their often negative experiences of care. Clinics run by consultants at a large District General Hospital were rated poor compared to smaller, more localised clinics run by women for women. To fully understand such empirically based conclusions, we must turn to the wider context.

The explanation for gender-based inequalities in health care can be found in the position of women as reproducers, and the fact that society takes this view to the exclusion of a more egalitarian position. Thus, women's consumption of care is essentially regulated by men and predicated towards the success of the birth process. This goal is seen in terms of scientific technocratic medicine; a medicalisation of the birth process takes place which excludes the more human elements. Reproduction is promoted and ambivalent attitudes towards non-birth are maintained. Childlessness is perceived as unusual, and family planning is a low prestige area of health care (Doyal, 1979). Access to abortion is controlled by a male-dominated Parliament. Even beyond the issue of reproduction such stereotypes as the overstressed housewife gulping tranquillisers prescribed by uninterested general practitioners, and the neurotic, undervalued mother whose children have left home predetermine a particular conception of health care for women (Barrett and Roberts, 1978). The view is of women as somehow imperfect and therefore of less functional use for the economy.

The numerically dominant position of women in the British National Health Service only serves to emphasise our point. With few exceptions, women are employed in the lower echelons of the service (Ehrenreich and English, 1976; Young, 1981) and are unlikely to gain promotion into senior decision-making positions. Even when this does occur, any attempts to change the nature of care for women or adopt different perspectives is unlikely to lead to real change. The entrenched power of the medical establishment and the functionality of current practice for capital accumulation ensure this.

We have set out a rather extended discussion of the determining factors in health care inequalities for women as it is a crucially important issue. We will conclude this section with a briefer look at two further areas of health care which have been similarly overlooked. Within the context of health care as a whole, occupational health care and preventative health care have been vastly neglected. They are undeniably cinderella services in terms of expenditure, organisation and manpower. To explain this inequality and neglect of interest we need only note that occupational health care can cast considerable aspersions on the safety of the capital accumulation process, and preventative health care does not accord with the interventionary biomedical model of illness to which the dominating class in contemporary medical practice subscribe and from which they gain their position and benefits.

Kinnersly (1973) estimates that there are 2,000 deaths from industrial injuries in Britain each year. A further ten million people have to receive first aid. These injuries and deaths are disproportionately visited upon lower status groups, yet the occupational health service is practically non-existent. Health care has been defined as being primarily the responsibility of the individual concerned, with doctors providing the means to get a person back to work. Ill-health produced by work does not enter the equation because it undermines the politico-economic *raison d'être*. It is essential that workers are fit for work; a service which ensures that they are not damaged by work and can enjoy their leisure is not essential and can accordingly be underfinanced. Of course, for higher status groups in less dangerous jobs this is less problematic. In this way a class-based inequality underlies the neglect of occupational health. Individual freedom of choice is one of the factors underlying the neglect of preventative health care; another is undoubtedly non-correspondence with the shibboleth of interventionary medicine. For example, attempts to promote health by curbing cigarette smoking may be construed as curtailing an individual's free choice. They also, however, ultimately undermine the rationale for high profile cancer surgery and the profits of the cigarette industry. There are vested capital interests eager to ensure neglect. A weak preventative sector means a continuing market for highly expensive medical technology and for the estimated £1.5 million weekly UK cigarette promotion budget (Doyal, 1983).

Local explanations?

So far the conclusions which we have reached concerning the explanation of inequality have been most obviously applicable to analyses conducted at a fairly broad level. They may seem adequate for studies at the level of the nation state and above, but rather hard to apply if we are trying to account for the local inequalities which are experienced during everyday life. We must now examine this problem, and consider the importance of the local in explanation, and the nature of those explanations which claim to account for local inequalities. To translate the general arguments which we set out earlier to a specific local situation we must make two points.

First, we must accept that, to a very great extent, society and its constituent parts in any particular locality operate as a small-scale reflection of the larger societal environment. The same contradictions, stresses and governing concepts are evident. Thus, the organisation, distribution and consumption of health care services are guided, at both the national and the local scale, by the broad politico-economic structures which we noted in the last section. This is not, however, to deny the importance of geography because our second point is that the local outcome of politico-economic structuration will always, to some extent, be unique. This is because of the way in which overall cultures, administrative processes, class relations and political organisation come together in a particular locality. Such locality effects are crucially mediated by the wider context. What we are suggesting is that, while the root of explanation may be in the wider politico-economic context, the nature of outcome is local. This perspective, which owes something to the work of the sociologist Anthony Giddens (1979, 1981), rests on the identification of the specificity of location and the way in which broader politico-economic structures come together at a particular place.

We can illustrate these contentions by looking critically at some of the theories which suggest a local role in explanation. First, we can consider the case of one of the many changes in the methods of health care delivery which have been devised in Britain since 1948. In the late 1970s health visitor work patterns were changed from a system based on geographical areas to one based on attachment to a general practitioner (Payne and Hall, 1981). A local spatial consequence of this change was that detailed parochial knowledge of health care needs was lost because general practitioner lists are less spatially concentrated. Responsiveness was also lost in areas with poor general practitioner services. These outcomes were consequences of the change as were the localised benefits in terms of integrated primary care which were expected to flow from the initiative; neither were explanations. Each was a locality or contextual effect. The initiative itself therefore constitutes only a partial explanation. For the root cause we need to look at the correspondence of the initiative with the perceived cost efficiency of the principle of centralisation which was popular with the medical establishment, and the way in which the initiative allowed for the maintenance of continuing domination of predominantly male doctors over predominantly female health visitors.

The example which we have just cited amply demonstrates that changes in health care organisation, although they have considerable impact on local inequalities in care, do not constitute an explanation for that inequality. We can apply the same analysis in other cases. A frequently attempted explanation for local health care consumption inequalities is the activity of partial and interested local decision makers. Such 'gatekeepers' or local managers, as they are termed, are perceived to use their influence to ration the consumption of care to the benefit of particular social groups. This has consequences for social access to care because consumption is constrained by the difficulty of negotiating the hurdles placed by the gatekeepers. Thus, general practitioners decide whether or not a person is ill and also whether or not they will be allowed access to the hallowed arena of consultant care. Receptionists are similarly important as they operate the front line of care and control non-price rationing systems like appointments, eligibility criteria and queueing systems (Williams, 1977). In both cases social class is crucially important; it determines when the decision is made to contact the doctor and the effectiveness with which symptoms are described (Magi and Allander, 1981). Behaviouralist attempts to account for local variations in the quality and provision of health care are also of interest. It can, for example, be argued that better doctors stand a greater likelihood of obtaining a post in a practice which is already of good quality or in a good neighbourhood. Doctors are also assumed to have a rational and free choice of where to practise and, finally, it appears that doctors equate success with practice at or close to a major teaching hospital. Thus Knox (1978) found clustering of general practitioner surgeries around hospitals and, in so far as we can measure quality of care by the distinction awards given out annually by the British National Health Service, the recipients of the highest awards cluster in the Regional Health Authorities containing the Oxford, Cambridge and London teaching hospitals (Cooper and Culyer, 1969).

The problem with the ideas of the gatekeeper and the behaviouralist approaches is that they are superficial and partial explanations; they both fail to place inequality within its wider environment. The important role played by professionalisation in medicine is not emphasised and still less is it considered in the context of a capitalist social system. Gatekeeping and the local factors affecting quality and provision of care both, in reality, reflect the need to maintain care as a scarce resource in the gift of the state, and illustrate the dominance of the expert medical practitioner over the sick individual.

Class conflict in its broadest sense is the key explanatory factor which underlies another putative local explanation for inequality. In Chapter 6 we outlined the way in which certain population sub-groups consistently appear to be provided with less and, effectively, consume less health care. Particularly where such groups are spatially concentrated a situation develops in which inequality may be clearly identified and equated with the social group concerned. Underclass explanations address this situation and suggest that the existence of the social group constitutes the explanation for inequality. This assumption is not, however, adequate. The underlying cause of the inequality is the position in society of the population sub-group. This position indicates the utility of the sub-group for the economy, the prestige which will flow to the doctor from treating them, and, in certain health care systems, the possibilities which they offer for financial reward to the doctor concerned. Thus, in Auckland and Wellington, New Zealand, there are proportionately fewer doctors in Maori and Polynesian areas (Barnett, 1978). This does not mean that the Maoris and Polynesians themselves are the cause of their own disadvantage; nor yet is the geography of health care disadvantage in New Zealand explained directly by the residential geography of these groups. What is indicated is that good health care for Maoris and Polynesians is not essential to the functioning of the economy and can be neglected.

A final potential local explanation for health care inequality is provided by democratic pluralism. Some would argue that certain aspects of health care inequality at the local level in Britain come about because there is no means by which effective pressure can be brought to bear on decision making. Administrative attempts to provide for a consumer input into decision making in the British National Health Service have been largely ineffectual and, despite the several reorganisations which have taken place since 1948 (see Chapter 8), there is now arguably less representative democracy in health care. Members of health authorities are appointed, not elected, and, through the vetoing power of the Secretary of State, chairpersons can be selected who comply with governmental views. The Community Health Councils, set up as watchdog bodies, have a consultative but not legislative function, and are also appointed rather than elected. There is no form of areal advocacy to draw attention to spatial inequalities and, indeed, it is specifically discouraged. Again, however, there is a neglected broader structural context to this situation. An assumption is made that local democracy could be effective, yet it would still be operating within the same political-economic context; the paramount goal would still be cost-effectiveness, value for money and efficiency rather than, specifically, equality. Indeed, it might be argued that the existing mechanism of consumer democracy is a legitimising smokescreen and its scope for action has been deliberately limited so as not to subvert the current status quo (Rose, 1975).

Conclusion

In this chapter we have been concerned to show that medical geographers must build upon the description of patterns of inequality in health care provision and consumption. It is not enough, for example, to know that the Flowers Report recommended the rationalisation of the teaching hospitals of London into six institutions (University of London, 1980). It is not even enough to know where these institutions were to be, how the rationalisation was to take place and what the consequent inequalities were likely to be. We must realise that, although the entire exercise was presented as one of rational planning and the creation of centres of excellence, it was, in the main, heavily indicative of the environment of fiscal constraint in which British health care is currently operating. It is necessary to identify the underlying structural determinants of what is under investigation. Furthermore, the researcher must try to get to the start of the causal chain of explanation and avoid superficiality.

To this end we have concentrated mainly on one single approach to explanation; one that draws extensively on the work of Vincente Navarro and Lesley Doyal. This approach sees inequalities in health care as the inevitable outcome of the shortcomings of a capitalist political economy. Processes such as reproduction and legitimation play a key role, and we might broadly say that the political economy constitutes the root of inequalities and other factors flow from it. Space and locality have a crucial role to play in explanation because the various processes concerned can be manifested in different ways at different places.

Guided reading

There are few texts explicitly dedicated to the task of explaining health care inequality, and very little material at all which specifically approaches this task from a non-marxist perspective.

This is unfortunate as it means that there is no real opportunity to consider other theoretical orientations.

The approach which we have developed in this chapter closely follows two authors: Doyal (1979) and Navarro (1976, 1978). We have only indicated the broad dimensions of the considerable potential offered by these texts. Anyone interested in attempting a comprehensive explanation of health care inequality should read them. Reading Doyal is perhaps the best way to begin.

For further study an examination of the *International Journal of Health Services* can be recommended.

References

Barnett, J. (1978), 'Race and physician location: trends in two New Zealand urban areas', *New Zealand Geographer*, vol. 34, pp. 2–12.

Barrett, M. and Roberts, H. (1978), 'Doctors and their patients: the social control of women in general practice', in C. Smart and B. Smart (eds), *Women, Sexuality and Social Control*, pp. 41–52, London, Routledge & Kegan Paul.

Cartwright, A. and O'Brien, M. (1976), 'Social class variations in health care', in M. Stacey (ed.), *The Sociology of the NHS*, Sociological Review Monograph 22, pp. 77–97, Keele, University of Keele.

Clarke, J. (1983), 'Sexism, feminism and medicalism: a decade review of literature on gender and illness', *Sociology of Health and Illness*, vol. 5, pp. 62–82.

Connel, J. and Engels, B. (1983), 'Indian doctors in Australia: costs and benefits of the brain drain', *Australian Geographer*, vol. 15, pp. 308–15.

Cooper, M. and Culyer, A. (1969), 'An economic assessment of some aspects of the operation of the National Health Service', in British Medical Association, *Health Service Financing*, pp. 147–57, London, BMA.

Coupland, V. (1982), 'Gender, class and space as accessibility constraints for women with young children', in Health Research Group (eds), *Contemporary Perspectives on Health and Health Care*, Occasional Paper 20, pp. 51–70, Department of Geography, Queen Mary College, University of London.

Department of Health and Social Security (1980), *Inequalities in Health*, Report of a Research Working Group chaired by Sir Douglas Black, London, DHSS.

Doyal, L. (1979), *The Political Economy of Health*, London, Pluto Press.

Doyal, L. and Epstein, S. S. (1983), *Cancer in Britain: the Politics of Prevention*, London, Pluto Press.

Ehrenreich, B. and English, D. (1976), 'Witches, midwives and nurses: a history of women healers', New York, Readers and Writers Publishing Co-operative.

Giddens, A. (1979), *Central Problems in Social Theory: Action, Structure and Contradiction in Social Analysis*, London, Macmillan.

Giddens, A. (1981), *A Contemporary Critique of Historical Materialism*, vol. I, London, Macmillan.

Hall, T. and Diaz, S. (1971), 'Social security and health care patterns in Chile', *International Journal of Health*, vol. 1, pp. 362–77.

Joseph, A. and Phillips, D. (1984), *Accessibility and Utilization: Geographical Perspectives on Health Care Delivery*, London, Harper & Row.

Keat, R. (1979), 'Positivism and Statistics in Social Science', in J. Irvine, I. Miles and J. Evans (eds), *Demystifying Social Statistics*, pp. 75–86, London, Pluto Press.

Kinnersly, P. (1973), *The Hazards of Work and How to Fight Them*, London, Pluto Press.

Knox, P. (1978), 'The intraurban ecology of primary medical care: patterns of accessibility and their policy implications', *Environment and Planning A*, vol. 10, pp. 415–35.

Knox, P. and Bohlund, J. (1983), 'The locational dynamics of medical care settings in American cities, 1860–1940', paper given at Annual Conference, Institute of British Geographers, University of Edinburgh.

London Health Planning Consortium (1979), *Acute Hospital Services in London: A Profile*, London, HMSO.

London Health Planning Consortium (1980), *Towards a Balance: A Framework for Acute Hospital Services in London*, London, HMSO.

Lupton, C., Moon, G. and Mountifield, I. (1985), *Women's Experiences of Antenatal Care*, Portsmouth, Social Services Research and Intelligence Unit.

Magi, M. and Allander, E. (1981), 'Towards a theory of perceived and medically defined need', *Sociology of Health and Illness*, vol. 3, pp. 49–72.

Navarro, V. (1976), *Medicine under Capitalism*, London, Croom Helm.

Navarro, V. (1978), *Class Struggle, the State and Medicine*, London, Martin Robertson.

Payne, L. and Hall, D. (1981), 'Resource allocation in urban services: spatial response to financial constraints with reference to health visiting in Kent, England', *Geoforum*, vol. 12, pp. 245–54.

Phillips, D. (1981), *Contemporary Issues in the Geography of Health Care*, Norwich, Geo Books.

Rathwell, T. and Phillips, D. (1986), *Health, Race and Ethnicity*, London, Croom Helm.

Rose, H. (1975), 'Participation: the icing on the welfare cake', in K. Jones (ed.), *Yearbook of Social Policy in Britain 1975*, pp. 63–77, London, Routledge & Kegan Paul.

Smith, D. (1977), *Human Geography: A Welfare Approach*, London, Edward Arnold.

Titmuss, R. (1971), *The Gift Relationship*, London, Allen & Unwin.

Townsend, P. (1979), *Poverty in Britain*, Harmondsworth, Penguin.

Townsend, P. and Davidson, N. (1982), *Inequalities in Health: The Black Report*, Harmondsworth, Penguin.
University of London (1980), *London Medical Education: A New Framework* (The Flowers Report), London, University of London.
Williams, W. (1977), 'The receptionist: barrier or link?', *Medical Secretary*, vol. 30, pp. 25–8.
Young, G. (1981), *Women, Health and Reproduction*, London, Routledge & Kegan Paul.

CHAPTER 8

Planning policy and the health services

Introduction

In the two preceding chapters we have moved from noting the considerable extent of inequalities in the levels of health care experienced in contemporary society, to a consideration of ways in which these inequalities may be explained. We turn now to an examination of some of the approaches which health care planners and policy makers have employed when intervening to remove, or at least ameliorate, these inequalities. Our attention will focus on case examples with an avowed welfarist orientation. The possibilities afforded by individualist and structuralist alternatives will be considered in the final chapter.

Health care planning and policy making is, of course, a vast field of study in its own right. In Britain as elsewhere whole books, for example Ham (1981, 1982), Butler and Vaile (1985), are devoted to the subject. The relevant governmental literature is similarly extensive and is exemplified by such recent studies as the Black Report (DHSS, 1980) on inequalities in health care, the Acheson Report (LHPC, 1981) on primary care in Inner London, the Bonham-Carter Report (DHSS, 1969) on district general hospitals, the Griffiths Report (DHSS, 1983) on management in the health services, and the Körner Report on health statistics and information (NHS/DHSS, 1982). The situation in the USA is not dissimilar. In the face of such a plethora of material we have inevitably had to be selective. We have chosen to focus on seven case study areas which possess clear spatial dimensions. The selected topics are the reorganisation of the health care administrative system, the direct allocation of financial resources, the incentives designed to direct general practitioner manpower to particular spatial areas, area workload assessment, location-allocation modelling, facility impact studies and health education.

Reorganising the system

In deciding to reorganise a health care administrative system, health care planners and their political masters are in some way assuming that the existing system is wrong. Inequality is viewed as a product of inefficiency stemming from the inadequacies of the current organisational structure. Reorganisation, it is argued, would promote a more efficient system under which resources would be released from bolstering the ineffective bureaucracy and become available for the promotion of health care rather than health administration. To some extent this assumption is true; however, it should be placed within a wider setting. Except in totally free-market situations, health care systems are fairly static entities, developed at a particular period in time in response to the contemporary structure of society and the interests of the dominant group in that society. Not surprisingly the milieu within which the system is designed to operate changes over time and as this dislocation increases so inequality is enhanced. Trends such as the ageing of the population or the suburbanisation of housing, fashions like health centres, private health care, performance indicators and primary care teams, and crises, such as the current economic depression and its associated cut-backs in caring services, all demand dynamic change in health care organisation. To promote reorganisation as a simple once-and-for-all panacea is to gloss over this important context.

Let us elaborate on these contentions with reference to changes in the organisation of the British National Health Service. The 1974 reorganisation can be seen as a response to the deficiencies of a system set up in 1947 to cope with the changed society of the 1970s. Between 1947 and 1974 health care in England and Wales was organised on a tripartite basis (Figure 8.1):

1) 134 independent Executive Councils conforming to local authority boundaries or groups of local authorities and providing general practitioner and dental care in England and Wales;
2) 174 local government authorities, responsible for preventative, community and ambulance services;
3) 15 regionally based Hospital Boards and sub-tier of 400 Hospital Management Committees governing one or more hospitals.

Teaching hospitals remained largely autonomous. Health care administration was bound closely to a local government system which, it was recognised, had itself become anachronistic. The passage of a baby boom and the longer-term ageing of the population meant that the administrative system, spatially based on jurisdictions derived in the 1930s at the latest, was not in tune with contemporary population dynamics or population distributions and was, thus, consequently inadequate for the task of facing new and extensive demands. The tripartite system had developed to cope with problems of acute ill-health and infectious disease; by the late 1960s the focus of medical care had had to shift to chronic long-term illnesses requiring a very different form of treatment.

It would seem, therefore, that a changing population distribution and changing population needs provided the structural context to the 1974 reorganisation. Health care planners attempted to meet these radical changes through the introduction of 'efficient' management and a radically different health care system. Brown (1979) argues that the management aspect was emphasised; perhaps even at the expense of the effective solution of health care needs. The reorganisation, accomplished in April 1974, aimed to provide an integrated service with all functions under the umbrella of the National Health Service (Figure 8.2). A clear account was to be taken of local needs and a hierarchical system of accountability established. These aims were not wholly successful; although in a strict sense the three previously separate medical services were brought together, the Executive Council services continued to maintain considerable autonomy, and environmental health (concerned with sanitation and public health) remained under the aegis of local government. In England and Wales, fourteen Regional Health Authorities, with associated teams of Regional officers, headed the hierarchy and

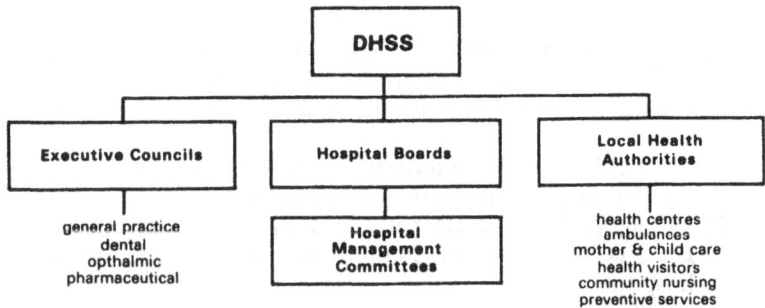

Figure 8.1 British National Health Service Organisation, 1948–1974

were given responsibility for strategic planning. Area Health Authorities provided the major administrative body and were involved with the co-ordination of services, liaison with local government and the assessment of local needs. They were specifically set up to be spatially congruent with local government areas to facilitate these goals. Beneath the area level, some 56 per cent of Area Health Authorities were subdivided into Districts from which the day-to-day delivery of care was to take place. The larger Districts were themselves divided into sectors for further administrative convenience.

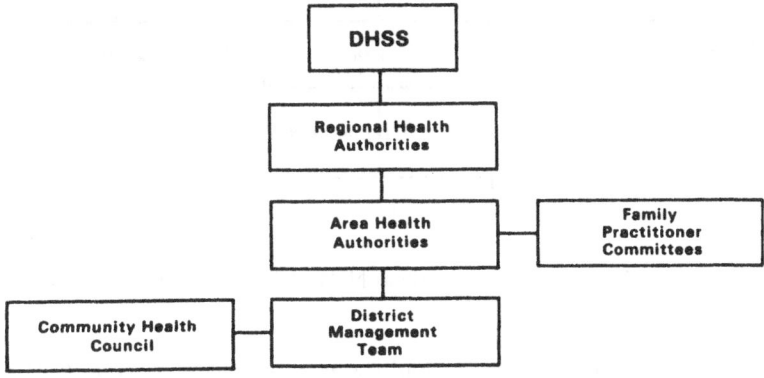

Figure 8.2 The reorganised British National Health Service, 1974

Following reorganisation, the rivalry and lack of communication between the different parts of the service, though still apparent, was much diminished. Facility and service planning in a more integrated fashion became possible, but other difficulties were manifested. For community health staff, the transfer from local authorities to health authorities meant a loss of contact with social service departments which continued under local government control. Integration within the health service was gained at the expense of external links, and the loss of co-ordination meant that the consumers of care arguably received a less effective service. In fact, in both medical and governmental circles, the opinion by the late 1970s was that an unwieldy administrative system had been set up (DHSS, 1979a). Too few resources were devoted to medicine and too many to administration. This opinion, although based on the questionable assumption that the production of good health flows from medical effort rather than the equitable distribution of resources, had considerable support.

Inequalities in health persisted while decision making remained slow and, of course, costs continued to rise, fuelled by the technological requirements of medicine and the changing needs of the population.

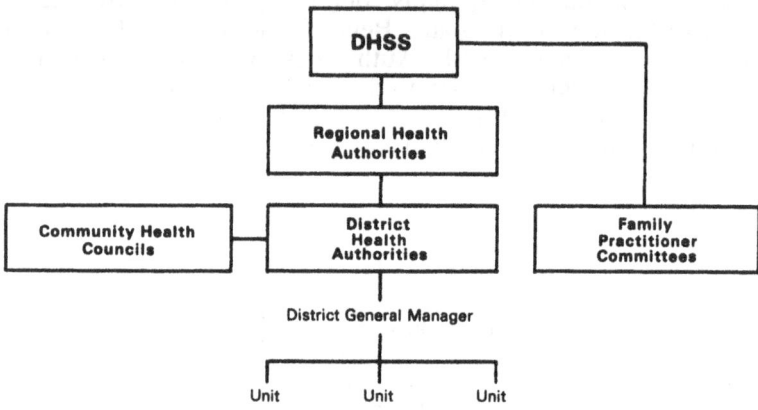

Figure 8.3 The reorganised British National Health Service, 1982

The outcome of the difficulties was a second reorganisation abolishing the Area Health Authorities (Figure 8.3). This was enacted in 1982 and is perhaps better seen as an incremental shift within the system (Walker, 1984). Area and District functions were combined and discharged at District level. Areas which had not been subdivided into Districts themselves became Districts. It was hoped that the new system, with its strategic Regional tier, and more responsive, localised District level would prove to be a greater success. Full evaluation of the effectiveness of the 1982 reorganisation remains to be made; however, the indications are that in governmental quarters there exists a view that there remains scope for a still greater input of efficient management-science-based approaches. Indeed further changes have taken place, and, in particular, the Griffiths Report introduced the appointment of District General Managers in addition to the existing District Medical Officers (DHSS, 1983). These managers were to bring business management expertise to the National Health Service and thus, it is claimed, contribute to enhanced efficiency. The crucial question is, of course, whether this more *efficient* system will be any more *equitable* than any of the earlier manifestations of National Health Service organisation.

Directed financial resource allocation

In the face of inequality, another policy which health care planners may adopt is to distribute their limited financial resources according to some criterion of need. The British Resource Allocation Working Party (RAWP), set up in 1975, provides a classic example of an attempt at this approach to health care planning.

The RAWP aimed:

> to review the arrangements for distributing NHS capital and revenue to RHAs, AHAs and Districts respectively with a view to establishing a method of securing as soon as practicable a pattern of distribution responsive objectively, equitably and efficiently to relative need. (DHSS, 1976, p. 5)

Need and equity were thus explicitly stated in the remit of the Working Party, and an official recognition of the persistence of health-based inequalities was established.

The evidence involved in reaching this conclusion was certainly not conclusive. Indeed, while Titmuss (1968), Hart (1971) and, later, the Black Report (DHSS, 1980) each argued that income-based inequality had remained a problem despite the provisions of the National Health Service, Rein (1969) and Cochrane (1972) suggested that, to a large extent, good health had become available to all. In Chapter 6 we set out some of the evidence for inequalities in health; the RAWP indicated the way by which initially Labour and later Conservative governments addressed the problem of redistributing care.

Prior to the RAWP, National Health Service financial allocation had been on an incremental basis (Walker, 1984). The budget of a Regional Health Authority in any one year would be the same as that which had been required to run the authority in the previous year, plus a percentage increase to allow for the notional growth of the client population. Little account was taken of changing needs between different authorities and, to a considerable extent, a budget would ultimately be reflective of the nineteenth-century philanthropy which had originally occasioned the development of still-existing hospital provision. Since RAWP there has undoubtedly been redistribution of health service monies, with the general movement being from regions which had been over-provided in practical terms towards those which were under-provided. All this movement of finance has, however, taken place within the context of little if any increased funding for the health service; thus the exercise has essentially involved the redistribution of an existing and inadequate financial base.

The RAWP has operated through distributing the amount of nationally available finance to the fourteen Regional Health Authorities in proportion to their needs as defined by the 'RAWP formula'. This formula, which is held to indicate relative need, facilitates subdivision of the national expenditure base between the Regional Health Authorities. In the case of revenue expenditure, the need-based apportionment which is thus arrived at is described as a 'revenue target'. The actual expenditure of an authority is then compared to this revenue target and under- and overfunded regions identified. The revenue target also serves as a functional goal towards which all regions are expected to move as speedily as financial and political circumstances permit. We might note, coincidentally, that the original RAWP document (DHSS, 1976) tried to express this operation as a formula; the formula used was mathematically incorrect.

The calculation of the revenue targets provide the key to the impact of the RAWP recommendations concerning health authority budgets (Eyles and Woods, 1983). The equalisation of resources which RAWP was meant to bring is a consequence of this operation. We may broadly summarise the build-up of the revenue target as a seven-stage process (Figure 8.4):

(i) As a first stage, health care planners make estimates of the mid-year populations for each region. If this were to be the only stage of the distributive process, *per capita* equality of expenditure would result.

(ii) The RAWP incorporates a weighting to reflect need; it approaches this task by, first of all, recognising that different sectors of medical care generate different needs. Seven sectors of care were originally each given separate consideration, but family practitioner administration was dropped in March 1985, leaving six. These are non-psychiatric in-patients, day- and out-patients, mental illness in-patients, community services, ambulances, and mental handicap in-patients. For all except ambulance services, national age/sex adjusted usage rates are then used to weight the crude regional populations. In essence this process weights the population of regions where there are large numbers of high-usage groups such as the elderly. In this way, it is argued, the different demands on the different sectors of the National Health Service are accounted for. In reality it is

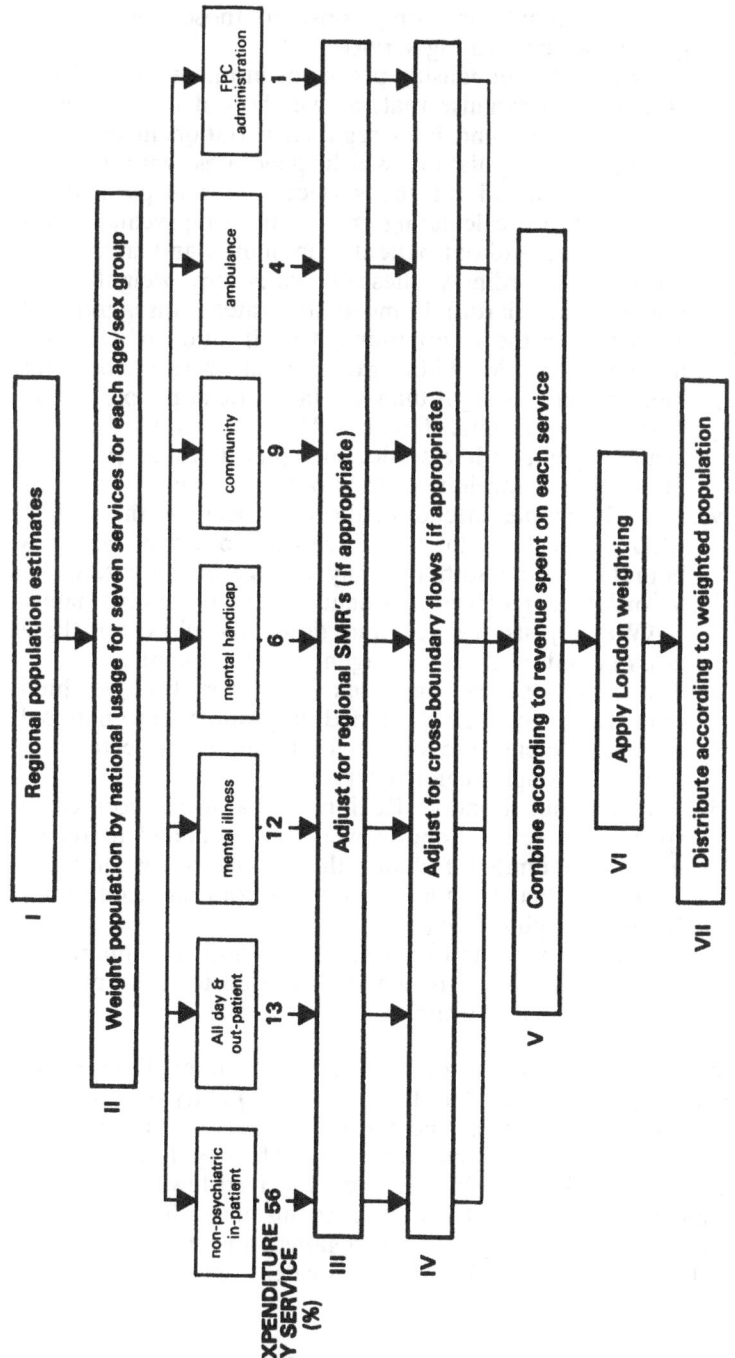

Figure 8.4 The RAWP process *Adapted from Woods (1982)*

a poor surrogate as it only considers those who actually get to use the existing service.

(iii) Further to the equalising processes in phase (ii), RAWP also has to recognise that, as we showed in Chapter 2, there is a profound inter-regional variation in ill-health among the population which generates variations in potential demand for the services. This is particularly important for calculating need for non-psychiatric in-patient, day- and out-patient, community and ambulance services; accordingly these services are weighted by regional standardised mortality rates, disaggregated according to the International Classification of Diseases in some cases. Morbidity rates would, in truth, be a far more effective weight than mortality; dead people are no burden on hospital finance. Mortality is used as an available, but rather crude, surrogate because (Chapter 2) adequate morbidity data is not available.

(iv) Not all people attend medical services in the region where they live; they may choose to travel across a boundary, or an authority may arrange for them to do so in order to receive treatment otherwise unobtainable. RAWP adjusts the revenue target to allow for these cross-boundary flows and agency arrangements.

(v) The sector-specific age, sex and mortality weighted populations are then combined, in proportion to national revenue expenditure on each of the seven sectors, to produce a single regional target.

(vi) For the four London Regional Health Authorities, a special subjective London weighting is added to recognise, in a limited fashion, the special needs of those regions particularly those stemming from the need to pay London-weighted salaries.

(vii) Finally the weighted regional population is compared to nationally available resources and used to make proportional regional allocations (Figure 8.5).

Through its derivation of the revenue target RAWP has certainly facilitated the identification of inequality – at least in terms of the limitations of the formula. Figure 8.6 outlines the general picture: Oxford and the four Thames Regions were 'overfunded' in 1977–8, whilst the rest received fewer funds than they apparently merited. To counteract this inequality, and move towards the revenue target, these regions had finances removed from them and given to the under-provided regions. The impact

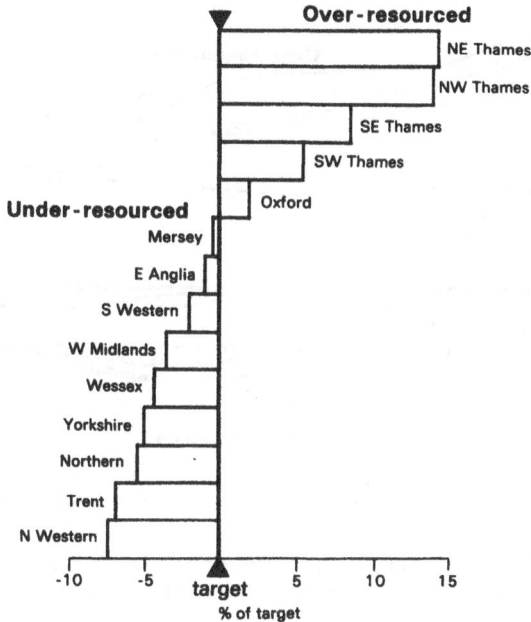

Figure 8.5   Regional resource allocation targets, 1977

of the general fiscal crisis of the late 1970s was such that there was no question of levelling-up the poor regions to the standard prevailing in the best regions. The outcome has been a gradual coming together of regions at a medium standard of funding rather than a high standard, and some regions, notably Oxford, have experienced widely fluctuating fortunes; Oxford is now underfunded. In relative terms losses have also occurred in three of the four Thames regions.

The revenue target approach has also been used to guide resource allocation in capital spending and joint funding (the latter facilitates projects on the health/social services interface). In the case of capital spending, community services are removed from the calculations as they are claimed to depend little on equipment and buildings. A capital allocation is then calculated by comparing weighted populations with the existing stock of buildings and equipment. Some marked changes in capital programmes have taken place. The Trent Region, for example, now claims to have an increased capital programme greater in real terms than that of 1969/70. This has allowed much-needed

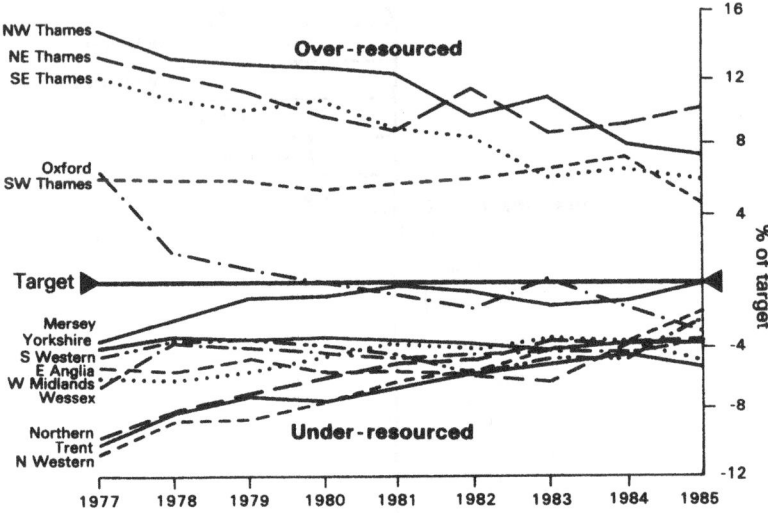

Figure 8.6 Changing regional finances since RAWP
Source: Annual Report of Health Services in England, 1985

renovation of older buildings and some flexibility in development programmes.

Criticism of the RAWP, on a general level, has focused on its financial background. The redistribution which has taken place should be seen in the context of high and increasing need such that one should expect a *rise* in expenditure – certainly over the period since 1970 when inflation has strongly affected costs of medical equipment and ageing and unemployment have had undeniable effects on health. In fact, as we have seen, the RAWP has levelled down rather than up.

A major criticism from medical circles has been the use of mortality rather than morbidity data as a measure of need. Geary (1977) argues that the latter would have provided a far more accurate measure. Mortality is comparatively cheap; illness, on the other hand, costs a great deal, particularly in the case of chronic illnesses, such as epilepsy, rheumatism and mental illness, requiring long-term treatment or hospitalisation. Though there were very good operational reasons, associated with the availability of suitable data (see Chapter 2), for choosing mortality, morbidity would have been a more accurate indicator of the pressures acting upon the services. It would probably have also produced a rather different resource distribution (Knox, 1978). In addition to this major criticism, there have also been suggestions that the adopted formulation of the standardised mortality ratio may, itself, constitute but one way in which the data could have been used. Palmer *et al.* (1979) recomputed the non-psychiatric in-patient target allocations, using four alternative indices of mortality, and indicated the considerable sensitivity of the RAWP figures to the mortality measure employed (Table 8.1). Some regions would have radically changed their position relative to the revenue target.

Another criticism levelled at the operation of the RAWP concerns its regional basis for allocation. The computation of the revenue targets is made for Regional Health Authorities, and the RAWP, as a whole, thus aims to equalise disparities in financial allocations between regions. Prior to the RAWP, however, variations in expenditure within the regions were far greater than

Table 8.1 The effect of mortality indices on RAWP targets (non-psychiatric in-patients)

Region	Age specific mortality ratio	Relative mortality index	Yerushalmy mortality index	Potential years of life lost index
Northern	+	+	+	+
Yorkshire	+	+	+	+
Trent	−	+	+	−
East Anglia	−	−	−	−
N.W. Thames	−	−	−	+
N.E. Thames	−	−	−	−
S.E. Thames	+	+	+	+
S.W. Thames	−	−	−	−
Wessex	−	+	+	+
Oxford	−	−	−	−
South West	−	−	−	−
W. Midlands	−	−	−	−
Mersey	+	−	+	+
North West	+	+	+	+

+ = percentage rise in revenue allocation
− = percentage fall in revenue allocation

For details of the formulae used to construct these indices see Palmer *et al.* (1979). The Standardised Mortality Ratio used in the actual RAWP formula is covered in our discussion of mortality data in Chapter 2.

those between regions. The RAWP recommendations ignore these large inequalities. Mersey Regional Health Authority, for example, was identified as underfunded although within its boundaries it contained the highest spending of the old Area Health Authorities (Figure 8.7).

We may further argue that the RAWP proposals were fundamentally conservative. The existing inequality between different sectors of the health service was maintained with no overt commitment to community services; general practitioner services remained largely outside the scheme and (Walker, 1984) considerable successful lobbying on behalf of prestigious hospital services took place. For example, a fixed addition to allocation was given in proportion to the number of medical students. The use of age/sex utilisation rates also fossilized health care

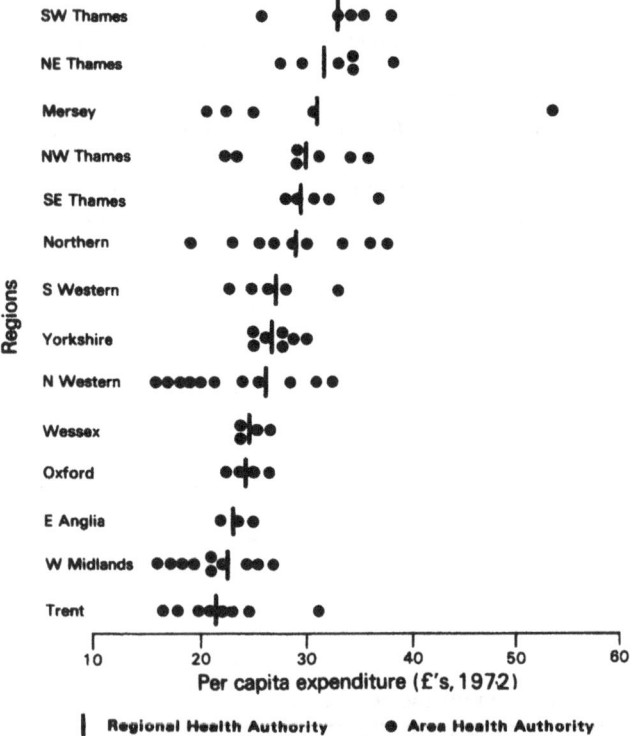

Figure 8.7 Intra- and inter-regional funding and RAWP
Source: original data in Buxton, M. (1976) *Health and Inequality*, Open University, Milton Keynes

consumption at its existing level, and made no attempt to encourage or even allow for increasing the utilisation of services for the chronically sick or elderly. Most importantly, in the context of our earlier discussions, the RAWP was conservative in ascribing the outcome of good health to the input of finance to hospital-based *medical* care; the social basis to inequalities in health was ignored.

Following the national RAWP exercise, each individual region was encouraged to produce its own mini-RAWP. Most regions initially followed broadly similar methodologies to the national RAWP operation, although each made modifications to confront local difficulties and some now use radically different approaches.

In London an attempt was made to introduce a deprivation factor reflecting the poor social conditions which health care was hoping to address in the inner city areas (DHSS, 1979b). The introduction of this factor reflected a realisation that, had the RAWP recommendations been introduced in unmodified form, the inner London areas and districts would have been revealed as heavily overfunded. Table 8.2 illustrates this point for North East Thames RHA.

Deprivation was calculated from census data on New Commonwealth immigrants, pensioners living alone and substandard housing. District Health Authorities, and, at the time of the commencement of the mini-RAWP, Area Health Authorities, were allocated combined deprivation scores and categorised as high, medium or low deprivation (Figure 8.8). High deprivation

Table 8.2 Inner city overfunding in North East Thames Regional Health Authority

Area Health Authority	Revenue allocation as percentage of RAWP target
Rural Essex	79.8
Barking and Havering	89.5
Enfield and Haringey	99.6
Redbridge and Waltham Forest	82.4
Camden and Islington	145.2
Inner City and East City London	125.1

Source: After Eyles *et al.* (1982).

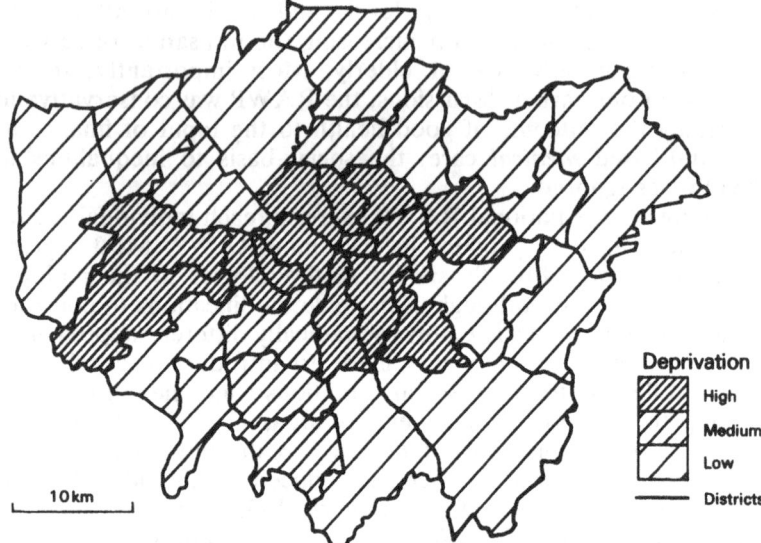

Figure 8.8 The deprivation factor in the Thames Regional Health Authorities
Source: Woods (1982)

was assigned a notional 5 per cent increase to their population within the revenue target build-up.

The London deprivation factor has been problematic in practice (Woods, 1982). Primarily this is because all deprived people do not live in places which are, on aggregate criteria, deprived. It is clear that for the deprived areas of London, there are usually more deprived living outside those zones than live within them (Table 8.3). Furthermore, because of the overfunding of the Thames Regions, London has still lost out. The deprivation factor has not cancelled out the effect of the transfer of finances away from the Thames Regions. For the consumer of medical care the problem has been exacerbated; as we showed in Chapter 6, the quality of inner city primary care is poor and, in fact, in such locations the hospitals often act as surrogate primary care providers. The reductions in finance have lessened their ability to perform this function and caused a strange coalition between community groups and the prestigious teaching hospitals.

Direct financial resource allocation policies, such as those of the RAWP, are seductively attractive in apparently allowing the development of equality through redistributive means. We should not be taken in by those claims unless empirical substantiation is

Table 8.3 Deprived people and deprived places in the Thames Regional Health Authorities

Indicator	Percentage of selected group in	
	high deprivation zone	*low deprivation zone*
Single pensioner households	23.5	56.6
New Commonwealth immigrants	43.7	31.9
Households lacking exclusive amenities	42.2	38.5
Economically active males 15–64 in Social Class V	33.5	50.9

Source: After Woods (1982).

available. In the case of the RAWP many drawbacks are evident; the methodological base is open to debate and the financial environment of the operation is such that the whole exercise could be described as a smokescreen for cut-backs. Perhaps the single most important failings are the implication that the quality and success of health care can somehow be equated simply with financial input, and the presumption that equality between people will flow from equalising allocations to regions.

Directed manpower allocation

In Chapter 7 we noted the part which the preferences, predilections and career-mindedness of medical professionals played in bringing about distributional inequality in the provision of health care. General practitioners, for example, may choose to settle in areas which are close to medical schools, or which have 'pleasant' residential environments. As a result some areas may become significantly under-doctored. An administrative response to this situation is to provide financial incentives to attract personnel to under-provided areas, and set up sanctions or quotas to limit recruitment elsewhere.

Let us consider the way in which this approach operates in Britain. England and Wales are divided up into ninety-eight Family Practitioner Areas corresponding to shire counties and amalgamations of metropolitan districts and London boroughs. Each Family Practitioner Area is further subdivided into Medical Practice Areas. A policy has been developed based on the number of patients registered with family doctors in each Medical Practice Area. Butler *et al.* (1973) have examined the operation of this policy in great detail. They note how, from almost the founding date of the National Health Service, some doctors, on finding themselves with small lists of patients, have been applying to have their local Medical Practice Areas declared closed to new doctors in order to protect themselves from competition. Others, in the opposite situation, have sought financial reward for large lists.

Table 8.4 outlines the present way in which Family Practitioner Areas are classified. In restricted areas, where the average list of family doctors is deemed to be small, any applications to set up a medical practice are usually refused – even when the application is as a replacement for a doctor who has left. Automatic registration of new family doctors remains problematic in intermediate areas, but in open areas it is usually granted. In designated areas financial inducements exist to encourage family doctors to set up practice and remain in the locality. These inducements are on a sliding scale reflecting average area list size, and comprise an initial practice allowance payable on setting up a practice and a designated area allowance which acts as a salary supplement.

There is little evidence that the financial incentives themselves have directly altered the inequality in the distribution of family doctors. The number of designated (that is under-doctored) areas had begun to fall slightly when Butler and his colleagues completed their work: 'Within the last year, however, there has been a slight drop in the number of these areas [Medical Practice Areas], although they still account for almost a fifth of all practice areas' (Butler *et al.*, 1973, p. 27). This process continued and, by April 1986, there was only one designated area in

Table 8.4 The classification of Medical Practitioner Areas

Type	*Average list size*	*Right to practise*
Restricted	less than 1,700	Normally refused
Intermediate	1,700–2,100	Problematic
Open	2,100–2,500	Automatic
Designated	above 2,500	Encouraged

England and Wales, located in County Durham. This does not mean, however, that there has been an improvement, or even that the policy has occasioned change.

A word of caution is necessary. Medical Practice Areas are dynamic creations, changing rapidly both in terms of their boundaries and their numbers. Over the years there has been a policy of amalgamating smaller practice areas into larger units. This may obscure the true extent of localised under-doctored areas. In addition the threshold list sizes defining the classification have been subject to changes. Furthermore, as we suggested earlier, it is by no means clear whether the decline in the number of designated areas can actually be linked to the financial inducements. The availability of family doctors to set up in practice is inextricably bound up with the numbers of general practitioners produced from medical schools and their employment prospects within the overall economy. Before the current financial crisis, these prospects were fairly good for newly trained general practitioners; 323 areas were designated in 1970. Now 75 per cent of England and Wales is classified as either restricted or intermediate and general practitioner unemployment is doubling annually.

Table 8.5 demonstrates the geographical dimension of the

Table 8.5 Medical Practitioner Area classification in England

Standard region	*No. of GP principals[a]*	*% in designated areas[a]*
Northern	1,315	57
Yorkshire & Humberside	1,919	50
East Midlands	1,351	44
East Anglia	702	17
South East	7,441	20
South West	1,723	8
West Midlands	2,004	66
North West	2,644	42

Source: Adapted from Butler, Bevan and Taylor (1973).

a. Figures for 1 October 1970

designated area policy at the time of the Butler research. Standard Regions in the South of England were clearly less under-doctored than those in the Midlands and North. Although one-fifth of GP principals (all family doctors with their own lists) in the South East were in designated areas, only 8 per cent of those in the South West were in similarly disadvantaged locations. In the West Midlands, in contrast, fully two-thirds of the principals were in designated areas.

Our verdict on the designated areas policy must be that it represents at least a recognition of the purely quantitative aspects of inequality in primary care. The qualitative aspects of inequality in care are, however, neglected. The policy does not consider, for example, the problems posed by elderly single-handed family doctors in poorly equipped inner city practices. Nor yet does it come to terms with the existence of very localised inequalities in care, such as those evidenced in the discussion of inverse care in Chapter 6. The size of Medical Practice Areas is such that they may, in some cases, encompass both an overcrowded inner city and an under-doctored suburban estate. The designated areas policy is at best, then, a partial measure the success of which is difficult to evaluate because of the problem of causally linking the policy to the outcome.

The issue of the quality of care, which we suggest was missing from the designated areas policy, was to a small extent the subject of the Inner City Group Practices Initiative, initially intended to run, in England and Wales, until April 1986. For a short trial period, cash incentives were made available in thirty-one inner city areas to facilitate the formation of group practices from former single-handed practices. The presumption, and it is questionable, was that group practices would provide better quality care. Participants in the scheme received up to £4,000 each; however, if grouping arrangements broke down they were expected to repay the sum involved. The scheme was small in scale, aiming only to create fifty new group practices, and could make little impact on national levels of single-handed practices – 13 per cent of doctors nationally and one in three in London are single-handed. It was primarily, like many policies, a cosmetic exercise awaiting a realistic finding and evaluation of questions like whether or not single-handed practices are necessarily all bad, and whether financial provision for the purchase of larger premises should be made.

Workload indices

Both the policy approaches that we have looked at so far have been essentially economically driven. The RAWP attempts to equalise finances, and the designated areas policy uses extra finance as an inducement. We will now shift our focus away from this emphasis towards a consideration of health and health care planning techniques which make a more explicit attempt to incorporate social and demographic factors. Our first case study considers workload indices.

The calculation and use of indices has a considerable pedigree, particularly in geography. In the simplest terms an index may be nothing more than a single variable from some data source such as the census. For example, through the social model of ill-health, we may equate certain social or demographic variables with an increased likelihood of ill-health. By extension we may link this, at least on a broad level, with an increased need, or perhaps demand, for health care. Mapping of such indicators can be used to show that the pressures on health care resources are not spatially equal: they provide an index of health care needs or provision.

If we wish to incorporate more than one variable into an index we are faced with a slightly more sophisticated but still relatively simple calculation. Multivariable indices, such as are inevitably needed to confront such concepts as need, are most easily constructed via the standard score transformation (Smith, 1975).

An interesting example of research using indices is provided by Jarman (1983, 1984), in two studies examining the location and characteristics of areas thought to generate high workloads for general practitioners. The 1983 Jarman study used the electoral wards of Inner London as its spatial basis. As an initial stage to the research, a sample of doctors were asked to list the population sub-groups which they felt, on the basis of their own experience, contributed most to the generation of work. This list was then rationalised to the ten most frequently mentioned variables. Each of the reduced list variables was assigned a weight according to its relative importance to the doctors. The final variables with their associated weights are set out in Table 8.6. The index calculation process involved firstly transforming the raw variables to an approximation of the normal distribution using an arc-sine transform, secondly converting these values to standard scores, and finally multiplying by the weights. A simple summation process then generated index scores for each ward; some wards contributed considerably more to hypothesised workload than others, illustrating the inequality in GP working environment and providing evidence for health care planning.

The construction of indices is undoubtedly very easy; indeed we might consider it to be an overtly simplistic approach particularly when applied to subjects as complex as need or workload identification. While on the one hand we might contend that indices provide a base line, incorporating social factors, from which some assessment of health care resource provision can be made, we should never lose sight of the dangers of poorly formulated indices. 'Garbage in; garbage out' certainly applies. The variables chosen to construct an index clearly constrain its final value. Jarman's index, for example, is heavily biased towards old people and makes the assumption that they contribute most to general practitioner workloads. Similar assumptions or interpretations of revealed relationships underlie the construction of many indices. As a consequence it is perfectly conceivable that an index could be constructed to 'prove' almost any contention which might be put forward. The implications of this possibility for the (mis)use of indices in resource planning exercises are obvious. Critics might point to the construction of measures used by the Resource Allocation Working Party (DHSS, 1976) as an example. Indices, furthermore, are static conceptions taking little account of future need patterns. The implication is that indices, like most quantitative techniques, should be treated with caution and informed circumspection rather than ignorance.

Table 8.6 Variables and associated weights in the Jarman workload index (1983)

Indicator	Weight
Over 65	2.5
Pensioners alone	2.6
Under 5	1.9
One-parent families	1.2
Unskilled workers	1.5
Unemployment	1.3
Lack of amenities	1.4
Overcrowding	1.2
Change of address	1.1
Ethnic minorities	1.0

In the 1984 version 'Over 65' and 'Lack of amenities' were removed.

Location-allocation modelling

Location-allocation modelling provides another example of a health care planning approach which is not explicitly grounded in financial considerations. Highly popular as a practical planning tool during the early 1970s, particularly in the USA (Morrill and Earickson, 1969; Morrill and Kelley, 1970), the approach has more recently been of largely academic interest (Clarke, 1984) despite a manifestly conservative ideology which would be expected to appeal in the political environment of Britain and the USA in the 1980s.

This last sentence needs some explanation. On the surface we might see location-allocation modelling as fulfilling two applications. Firstly, as the name suggests, it can be used as a practical planning tool to inform decisions concerning the location of a finite number of facilities within a region (Figure 8.9). Secondly it may be used in a comparative fashion to examine variation between an observed facility distribution and an optimum, theoretically derived distribution (Figure 8.10). In the latter case it is rather similar to accessibility modelling (Chapter 6); however, the existing facility distribution is assessed by reference to external rather than internal comparison. On a deeper, more philosophical level we should note however that generally, but not always, location-allocation modelling is predicated to a goal of efficiency. The derived optimum locations maximise efficiency with regard to the inputs to the model. These inputs are, in very basic terms, social and demographic data (disaggregated by sex and age and ascribed to particular locations) and the spatial separation of these locations. The approach allocates facilities to locations which will serve the most people yet minimise the distances to be travelled. The efficiency of this solution is achieved at the expense of reality and equality. Specifically, although social data is used, the social context is neglected; there is no consideration of political factors, land availability, locational conflict or planning regulations.

Given such criticism, it would seem difficult to marshal much support for location-allocation modelling. Yet, paradoxically, when used in a comparative fashion, it has considerable utility particularly when an existing facility distribution is compared with a set of alternatives illustrating different assumptions within the model. For example, let us consider the simplest allocation problem: that of allocating a single facility within a defined region.

We could make a very simple assessment of the problem via

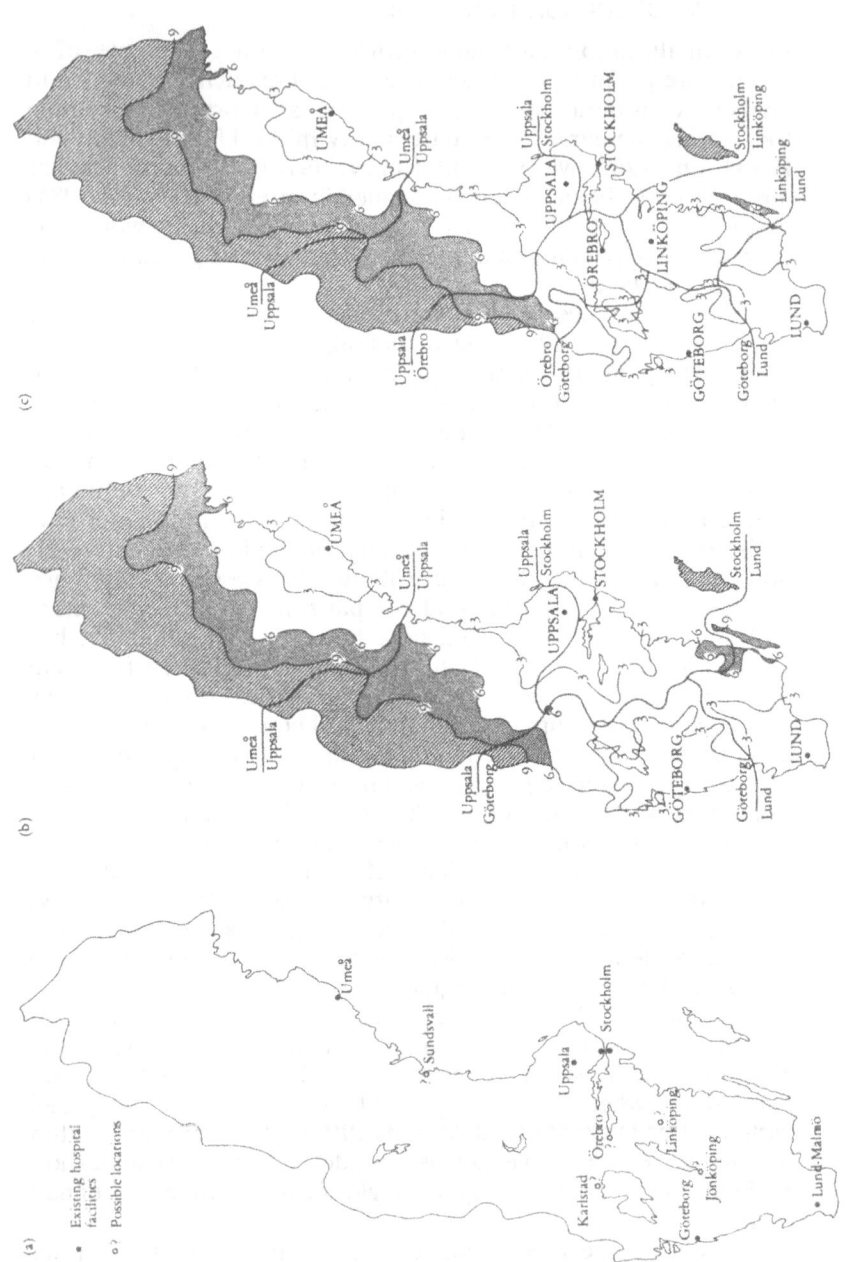

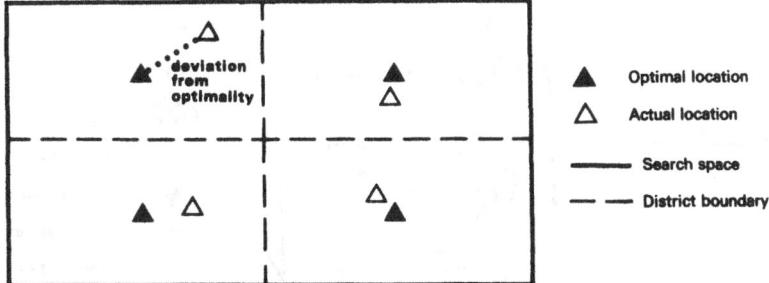

Figure 8.10 Actual and optimal facility distributions

nothing more complex than the computation of the mean centre of the region weighted according to the population of the constituent sub-regions. Ebdon (1977) provides a simple exposition of this technique requiring the investigator to know only the grid coordinates of constituent sub-regions and their respective populations. To render the example slightly more complex, the weighting variable might be an index representing the social or demographic make-up of the sub-regions. Figure 8.11 shows a simple example. The task was to suggest a highly idealised distance-minimising location for a child health clinic and compare that location with the existing situation. The catchment of the clinic was defined to include enumeration districts from the 1981 British Census and the centroids of the enumeration districts together with the child population of each enumeration district used to compute the new location; a similar approach could be used in the USA. The indications from the exercise were that, admittedly on limited criteria, the existing clinic was poorly sited. Any recommendation for a new clinic would however demand a full survey of needs, transportation availability and site suitability. Moreover, geographic accessibility may not lead to consumers necessarily using a facility.

Impact studies

For our fifth case study we are going to change our focus from looking at the derivation and drawbacks of policies and planning techniques to a consideration of what happens on a very local scale when a policy is enacted. We will take, as our example, the

Figure 8.9 Optimal location of health facilities, specialised hospitals. Sweden (a) existing facilities, (b) with isochrones, (c) optimal solution, two extra facilities, revised isochrones
Source: Godlund, S. (1961) Population, Regional Hospitals, Transport Facilities and Regions, *Lund Studies in Geography*, Series B, No. 21

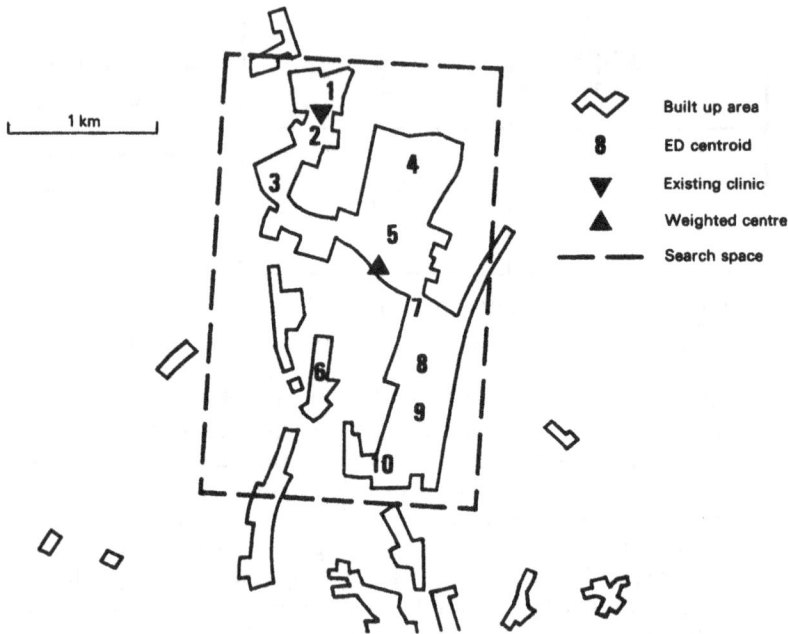

Figure 8.11 The facility location problem: child clinic, Horndean, England

policy of care in the community under which, as we saw in Chapter 5, large numbers of mental patients now live 'in the community' rather than in hospitals. They reside in small-scale hostels or lodging houses and receive treatment in day clinics.

In both Britain and North America, proximity to the facilities which service the community care policy is problematic. These problems stem from what we may term 'negative externality'. Despite their overall benefit to the community at large, the facilities confer local disbenefits in the form of generation of traffic noise and movement, parking difficulties, unsociable business hours and people perceived to be 'abnormal'. Thus, as we showed in Chapter 5, a policy which has support may still, on a local level, be offensive to social norms.

Most medical facilities experience these negative externalities to some extent. In most cases the benefits outweigh the problems (Figure 8.10). For 'noxious' facilities, like those catering for the mentally ill, which hold deviants from society and are therefore regarded with some fear by society, this is emphatically not the case. For these facilities those who live nearby might even be considered to be suffering a form of inequality by absorbing the localised negative impact of a facility which is, on a wider scale, a benefit to society as a whole. It is a task of health care planners to overcome or allay these fears in order that the continued operation of the policy can be ensured. The planner acts to facilitate social policy and, effectively, engineer supportive socially derived attitudes. Dear *et al.* (1977) working in North America, have suggested that a crucial distinction to be made in discussing facility impact is that between tangible and intangible effects. The former constitute measurable or otherwise quantitative impacts, such as changing levels of complaints concerning noise, increasing rates of petty crime in facility neighbourhoods, falls in local property prices, and experiences associated with abnormal behaviour by facility patients. Intangible effects refer to unquantifiable feelings and fears – for example those concerned with the perceived status of a locality. Burnett and Moon (1983), while largely supporting this view, contended that a further distinction can be made relating to the changing assessment of externality effects before and after a facility opening. Prior to opening, reaction is usually severe as residents

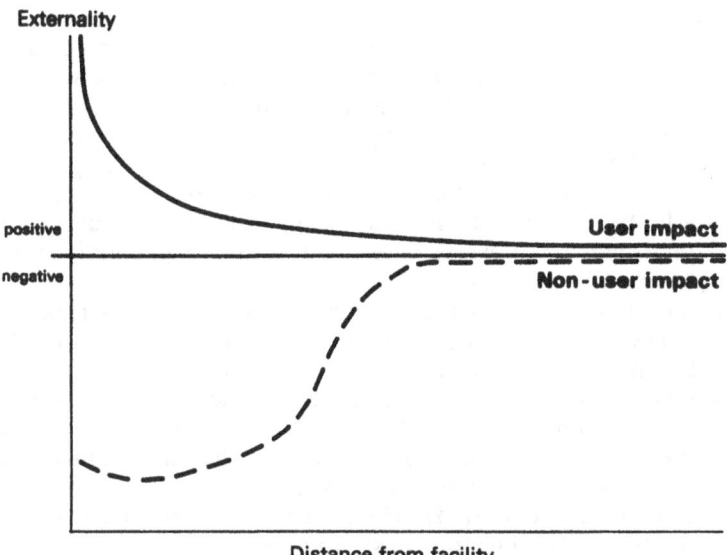

Figure 8.12 Positive and negative externality

in the vicinity of the projected facility allow their imagination to play on the most extreme possible impacts. After opening, those perceptual fears are mitigated by the experienced reality of impact. Indeed, in the case of facilities in Portsmouth, England, there is evidence that this mitigation can itself be extreme. A sample survey of residents living within 400 metres of community-based facilities housing patients with a history of mental disorder indicated that less than 10 per cent of respondents were actually aware of the presence of the facility (Moon and Burnett, 1984).

Once the dimensions of facility impact are known, the health care planner can begin to try to understand why negative reaction may occur in certain cases. This is not only a case of identifying the factors which superficially relate to reaction – such as a fear of falling property prices – but also an indication of a need for an understanding of why some individuals rather than others should react negatively. Dear and Taylor (1982) have considered this latter question in some depth (Figure 8.13). They hold that individual attitudes towards the mentally ill and the treatment of mental illness are of considerable importance. These attitudes are to a large extent a function of socio-economic status, with prosperous owner-occupiers who have most to lose being more likely to reject a facility. Planning for the acceptance of a facility must, however, go beyond this simplistic analysis and consider interpersonal relations and the ways in which the mentally ill should be treated. Thus, the health care planner should consider the neighbourhood characteristics of the environments in which it is proposed facilities should be located, and the characteristics of the facilities themselves. It appears that transient, low status neighbourhoods in decayed inner cities have low political cohesion and are less likely to muster objections to facilities. Similarly, facilities which are small scale, secluded, well designed, well managed and possess a resident staff are generally more acceptable (Dear, 1976).

When opposition to a facility does occur, Smith and Hanham (1981) have suggested that we can consider the actions of the population in the impacted neighbourhood in terms of three concepts drawn from political science. Residents may choose to remain in the impacted neighbourhood and resign themselves to negative externalities (the loyalty option). They may remain but enter into vocal opposition (the voice option), or they may move elsewhere (the exit option). Health care planners must aim to convert these differing strategies into a fourth outcome: support.

Research and public relations are one way of achieving this goal. The former usually acts as a basis for the latter, for example

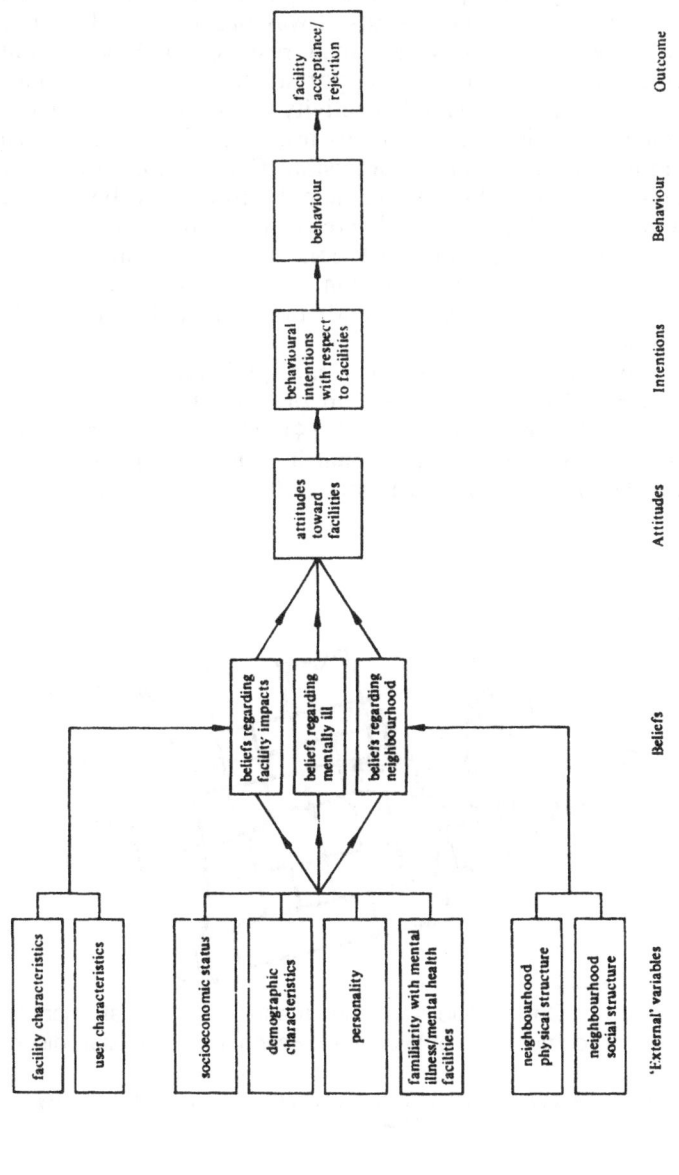

Figure 8.13 Dear and Taylor's model of facility impact

Source: Dear and Taylor (1982)

research in Philadelphia and Toronto (Dear, 1977; Boeckh *et al.*, 1980) has considered the putative impact of hostels on property prices. No evidence of such an effect could be found, and a far greater importance was attached to the influence of 'extras' such as heating. A greater proportion of people, however, tended to move from an area where a hostel was proposed. If health care planners can successfully use such research findings at public enquiries the groundwork for support at least will have been done.

It is important that planning activity of this form is carried out; continued opposition can lead to fewer facilities and a greater probability of inequalities in provision. Furthermore, the successful integration of facilities into a neighbourhood has been claimed to be vital in aiding successful care of the mentally ill (Trute and Segal, 1976). Indeed this last conclusion is applicable to all medical facilities as the opposition of local residents may inhibit the effectiveness of carers and may dissuade clients from attending.

The one situation which health care planners must particularly seek to avoid is the ghettoisation of patients into areas where they can find acceptance (Figure 8.14). This has been a particular problem for the mentally ill in that it confers inequality both for the residents of the limited number of areas into which facilities

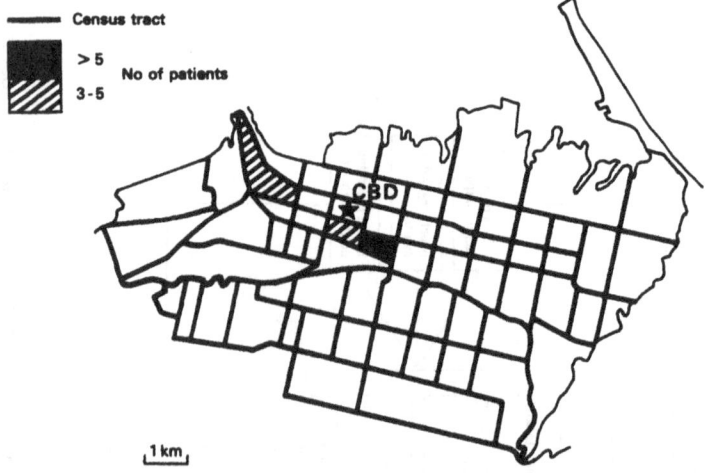

Figure 8.14 Ghettoisation of the mentally ill: initial location of institutional component of Hamilton Psychiatric Hospital discharge cohort

Source: Dear, M. (1977) Psychiatric patients in the inner city, *Annals, Association of American Geographers*, 67, 564–88

are concentrated, and also for the residents of the consequentially limited number of homes. This aim of avoiding ghettoisation is now being realised and facilities for the mentally ill are now the subject of policy recommendations designed to aid acceptance and restrict the problem of negative externality (Wolpert and Wolpert, 1976; Department of Environment, 1975).

Health education

In the last example, we broadened our discussion and looked at the issues which policy implementation raises for its practitioners. In this final case study we are going to continue this broadening process with a consideration of the wider, more philosophical questions which can be raised in the everyday operations of health care planning. We will take as our example health education.

Health education is the strategy adopted to try to facilitate the long-term shifts in culture and attitude which are required to counter inequalities in preventable illness. The central aim of the strategy is the education of the general public concerning health and health problems and the prerequisites of a more healthy lifestyle. This aim involves transmitting information to the general public via courses, leaflets, poster campaigns and other forms of publicity.

In considering planning for health education, it is crucially necessary to recognise that health education, particularly as it is practised within the British National Health Service, is in a position of considerable paradox. The rationale underlying the concern with preventable illness is, in itself, clear enough. Many childhood communicable diseases, for example measles and whooping cough, are eminently preventable yet inequalities remain (see Chapters 4 and 6). Similarly, evidence is widely available linking preventable factors with specific illnesses: smoking to chronic bronchitis and lung cancer, high carbohydrate diets to obesity. The paradox of health education lies in the solution which it proposes to this situation. Unlike other sectors of the National Health Service, health education is dedicated to a long-term, societally based approach rather than one of medical intervention. Thus, educational inputs emphasising the dangers of smoking or a poor diet are promoted rather than mainstream chemotherapy or surgery which intervene in pre-existing illnesses and do not seek to prevent the initial onset of a condition.

It should not be suggested that medical science falls completely outside the interests of health education. Where medical approaches have a demonstrable impact in preventing ill-health, they are encouraged. The case of immunisation illustrates this point, yet, again, the emphasis is on educating people about the effectiveness of the strategy rather than on immunisation in itself. Notwithstanding this caveat, it is clear that health education is one sector of the health services which does not conform to the stereotype of interventionist medicine. The consequence of this paradox is that funding and resources for health education are restricted. At least as a partial result of this, health education has often had only a limited effect and its potential has been poorly recognised. The Black Report noted this unfulfilled potential in its recommendation that 'An enlarged programme of health education should be sponsored by the government and necessary arrangements made for optimal use of the mass media, especially television' (Townsend and Davidson, 1982, p. 213). Until this recommendation is adopted by central government, health education remains faced with the problem of delivering an effective service within a resource-poor environment. This inevitability indicates a philosophical bind because it is necessary for health education professionals to recognise inequalities in preventable health and use these inequalities to discharge their work more effectively. The philosophical bind comes about because what amounts to 'targeting' health education material at those, for example, known to eat fatty food, may on the one hand be construed as efficient and cost effective, and on the other hand be seen as a form of victim-blaming (see Chapter 1).

We can illustrate these broad issues by considering the case of a campaign warning against the dangers of hypothermia. Health educators are concerned about hypothermia – a lowering of the body's core temperature to below 35°C/95°F (Collins, 1981) – because even in a mild winter 35,000 people over the age of 65 are in danger of becoming hypothermic (Fox *et al.*, 1973). In its more extreme form hypothermia kills, and it is a particularly severe problem for the elderly, living alone in poorly heated, inadequately insulated houses (Wicks, 1983). Furthermore, the elderly are notably susceptible because there is a tendency among the over-70s to have a decreasing ability to sense a need for warmth, whilst at the same time facing the competing demands of food and heating bills on a limited income.

Clearly hypothermia has a fairly precise social context. It is this which health education planners have increasingly begun to use in an effort to operate in a cost-effective fashion. The approach which has been used has been termed 'campaign targeting'

(Moon and Jones, 1985) and draws on work in market research. The basic premise is that, if the social profile of those whom it is wished to contact is known, then the Population Census can be used to identify areas where such people live. Saturation mailing can then be performed in the places where it would be presumed to be most effective (CACI, n.d.). In the case of a hypothermia campaign, this would be where there are high numbers of old people living in substandard housing (that is housing which would reasonably be expected to lack adequate heating and insulation).

The targeting strategy can be extremely effective. In Portsmouth and South East Hampshire District Health Authority three targeting exercises were performed using progressively more detailed data for the 1,007 census enumeration districts within the Health District. The enumeration districts were ranked according to the number of people associated with each of the selected indicators, and the cumulative percentage of people within the top 5, 10, 25 and 50 per cent of the enumeration districts was noted. The results of the exercise indicated considerable statistical concentration (Table 8.7). In the broad group of the elderly there were over 26,000 people; even if we targeted half the enumeration districts we would still reach less than three-quarters of our target population. We would do rather better if we focused on the elderly living alone, but the best targeting

Table 8.7 Area targeting for hypothermia in the Portsmouth District Health Authority

| Target population | *Percentage of target populated contacted in | | | |
	5% of EDs	*10% of EDs*	*25% of EDs*	*50% of EDs*
People aged over 75 (base = 26,619)	12	21	44	72
People aged over 75 living alone (base = 10,998)	17	28.	50	78
People aged over 75 lacking inside WC (base = 1,629)	40	62	91	+100

*base total of 1007 Enumeration Districts

+100% contact rate reached with 36% of Enumeration Districts

strategy is yet more detailed. All the elderly living in poor housing could be contacted by a campaign based on just over one-third of the enumeration districts.

The statistical concentration which the targeting strategy identified was paralleled, in the Portsmouth case, by a spatial concentration. The target population was located predominantly in the inner city areas of Portsmouth (Figure 8.15). By targeting their publicity onto those areas, health educators were able to run a highly successful campaign. Using our knowledge of epidemiology and attitudes to health, it should be possible to investigate the effectiveness of this approach in other cases, for example rubella screening or diet campaigning.

Figure 8.15 Spatial targeting for hypothermia

As we suggested above, targeting is not, however, without philosophical drawbacks. Indeed it may be argued to constitute a clear case of victim-blaming. The elderly are stereotyped as potential victims of hypothermia and solutions to the problem are offered suggesting that the condition is the result of individual irresponsibility. To avoid becoming hypothermic you should wrap up warmly and budget adequately for food and heat; if you do not then it will all be your own fault! Issues of fuel poverty and low income vanish, and the onus for change is placed on the individual.

In the search for cost-effective planning methods, health educators and others therefore lay themselves open to the charge of victim-blaming. A particularly clear example concerns the monitoring of new-born babies; designed to prevent sudden infant death syndrome, this approach involves identifying 'at risk mothers' on the basis of individual social and medical data (Carpenter *et al.*, 1977). As with the hypothermia case study, it has philosophical drawbacks but works very well. The problem of victim-blaming is thus peculiarly difficult; does the effectiveness outweigh the ethics? Clearly a far more effective long-term strategy would be to address the root causes of ill-health and health care poverty which we began to explore in Chapter 7 and to which we will return in the following chapter. For example, action against hypothermia should be combating fuel and financial poverty and the loneliness of old age. Such social engineering is currently politically beyond the scope of health education or indeed health intervention in its wider context. The virtual dismissal of the Black Report illustrates this point. It is for this reason that health education, potentially the source of a long-term attack on ill-health, can currently only offer short-term, temporary palliatives.

Conclusion

We have explored several ways in which policy makers, planners and researchers have confronted the problems of health care delivery. The way in which the health service is organised has been altered in response to the changing requirements within the service – and less clearly in response to the changing needs of the consumer. Financial and manpower policy and planning measures have been enacted to redirect resources to areas where needs are higher. Planning techniques, such as workload indices and location-allocation modelling have been adopted. New policies for the location of care facilities have been developed with, at least partially, unforeseen consequences for the health care planner involving the need to recognise how both the consumer of care and the resident of the neighbourhood in which a facility is sited must be considered when planning a new facility. Finally we outlined paradoxes within the role of health education in the fight against inequalities in preventable illness.

We would site all these measures, which are only a selection of what we could have covered, within a threefold framework. First, they are financially dependent; they cannot operate, even within their own limitations, without the necessary financial resource base. Whether or not this is forthcoming depends on the attitudes of the current government towards health care. Thus some policies receive only limited funding, whilst other measures develop as specific responses to poor funding. For example, preventative care is poorly funded because of ideological opposition to caring in situations which are not interventionary, high technology and high prestige. Health education falls within this underfunded arena and so is forced to adopt partial strategies such as targeting in order to make good use of inadequate resources. Other policies are developed with aims which may appear laudably redistributive but which are, in reality, covert recognitions of the low value which a government attaches to health care. Thus, attempts at reorganising the health service, and the work of the Resource Allocation Working Party, may, in reality, be seen as measures designed to cut the British National Health Service back to the minimum levels commensurate with ensuring adequate reproduction of labour power while still placating the wishes of the influential medical establishment.

The second element in our framework is a recognition that all policy and planning measures can only partially intervene in the fight against inequality. As we showed in Chapter 7, inequality has its roots in macro politico-economic structures which go beyond even the nation-state. Any approaches developed within a country cannot hope for complete success as they treat symptoms but not causes.

Finally, and in conclusion, we should note that the role of geography has been grossly neglected in the analysis of health care planning. A theme running through each policy or planning measure which we have discussed is that of spatial scale. Reorganisation was predicated, at least in part, on trying to find a common spatial scale at which health care could be delivered; it has failed. The Resource Allocation Working Party similarly failed to take into account the problem of local geographical disparities as did the policy of general practitioner area classification. Planning techniques, such as location-allocation modelling or workload indices, attempted to confront these problems but suffered from other drawbacks. Community care of the mentally ill provides a superb example of a policy derived in all good faith but failing to recognise the inevitability of local, geographically based, opposition. Finally our case study of health education shows that the pursuit of an overtly geographical approach can raise difficult philosophical problems in its own right. Geography, firmly sited in the social realm, is not without relevance.

Guided reading

Any reader interested in the subject of British health care planning and policy making would be well advised to read one of the standard texts on the subject. Ham (1982) and Klein (1983), Iliffe (1983) or Thunhurst (1982) are good examples and provide critiques and legislative outlines of the British National Health Service. Similar texts exist covering countries other than the UK, for example Oatman (1978) on the USA and Kaser (1976) on Eastern European countries.

The reorganisation of the British National Health Service is well covered in Levitt and Wall (1985) and the Resource Allocation Working Party has been the subject of detailed research by Eyles *et al.* (1982), Woods (1982) and Eyles and Woods (1983). Butler *et al.* (1973) and Butler (1980) are the key texts on the direction of medical manpower.

The references cited in the text provide the best source material on workload indices and location-allocation modelling whilst Dear and Taylor (1982) is the best introduction to the complex field of facility impact studies. A useful overarching view of policy making is provided by Walker (1984, Chapter 7). Health education is given a general coverage by Sutherland (1979) and its current practice is detailed in *Health Education Journal*.

References

Boeckh, J., Dear, M. and Taylor, S. (1980), 'Property value effects of mental health facilities', *Canadian Geographer*, vol. 24, pp. 270–85.

Brown, R. (1979), *Reorganising the National Health Service*, Oxford, Blackwell.

Burnett, A. and Moon, G. (1983), 'The effects of community based residential facilities on neighbourhoods', End-of-grant Report to the Economic and Social Research Council, London, ESRC.

Butler, J. (1980), 'How many patients?', Occasional Papers on Social Administration no. 64, London, Bedford Square Press.

Butler, J., Bevan, J. and Taylor, R. (1973), *Family Doctors and Public Policy*, London, Routledge & Kegan Paul.

Butler, J. and Vaile, M. (1985), *Health and Health Services: An Introduction to Health Care in Britain*, London, Routledge & Kegan Paul.

CACI (n.d.), *Target Marketing from CACI*, London, CACI.

Carpenter, R., Gardner, A., McWeeny, P. and Emery, J. (1977), 'A multistage scoring system for identifying infants at risk of unexpected death', *Archives of Disease in Childhood*, vol. 52, pp. 606–12.

Clarke, M. (ed.) (1984), *Planning and Analysis in Health Care Systems*, London, Pion.

Cochrane, A. (1972), *Effectiveness and Efficiency: Random Reflections on Health Services*, London, Nuffield Hospitals.

Collins, K. (1981), 'Thermal comfort and hypothermia', *Journal of the Royal Society of Health*, vol. 1, pp. 16–18.

Dear, M. (1976), 'Spatial externalities and locational conflict', in D. Massey and P. Batey (eds), *London Papers in Regional Science, 7: Alternative Frameworks for Analysis*, pp. 152–67, London, Pion.

Dear, M. (1977), 'Impact of mental health facilities upon property values', *Community Health Journal*, vol. 13, pp. 150–7.

Dear, M., Fincher, R. and Currie, L. (1977), 'Measuring the external effects of public programmes', *Environment and Planning A*, vol. 9, pp. 137–47.

Dear, M. and Taylor, S. (1982), *Not on Our Street*, London, Pion.

Department of Health and Social Security (1969), *The Functions of the District General Hospital* (The Bonham-Carter Report), London, HMSO.

Department of Health and Social Security (1976), *Sharing Resources for Health in England: Report of the Resource Allocation Working Party*, London, HMSO.

Department of Health and Social Security (1979a), *Patients First*, London, HMSO.

Department of Health and Social Security (1979b), *Assessing Target Allocations within the Thames Regions*, London, Joint London Regional Working Group, DHSS.

Department of Health and Social Security (1980), *Inequalities in Health*, Report of a Research Working Group chaired by Sir Douglas Black, London, DHSS.

Department of Health and Social Security (1983), *NHS Management Inquiry* (The Griffiths Report), London, DHSS.

Department of the Environment (1975), *Hostels and Homes*, Development Control Policy Note 15, London, HMSO.

Ebdon, D. (1977), *Statistics in Geography: A Practical Approach*, Oxford, Blackwell.

Eyles, J., Smith, D. and Woods, K. (1982), 'Spatial resource allocation and state practice: the case of health service planning in London',

Regional Studies, vol. 16, pp. 239–53.

Eyles, J. and Woods, K. (1983), *The Social Geography of Medicine and Health*, London, Croom Helm.

Fox, R., Woodward, P., Exton-Smith, A., Green, M., Donnison, D. and Wicks, M. (1973), 'Body temperatures in the elderly: a national study of physiological, social and environmental conditions', *British Medical Journal*, vol. 1, pp. 200–6.

Geary, K. (1977), 'Technical deficiencies in RAWP', *British Medical Journal*, vol. 1, p. 1367.

Ham, C. (1981), *Policy Making in the National Health Services: A Case Study of the Leeds Regional Hospital Board*, London, Macmillan.

Ham, C. (1982), *Health Policy in Britain*, London, Macmillan.

Hart, J. T. (1971), 'The inverse care law', *Lancet*, vol. 1, pp. 405–12.

Iliffe, S. (1983), *The NHS: a Picture of Health?*, London, Lawrence & Wishart.

Jarman, B. (1983), 'Identification of underprivileged areas', *British Medical Journal*, vol. 286, pp. 1705–9.

Jarman, B. (1984), 'Underprivileged areas: validation and distribution of scores', *British Medical Journal*, vol. 289, p. 1587.

Kaser, M. (1976), *Health Care in the Soviet Union and Eastern Europe*, London, Croom Helm.

Klein, R. (1983), *The Politics of the NHS*, London, Longman.

Knox, E. (1978), 'Principles of allocation of health care resources', *Journal of Epidemiology and Community Health*, vol. 32, pp. 3–9.

Levitt, R. and Wall, A. (1985), *The Reorganised National Health Service*, London, Croom Helm.

London Health Planning Consortium (1981), *Report of the Primary Health Care Study Group* (The Acheson Report), London, DHSS.

Moon, G. and Burnett, A. (1984), 'The neighbourhood impact of mental health hostels', Hostels Impact Project Working Paper 5, Geography Department, Portsmouth Polytechnic.

Moon, G. and Jones, K. (1985), 'Targeting resources for health education', paper given at the Institute of British Geographers Annual Conference, University of Leeds

Morrill, R. and Earickson, R. (1969), 'Variations in the character and use of hospital services', *Health Services Research*, vol. 3, pp. 224–38.

Morrill, R. and Kelley, M. (1970), 'The simulation of hospital use and the estimation of location efficiency', *Geographical Analysis*, vol. 2, pp. 283–300.

NHS/DHSS (1982), *Steering Group on Health Services Information First Report* (The Korner Report), London, DHSS.

Oatman, E. (1978), *Medical Care in the United States*, New York, Wilson.

Palmer, S., West, P., Patrick, D. and Glynn, M. (1979), 'Mortality indices in resource allocation', *Community Medicine*, vol. 1, pp. 275–81.

Rein, M. (1969), 'Social class and the health service', *New Society*, no. 14, pp. 807–10.

Smith, C. and Hanham, R. (1981), 'Proximity and the formation of public attitudes towards mental illness', *Environment and Planning A*, vol. 13, pp. 147–65.

Smith, D. (1975), *Patterns in Human Geography*, Harmondsworth, Penguin.

Sutherland, I. (1979), *Health Education: Perspectives and Choices*, London, Allen & Unwin.

Thunhurst, C. (1982), *It Makes You Sick: The Politics of the NHS*, London, Pluto Press.

Titmuss, R. (1968), *Commitment to Welfare*, London, Allen & Unwin.

Townsend, P. and Davidson, N. (1982), *Inequalities in Health: The Black Report*, Harmondsworth, Penguin.

Trute, B. and Segal, S. (1976), 'Census tract predictors and the social integration of sheltered care residents', *Social Psychiatry*, vol. 11, pp. 153–61.

Walker, A. (1984), *Social Planning: A Strategy for Socialist Welfare*, Oxford, Blackwell.

Wicks, M. (1983), 'Cold condition, hypothermia and health', in J. Bradshaw and T. Harris (eds), *Energy and Social Policy*, pp. 85–104, London, Routledge & Kegan Paul.

Wolpert, J. and Wolpert, E. (1976), 'The relocation of released mental hospital patients into residential communities', *Policy Sciences*, vol. 7, pp. 31–51.

Woods, K. (1982), 'Social deprivation and resource allocation in the Thames Regional Health Authorities', in Health Care Research Group (eds), *Contemporary Perspectives on Health and Health Care*, Occasional Paper 20, Department of Geography, Queen Mary College, University of London.

CHAPTER 9

Critical perspectives

The reader who has followed the previous chapters sequentially will know that there are many 'loose ends', as we have given many signposts to matters to be discussed in this final chapter. There have also been major contradictions in our arguments which we must now confront. In particular, in discussing explanations of health care inequality in Chapter 7, we characterise quantitative approaches as providing only partial explanations. Yet this methodology is central to the chapters on disease ecology, especially Chapter 3. Moreover, in Chapters 7 and 8 we have been exclusively concerned with the provision of medical care as a means of achieving better health. Yet, this argument is contrary to the implicit assumption of earlier chapters that we need to examine the production of ill-health and not just the consumption of medical care. These contradictions in our presentation reflect similar characteristics in the work of medical geographers in general. Societal explanations have begun to permeate the geography of medical care, while leaving disease ecology largely untouched. There has been very little radical work on the social production of ill-health, and the positivist epidemiological approach of previous chapters is undoubtedly hegemonic. There are, however, signs of change.

While mainstream or 'positivist' epidemiology has remained resolutely quantitative, biological and concerned with individual-level explanations, there has been some recent development of what is being called a 'critical' epidemiology (Davies, 1982). Such critics have argued that the major causes of health inequality are not to be found internal to the body (genetic disposition) nor in individuals (personal behaviour) but in the organisation of society. They have argued that an epidemiology needs to be developed that regards material production as the ultimate determinant of social life and health. Moreover, there are now frequent calls from all parts of the political spectrum for changes in our whole approach to health. On the 'right' there is the claim that we cannot afford the increasing cost of technological medicine and, therefore, the onus of responsibility needs to be shifted to the individual to do things for himself. On the 'left' there is a realisation that equality of access to medicine as currently practised does not guarantee an improvement in health. In both cases, it is argued that medical care does not equal health care, and that the future does not simply mean more, or less, of the same.

This chapter offers critical perspectives in the sense of Habermas (1974), the aim being to uncover the real mechanisms of how society operates. Instead of reflecting and reproducing these mechanisms, the goal is to change them to achieve a better society. This requires a different approach to causal explanation from that developed in Chapter 3, and we begin with a consideration of the positivist view of science that is implicit in current epidemiology (Cameron and Jones, 1982), a realist critique is developed and applied to epidemiology, and then some illustrations of a materialist approach are given. Finally, in the light of these arguments, three alternative political strategies (free-market, welfarist, marxist) are outlined and evaluated.

Positivist epidemiology

Any critique of epidemiology needs to be set within the broader framework of a critique of the concepts of positivism, for most 'orthodox' epidemiologists are supposed to believe implicitly in the unity of science and to try to emulate the particular form of scientific method that has been used to great effect in the natural sciences. A positivist approach, as exemplified in Chapter 3, aims to account for observed phenomena by the development of laws based on empirical regularities. Following the familiar sequence, theory generates a set of hypotheses, valid data are collected, and the scientist as disinterested observer makes inferences on the basis of this information, either rejecting a hypothesis or accepting it as a tentative explanation which has to be submitted to further empirical tests. Scientific knowledge can only be gained by the judicious use of the senses to observe and measure 'facts' that are used to refute and test hypotheses and theories. Having developed tentative explanations, the empirical regularities can be used to predict what would happen if certain changes are made.

As it is only unbiased observation that allows the critical tests to be made there is no place for emotions and values. Events are separated into facts in one group and their evaluation, context and subjective nature in another. The objective is cut off from the subjective, facts are distinguished from values, body from mind, nature from culture, science from society; knowledge that cannot be empirically tested is irrelevant knowledge. Not only are observations supposed to be objective but the scientific method is also presumed to be value-free and to support no political or ideological position but merely to give the facts. In terms of positivist epidemiology, human disease is an observable, biological fact and is subject to cause-and-effect regularities which involve other observable phenomena. It is the job of the positivist epidemiologist to develop and use technical apparatus to discover the empirical evidence for the existence of particular regularities, the ultimate goal being a universally applicable model or explanation.

Realist science

In contrast realist science is not searching for regularity and order, but rather for an explanation of how things operate in terms of the underlying structures and mechanisms that are promoting or inhibiting change. Sayer (1984, p. 181) has noticed certain affinities in the positivist approach in terms of philo- sophical position, social theory and analysis; much social science has been concerned with positivism, the techniques of quantita- tive analysis and explanations based on individual-level data. Similar affinities can be discerned in critical epidemiology, and we can recognise the importance of a realist philosophy, a marxist social theory, an emphasis on qualitative change, and the contention that the underlying structures of society are the fundamental determinants of ill-health. As Chouinard *et al.* (1984) have argued, an epistemological home for marxist science has been found in realism and to underpin subsequent discussion we need to consider three elements of this approach as developed by Sayer (1984).

1) The essential difference between the objects of study in natural and social sciences.
2) The need to replace the concept of causation as empirical regularity by the retroductive method which seeks to uncover the real mechanisms that connect phenomena in a causal manner. Figure 9.1 summarises the differences between the two approaches with the positivist, empiricist method working 'horizontally' seeking regularities between observable outcomes and events (E); the retroductive

method working 'vertically', seeking the underlying structures (S) and mechanisms (M) that produce the observed
events.
3) The need to recognise that observations are not theory-free
but are theory-laden.

The positivist position is that there is no essential difference
between the objects of study in the natural and social sciences so
that social objects are considered discrete, separable and
observable by a human subject. But in the social sciences,
knowing subjects are part of the objects of study and the subjects
are related to the objects through a common culture. Moreover,
in the physical sciences, the object of study remains unchanged
by the concept of itself, so that calling moving ice a 'glacier' does
not affect it, while defining a person as ill certainly does and the
effect of this label depends on the concepts and beliefs of society
and who makes the definition. For the positivist, disease as an
object pre-exists in nature and is discovered as a scientific fact.
While not denying that symptoms are genuine, we must ask why
medical science has chosen to constitute disease in the particular
form (biological abnormality within the body) that it has.
In positivist epidemiology a relationship is deemed causal if
one set of discrete events (causes) in the social or biological
environment precedes a discrete disease entity (effect) in regular,
repeated succession, and the associations are replicated under

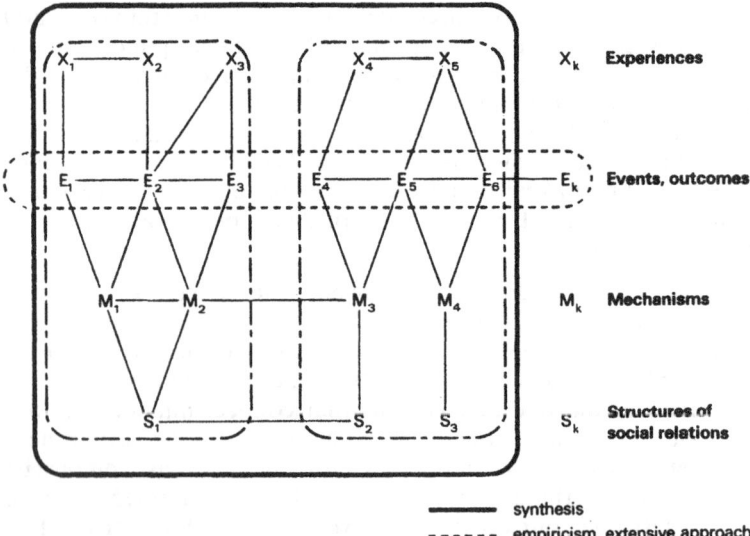

X_k **Experiences**

E_k **Events, outcomes**

M_k **Mechanisms**

S_k **Structures of social relations**

—————— synthesis
- - - - - empiricism, extensive approach
— · — intensive approach

Figure 9.1   Understanding reality

different conditions. Crucially, such notions make no reference whatsoever to causal mechanisms. There can be regularity without causality, as in the drunk who imbibed whisky and soda, gin and soda, vodka and soda believing that it was soda that was making him drunk. In the retroductive method, however, events are explained by identifying mechanisms which are capable of producing them. A fundamental distinction is made in realist science between two types of relationships: (1) external or contingent relationships where either object can exist without the other, and it is neither necessary nor impossible that they are involved in any particular relationship; (2) internal or necessary relationships where both objects are dependent on each other for their existence, so that a doctor cannot exist without a patient, an organism without an environment, a capitalist without a worker. Causation is not seen as regular association between externally related objects but as 'causal powers', mechanisms or 'ways of acting' that exist of necessity and internally in an object. To take Sayer's (1979) example of why gunpowder explodes, it is no good concentrating on the contingent conditions such as the presence of oxygen and a spark (for these often occur without an explosion) without considering the internal unstable nature of gunpowder. When we ask for an explanation, we usually already know that gunpowder and certain conditions (causes) produce explosions (effects) in regular association, but we need to know how and why this occurs. Similarly in epidemiology we need not just to measure the regularity between low social class and ill-health but we need to reveal the mechanisms and structures that explain such outcomes.

The realist view of causation accepts that there can be causation without regularity; what causes an event has nothing to do with the number of times it has been observed, for counteracting forces and changing conditions can override and conceal the operation of a particular mechanism. Unique events are caused no less than repeated ones. Regularities do exist, of course, and science has used them to gain an understanding of the causal mechanisms involved. Bhaskar (1975), however, recognises that the effects of a causal mechanism will only produce regularities under the particular conditions of a closed system in which two requirements are fulfilled:

1) intrinsic condition: there must be no change in the object possessing a causal mechanism;
2) extrinsic condition: there must be no change in the relation between the causal mechanism and its external conditions.

Thus a clockwork mechanism will not produce regularities if

there is metal fatigue in the spring (breaking the intrinsic condition) or if it is thrown into water (breaking the extrinsic condition). In the natural sciences, a closed system may occur naturally or be achieved by experimental control so that it becomes possible to use the regular sequences to elucidate the causal mechanisms involved. But, crucially, it is not regularity that is causal for regularity is merely being used to help to reveal causal powers.

In an open system, an observed regularity can only be transitory and irregular and it may be fundamentally misleading in terms of revealing the causal mechanisms operating in the system. All human and social affairs are open systems for we are all capable of self-change and learning new ways of acting (breaking the intrinsic condition) and human action can change the 'configuration' of the system (breaking the extrinsic condition). 'Everything is related to everything else' (Tobler, 1970) and 'everything is perpetual motion' (Mandel, 1979), and the insistence on the control or removal of this interconnectedness and the freezing of change may be detrimental to our study, for these may be the very conditions of an open social system that we should examine most fully. In particular in an open system, we must pay particular attention to qualitative change with 'emergent' powers.

Emergent powers are produced when previously separate, externally related objects undergo a qualitative change so that they become a set of internally related objects, that is a 'structure'; such structures may have new causal powers that do not exist in the original objects. For example, both oxygen and hydrogen can fuel a flame, but if they are combined to form water, they have the new power of being able to extinguish the flame (Sayer, 1979). The presence of emergent powers and structures has particular implications for methods of analysis and the scale or unit of analysis. The common method of causal disaggregation in which 50 per cent of the variations in a dependent variable are accounted for by this independent variable, and 20 per cent by that one, simply cannot be applied when emergent powers are present. To continue the example, it is plainly ridiculous to explain the power of water to extinguish flame by calculating the percentage variance explained by oxygen atoms and the percentage explained by hydrogen atoms; the structure of water cannot be reduced to its constituent parts in this explanatory context.

It is presumed in positivist social science that by understanding individual behaviour it becomes possible to understand society, that is the notion of methodological individualism. This of course ignores that society is an open system with structures and emergent powers and therefore cannot be fully understood by merely disaggregating it into its component parts. Human activity must be seen in the context of structures which are produced by people but which have a life of their own with their own constraints and powers; structures are made up of people but their powers are irreducible to the individuals who make them up. Explaining human relationships at the individual level misses the role of structures in producing those mechanisms that maintain the world as it is; radical change is to be achieved by altering the structures and not by mediating the effects of mechanisms. While structures can be changed (for they are being continually reproduced by people) it is often not easy to do so.

The final aspect of Sayer's (1984) realist conception of social science to be examined here is the notion that while data and observations are not completely determined by theory, they are to some extent theory-laden. Positivism presumes that facts are out there in the real world waiting to be measured and used to test a theory, but Sayer argues that data are not neutral, 'pure' observation but to some extent formed by theory. As Saunders (1981, p. 280) puts it, 'the point is not simply that theory determines where we look, but that to some extent it governs what we find'. If this is the case, data alone cannot be used for the crucial tests of a theory, because, while testing a theory against data may reveal a poor fit, the definition of the phenomena involved is unlikely to be challenged. Moreover, because we can have causation without regularity, not finding evidence for a causal mechanism in a particular empirical check does not mean that the mechanism never operates. As Elson (1979, p. 163) has stated

> the question of when we have sufficiently grasped the real relations under investigation, when we know enough about them to proceed to practical action is not one that can ever be finally decided by an empirical test. It must always be a matter of judgement.

Chouinard et al. (1984) have argued that realist research is not any less strong in adopting such a viewpoint for positivist research actually operates like this in practice.

It can also be argued that theory and science itself is neither neutral nor value-free but embraces a particular ideology or way of looking at the world. Chalmers (1982) argues that all knowledge is conditioned by the society in which it is produced.

Thus, positivist science and positivist epidemiology are inextricably bound up with the capitalist view of the world. This is not to argue that researchers are colluding in a gigantic plot, but that the development of science is constrained, funded, and allowed by existing societal norms and structures. For example, as Moss (1982) has shown, the vast bulk of cancer research is at the cellular level and concentrates on what is the biological cause of cancer; for Navarro (1980a) these research priorities reflect class dominance for if research was promoted into carcinogens in the workplace this would be a direct threat to those groups which play a major role in funding institutions for cancer research. A striking example in the British context is the Health Promotion Research Trust which is funded by the tobacco companies to do fundamental research provided it does not involve cigarette consumption!

Habermas (1970) has argued that by defining an ever-growing number of problems as amenable to positivist scientific intervention, social issues are removed from critical scrutiny and depoliticised. At the same time, science legitimises the current pattern of the dominant and dominated. In this context, positivist scientists masquerade as humble seekers of the truth while in fact perpetuating and promoting a capitalist society. In particular an epidemiology that concentrates on individual level explanations and does not consider the basic structure of society as in any way problematic, mystifies the social origins of disease and legitimates the status quo.

Critical epidemiology

The majority of epidemiological work is concerned with developing a biological explanation of disease. Information is collected on individuals which is then aggregated to allow statistical analysis to reveal regular associations between certain 'risk' factors and the disease. However, unlike the purely biomedical model of disease where disease is explained as a biological abnormality which is internal to the body, the epidemiological approach does take account of external influences, but these social and environmental factors are included in a narrow and selective manner, usually by stopping at the level of the individual person. thus while the biomedical model has been extended backwards, it has not been extended far enough to include societal, structural variables.

Modifying Armstrong (1980) there is a biomedical model:

High serum ─────────→ Ischaemic
cholesterol   heart disease

in which treatment is achieved by medical intervention; a behavioural model:

Poor ─────────→ High serum ─────────→ Ischaemic
diet   cholesterol   heart disease

in which the natural course of action is exhortation and education to modify behaviour, and a structural model:

Capitalist   High   Ischaemic
mode of → Working → Poor → serum   → heart
production   class   diet   cholesterol   disease

in which explanation and action lies at the level of society. But from a critical position, this extension of the causal chain is not enough and explanation still remains too closely attached to the biomedical model. As Navarro (1980a) put it 'the addition of new causes – social and environmental – to old ones does not constitute a break with the bourgeois vision of science'. Following Davies and Roche (1980) we must ask three fundamental questions: 'what is to be explained?', 'what should the unit of analysis be?', and 'how will the explanation proceed?' These questions will allow us to discuss the limitations of epidemiology as currently practised and to develop a critical perspective.

What is to be explained?

Much of epidemiology does not regard the definition of disease as in any way problematic. Following positivist science, technical biomedical criteria are used to define disease in a reliable, universal fashion. But while disease can be seen as a clinical syndrome, it is not merely so, always so, and naturally so. From a critical perspective the clinical view fails to consider the social construction of illness (Chapter 1) in which disease is seen not as fixed and universal but rather as a concept which is moulded by changing power relations in society.

Smith's (1981) social construction of 'black lung' is a very clear illustration of how disease definition is related to social and economic change. The first medical construction of the disease developed in the Pennsylvanian coalfields during the 1860s and 70s when the 'scientific' view of medicine had not yet achieved hegemony (Chapter 1) and miners were beginning to organise their own health care through mutual-aid associations. There was a growing recognition of miners' respiratory problems which were linked to breathing dust and gas at work. However, with the development of the Appalachian coalfield from the beginning of this century, the disease, or at least its definition, disappeared. Health care was controlled by the coal-owners and delivered via the company doctor, and while 'miners' asthma' was recognised, such conditions were considered natural and inevitable and not requiring scientific investigation. The mine itself was not seen as an inherently unhealthy place to work and the explanation of ill-health was couched in terms of lifestyle, particularly drinking.

Such views remained dominant until recently, and this was despite a minority of physicians (usually funded by the miners' union) who argued that black lung is an occupational disease. However, in the late 1960s the older generation of miners, who were facing retirement on relatively low wages, perceived that the industry was prosperous and started a campaign for compensation. In West Virginia a strike of 40,000 workers in 1969 compelled the state legislature to pass a compensation bill and by the end of the year Congress had passed a similar bill. Compensation was based on the diagnosis of a particular clinical entity called coal miners' pneumoconiosis (CMP) which was recognised by pathological changes of the lung, visible on x-ray. In this advanced form, the disease had a prevalence of 3 per cent among miners, while simple CMP, which was more widespread, remained attributed to lifestyle factors (especially smoking) and was non-compensatable.

This disease definition did not satisfy the mining communities for their knowledge of the disease did not coincide with the physicians' clinical entity. Miners with advanced CMP received compensation even if experiencing little disability, while others who were severely disabled did not because the disease remained unconfirmed by x-ray. The miners set out to redefine the disease not in terms of the individualised clinical diagnosis, but in terms of the experience that affected them as a group, working in a mine. They wished to reject the notion of a specific disease entity (advanced CMP) and specific causal agent (dust) and replace them with a general illness and disability produced by the total environment of the workplace, miners being compensated according to the number of years that they had worked. While a Federal legislative motion was defeated in 1976, by 1978 the medical requirements were changed so that ex-miners who had worked a substantial period underground were receiving compensation.

Table 9.1   A social classification of disease

	Related risks	*Resulting diseases*
How we produce	the use of various chemical and other toxic materials in mining, industry and agriculture.	occupational diseases and injuries e.g. asbestos diseases; numerous skin, lung, bladder and other cancers; radiation diseases.
	the careless use of capital-intensive productive methods.	industrial injuries and deaths; capital-labour substitution leading to unemployment and thereby to anxiety states, depression, alcoholism and the cigarette diseases such as bronchitis and lung cancer.
	increased use of human beings in passive, repetitive or machine-like roles.	obesity; industrial accidents; cigarette diseases; alcoholism; boredom and stress-related diseases and conditions.
	industrial pollution.	affects not only workers but also rest of our society, and other societies (e.g. lead pollution locally; sulphur dioxide and other pollution problems in Norway which are created in the UK).
How much we produce	pressures leading to damaging rapidity in the production process.	increased risks of accidents e.g. diving accidents; 'executive stress' leading to cigarette diseases, road accidents, alcoholism and over-eating (obesity).
	pressures related to frenzied and damaging marketing.	results in various conditions as indicated in 'executive stress' above and where domestic life significantly disrupted, to increased risk of mental illness.
	pressures to utilise new and inadequately tested forms of energy inputs.	nuclear power radiation hazards and deaths.

continued

Table 9.1 continued

	Related risks	*Resulting diseases*
	pressures to adopt damaging levels of goods, transport and labour mobility.	road traffic accidents affecting not only lorries but involving cars, coaches and buses; disrupted domestic life as above.
What we consume	the consumption of disease and accident linked products.	cigarette diseases; dental caries and the sweets and chocolate (and other sugar) diseases – including obesity and some diabetes; road accidents secondary to alcohol, hypnotic or tranquiliser consumption; poisoning from weed-killers and pesticides; aerosol sprays.
	the consumption of nutritionally deficient products.	refined flour and sugar i.e. fibre-deficient carbohydrates leading to diverticulitis, some cancer of the colon, etc.
	waste pollution hazards.	poisoning from heavy metal or other chemical and radioactive wastes e.g. to workers on waste tips or to others through water contamination etc.
How much we consume	pressures to consume more in an absolute sense e.g. advertising of the form 'eat more, drink more'.	advertising which contributes to over-eating and therefore to our major nutritional problem, obesity and associated diseases e.g. heart disease.
	pressures to replace/update consumer durables and other products at an ever-increasing pace (including 'planned obsolescence').	anxiety states and depression which arise from financial and other pressures to 'keep up with the Jones's'.
How we share	chronic persistence of shortages and inadequacies in housing and basic amenities despite ever	hypothermia, respiratory and gastro-intestinal conditons which arise from grossly inadequate housing and

Related risks	Resulting diseases
increasing levels of productive output and energy consumption.	sanitation, overcrowding and homelessness; accidents to children from the lack of safe and attractive play facilities e.g. the special problems of high-rise flats.
chronic problems of unemployment and poverty amongst specific sub-groups of the population.	many single-parent families; immigrants living in over-crowded and decaying urban areas with high unemployment; middle-aged and older unskilled workers whose physical fitness has been lost; agricultural workers who have little or no land of their own for vegetables, chickens etc. Generally, poverty and unemployment effects such as malnutrition or subnutrition, anxiety and depression and associated cigarette diseases, methyl alcohol drinking etc.

Source: Draper *et al.* (1977)

As this example makes clear, there is more to disease definition than simply a clinical perspective; indeed if epidemiology merely accepts biomedical diagnosis then 'society is epistemologically eliminated as an element in the etiology of disease' (Renaud, 1978). While medical science views disease as a deviation from a statistical norm, a critical approach views people as 'unhealthy when subjected to a social assault' (Berliner and Salmon, 1979). From this perspective it makes little sense to define disease in an individualistic way; diseases could be defined by their common social origins. The beginning of such a classification is given in Table 9.1, while Scott-Samuel (1981) goes further in suggesting that it would be appropriate not to have specialists studying specific organ systems but specialists in the illness of alienation, racism and inequality of opportunity. Thus, a critical epidemiology must adopt a viewpoint that considers dynamic relationships between social people rather than static relationships between natural objects. Diseases are not natural immutable categories waiting to be discovered. They must be examined as contigent and historically specific struggles over who, and for what purpose, provides the definition and makes the diagnosis.

What should the unit of analysis be?

While epidemiology makes use of aggregates, it is the individual that is the basic unit of analysis. For Kleinbaum *et al.* (1982) 'the level of primary biological interest to the epidemiologist is the individual'. Our criticism of this takes two forms: firstly, the aggregations that are used may be unsuitable for causal analysis and secondly, an explanation should not automatically begin with the individual, then ascend to the aggregate and then descend to the biological, but rather individuals must be seen in their social context. A critical epidemiology must be based on a macro- and not a micro-causality.

Aggregates of individuals are frequently used in epidemiology in the hope of revealing common properties of the diseased as compared to the non-diseased. Thus, groups such as 'men over 65' and Japanese women' are formed so that common risk factors (the regularities) can be identified. From a realist perspective these groups are 'taxonomic collectives' (Harré, 1979), in that they share similar (formal) attributes but they may not actually interact or have functioning, actual connections. The individuals may only exist as a group in the mind of the classifier. While such groupings may be adequate for description, they are unlikely to be suitable for explanation, for as the groups are internally heterogeneous they cannot be expected to behave consistently, that is they are distributively unreliable (Harré, 1979). As Sayer (1984) has noted, in the natural sciences it is often possible in closed systems to classify objects simultaneously in terms of taxonomy (shared attributes) and causality (functioning relationships). Under such conditions, statistical analysis can be used to assess the relative contribution of separate and additive causes, but when one heterogeneous group is correlated with another there may be little chance of revealing causal mechanisms.

To take an example, epidemiology frequently uses the taxonomic collective of social class whereby individuals are grouped by their occupation and other shared attributes. Such classes are qualitatively heterogeneous and distributively unreliable with the Registrar General's socio-economic Class II (employers and managers) containing individuals ranging from a corner-store shopkeeper, to the head of ICI, to a mid-Wales farm manager. Such occupational aggregations have little meaning for they do not correspond to real social situations. Such groups cannot be expected to behave in anything like a consistent manner and it is not surprising that epidemiology does not use social class as an explanation but only as an ordering framework to search for more 'specific factors'; a point we shall return to later.

The dominant approach in epidemiology is reductionist. The researcher tries to explain properties of complex wholes in terms of the parts or units of which they are composed and this is closely allied with the political position that the individual has priority over the collective. Accordingly, society is nothing more than the sum of individuals that make it up, while individuals are nothing more than the sum of the biological parts that make them up. But this ignores the notion that reality is stratified and that structures can exist with emergent powers which are irreducible to their constituent parts. We need the macro-gaze of social structure and not just the micro-gaze of biological individualism to provide an adequate explanation of the social phenomenon of ill-health. To take an example, the health of different workers may not be solely explained by their genetic susceptibility (biology) to toxins but by the nature of the organisation of the production process (society) that exposes workers to such dangers. Our aim should be not to reduce the social to the individual and explain by biological variations, but the reverse, that is to place and relate the individual and their biology to the social and political context. The irreducible unit of analysis becomes the relations and structures in which individuals are located.

How will the explanation proceed?

There are essentially two different methods of developing explanations of social phenomena; one can be called the extensive, and the other, the intensive. Extensive designs are able to discover how extensive events are, while intensive designs aim to discover the underlying causes of events (see Figure 9.1). It is the former methodology that is dominant in contemporary epidemiology. Almost as a matter of routine, the researcher does an extensive study based on a large-scale survey of the population using formal questionnaires, which are then analysed statistically. The aim is to identify distinguishing features that those with a disease have in common. This, of course, is the approach that has been considered in Chapters 2 and 3. The intensive design is different both in its aims and its methods and is sometimes known as the ethnographic approach. Typically, the research will concentrate on a few individuals or groups using long interactive interviews and not standardised questionnaires. The people in the study are not objects about which standard measurements are to be collected but subjects who are active agents in particular causal contexts. The intensive approach is usually considered worthless by the positivist epidemiologist, being dismissed as mere anecdotal evidence. The purpose of this section is to discuss some of the limitations of the extensive approach, to illustrate the intensive approach, and to argue that both methodologies are required for a full understanding of the social world.

The extensive approach faces a number of problems and, while we have already considered the difficulties of distributive unreliability and the theory-laden nature of data, the discussion here covers the inability of quantitative methods to reveal causal mechanisms and to accommodate structure, and the contention that these methods reify and compartmentalise the social world. Sayer (1984) and Pawson (1978) have argued that the use of causal models of the multiple regression or Simon-Blalock type is problematic because they cannot distinguish between formal, accidental associations and causal, functional ones. These techniques reveal how much variation variables have in common, but they do not necessarily deal with causal mechanisms for the 'forcing' or 'producing' element is not reflected in the mathematics. When we criticised Gardner (1973) in Chapter 3 for including latitude and longitude in his statistically successful model, we were making the same point. Even if we do find a non-spurious association, statistical methods say nothing about what is producing the regularities. Moreover, because these models cannot incorporate necessary, functional relationships, their calibrations merely represent temporary, superficial regularities occurring in open systems; it is not surprising that their predictions are often poor. Pawson contends that the high unexplained variation for such models represents not omitted variables (as usually assumed) but changes that are occurring in open, human systems. For him these techniques when used in social science do not measure law-like regularities but

> the description of momentary regularities literally plucked out of a much larger explantory structure . . . , a snapshot of the interrelationships . . . in the exposure time of the observation.

They are instantial relationships because there is no reason to expect structural relationships to be fixed. (Pawson, 1978, p. 624)

Reification is the action of turning a changing, changeable social process or relationship between people into a fixed, immutable relationship between things. Reification is inherent in much quantitative analysis for if objects are to be measured, they must be qualitatively invariant and not change their nature. In essence, social processes are 'frozen' and turned into 'things', existing relationships take on the appearance of being 'natural' and beyond human control, and attention is deflected from social processes to individual attributes. To illustrate this legerdemain we will consider the example of class. For the marxist, class is a relationship consisting of conflicting interests and powers reflecting the necessary and internal relationship between capital and wage-labour. Class is seen as a social position occupied by individuals in a structure whereby one class exploits another. Some elements of this position have been adopted by the Black Report (DHSS, 1980, p. 194) in its explanation of health inequalities: 'there is undoubtedly much which cannot be understood in terms of specific factors, but only in terms of the more diffuse consequences of class structure: poverty, working conditions, and deprivation in its various forms'.

While we will return to this materialist position later, it is crucial to grasp that for the marxist, class differences are not the pattern of inequalities but the explanation of such inequalities. In orthodox accounts, however, the social process of class exploitation becomes reified into the categories of social class whereby individuals are classified on the basis of a number of shared, but externally related, contingent attributes such as similar occupation, income, education and standing. Class is not an explanation but an ordering framework for describing health inequalities which is then used to search for an explanation. According to Susser (1973, p. 51), 'we do not find it comfortable to say that circumstances of social class are a direct cause of the death of an individual . . . additional factors must intervene between social positions and causes of death'. Such factors are usually conceived at the individual level and in particular genetic and behavioural lifestyle explanations are developed to account for the observed inequalities. For example, Reid *et al.* (1979) claimed that the higher incidence of uterine cancer among working-class women was due to a factor carried in the sperm of working-class males which had a simpler, more repetitive structure to its DNA than did middle-class sperm! While most epidemiology does not go this far, it is common for social processes to disappear as categories are produced and variables are measured, and a biological explanation developed.

An important aspect of reification is the compartmentalisation it involves. The social world is seen not only as things but as compartmentalised and separate things. For example, it is assumed that it is not the process of work that causes disease but rather certain factors in the work environment and that by using causal disaggregation it is possible to assess the independent influence (percentage variance explained) of each of these different factors. This presumes, of course, that causes can be varied independently of each other and that a factor when isolated operates identically when all conditions vary simultaneously. But if we accept that social actions represent structures of internally related elements (with emergent powers) then the 'control' or removal of these elements will fundamentally change the relationship; internal elements cannot survive extraction from a structure. Like Humpty Dumpty, once reality is broken up into externally related parts it becomes impossible to put it together again. Interconnection, what the realist researcher is looking for, becomes a severe technical problem for the positivist researcher in the form of multicollinearity (Chapter 3). There is again a strong political element here and Day (1980) has characterised such compartmentalisation as capitalism's way of viewing the world. Thus, by concentrating on the superficial associations between separated 'things' in the social and physical environment, we miss the underlying power relations that are shaping and structuring these things; existing relationships go unchallenged by default. Compartmentalisation is a major failing of orthodox epidemiology and this has been recognised by epidemiologists themselves, as Bridford, in a governmental report (1978, p. 3), concludes: 'a major fault in most cancer research in the western world is that most cancer research has been based on looking for a single or multiple cause, ignoring the inter-relationships among the assumed causes'. In short, extensive research and quantitative analysis has particular problems with structures, interdependence, emergence and distributive unreliability and these are the very areas in which a qualitative, intensive analysis comes into its own.

The basic difference between extensive and intensive approaches is that the former examines a few properties of a large number of individuals, while the latter considers a large number of properties of a few individuals. But as Willis (1980,

p. 89) writes: 'if the techniques of qualitative methodology make a decisive break from the quantitative, the way in which they are usually applied makes a secret compact with positivism to preserve the subject finally as an object'. That is, there should be also an important change in the social relationship between the researcher and the respondent with the two entering into interactive, non-hierarchical conversation (Oakley, 1981) rather than the objective interrogation of the extensive approach. Paraphrasing Harré (1979, p. 118), the interviewee is no longer an object of contempt to be probed, but a person to be co-operated with. Our attention here is not to discuss the methods and techniques of intensive research (Open University, 1979) but to illustrate with two examples the potential of this neglected approach.

Our first example is the work of Cornwell (1984) who provides accounts of health and illness in the working-class community of Bethnal Green in London. Believing that lay theories of health are not inferior, partial or distorted versions of medical truths but meanings used by actors making sense of their experiences of life, she employed the ethnographic method to try to analyse the life of a person in its entirety, in work, at home, and in the context of family and community. This involved intensive interviewing of a small number of people; the aim was to get people to tell their own story in their own way. The striking thing to come out of this work is that the initial accounts differed quite markedly from later interviews. The initial responses were seen as 'polite', morally correct replies, but when the respondents began to tell stories about themselves and their friends and families a very different picture emerged. Cornwell theorised that individuals hold two accounts of their attitudes and life events; one is for public consumption, the other is the private account. For example, in a public account a respondent would not even mention a major illness involving hospitalisation for he or she did not want to be classed as a 'moaner'. It was not the case of forgetfulness or 'recall-bias' but of deliberately choosing not to reveal. In the public account, working hard equates with a healthy life but in private they realised that work was crippling and disabling. The right attitude to be adopted in public was to be cheerful and hard-working; a great deal of emphasis was attached to the individual's ability and role in determining their own lot in life. In private, however, they realised that there was in fact little room to take individual action and that their lives were constrained by the practicalities and meanings of work and family life. This research well illustrates the ability of the intensive approach to penetrate beneath superficial appearances.

Our second example of the intensive approach comes from eight interviews conducted by Mitchell (1984) with active socialists and trade unionists. The interviews covered the topic of how health is determined by life experiences. She found a complex interconnectedness between physical and emotional aspects of life. In these people's stories it was not only physical agents such as using toxic chemicals at work, living in damp housing in areas of high pollution, and being forced to eat a poor diet due to inadequate income that were causal but it was also the loneliness, the powerlessness, the monotony, the demoralisation and the alienation of everyday life that were 'grinding them down'. Based on these interviews, Mitchell was able to construct eight key dimensions of what damages our health and she argues that these dimensions can all be interrelated for an individual and 'are far more complex and subtle than the crude correlations of epidemiology would have us suspect' (p. 101). The eight dimensions she identifies are:

> the degree of exposure to a hazardous environment both inside and outside of work;
> the extent to which work inside and outside the home is exhausting and the time and space that is available for recuperation;
> the amount of money that is available;
> the extent and severity of worries;
> whether an individual can feel hopeful and optimistic;
> whether an individual is powerless;
> whether an individual is bored and alienated;
> whether an individual is lonely and not loved.

Mitchell sees these dimensions as not being determined exclusively by the individual's attitude to life but rather by the constraints and expectations that are imposed by society.

A materialist epidemiology

Although we have now considered the matter and method of a critical epidemiology, we need to see how health and ill-health can be related to the broader social structure of society, that is to develop an 'Historical Materialist Epidemiology'. While this approach has its antecedents in the work of Engels in Manchester and Virchow in Silesia in the last century it is only in the last few years that interest has been renewed. The basic feature of a materialist approach is the emphasis on social relations between

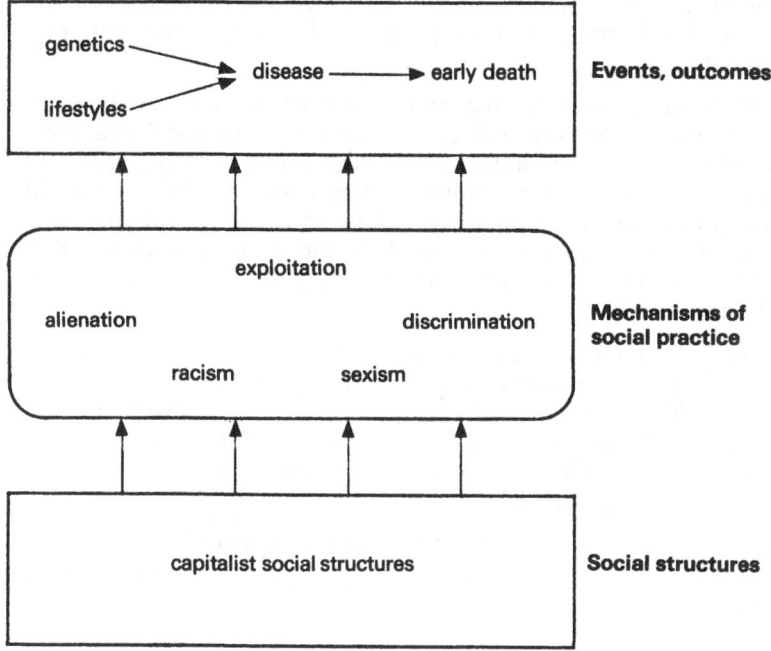

Figure 9.2 Understanding ill-health

groups which are defined primarily in terms of the ownership or effective control of the material conditions in which they live. This materialist approach also requires historical specificity in that societal relationships are not viewed as yet more static factors but as dynamic processes. Figure 9.2 contrasts the empiricist, individual-level explanation which is portrayed 'horizontally' with the realist, materialist explanation which is shown 'vertically'. In the empiricist strategy, the event of early death is associated and accounted for by earlier events in the life of the individual. Cause and explanation is at the individual level, people develop disease either due to genetic inheritance, about which little can be done, or due to self-damaging lifestyles. Crombie (1984, p. 15) manages to combine these explanations to provide an utterly reified, despairing view:

if differential ability to cope with life and its problems lies at the basis of most if not all of the other (secondary) factors (such as education, smoking, heavy drinking, socio-economic and occupational status, dietary and environmental inadequacy), and if coping has a large genetic element, then disparity, if not relative inequality, will always be the order of the day.

In marked contrast, the materialist attempts to go beneath these surface events to seek causes in the processes and structures of capitalist social organisation. While it would be silly to contend that individuals have no choice in their actions, what is required from a materialist epidemiology is how and in what ways personal choice is constrained by the structure and power relations within capitalism. According to Mitchell (1984, p. 105), in the search for a better health

> we have to look at the way work is organised, the hours of work, the division of labour, how caring for children, sick and very elderly people, making food and doing housework is organised, how houses are built, how cities are planned and who decides, who is in control and where the power lies.

While there is still a pitifully small literature on such a materialist approach, we have tried to exemplify the approach by considering three examples: cigarette smoking, hypertension and teenage drinking.

Cigarette consumption

Extensive, empirical epidemiological research has found regular associations between cigarette consumption and the development of disease, particularly cancer of the lung. An explanation that is often put forward for this consumption is 'lifestyle factors' and individual irresponsibility, for smoking is usually seen as a matter of free choice. But such an explanation ignores the social pressures, commercial inducements and the role of capitalism and the state in creating and promoting such consumption. Cigarette smoking can be seen as the consumption of a habit-forming drug which is produced and distributed in a social context in which it is more encouraged than discouraged. The government can be seen in a position of dependence on the tobacco companies, and while it receives over £4,000 million per year in tax revenue, only one million is spent by the DHSS on prevention (as discussed in Chapter 7); this mostly takes the narrow form of health education and 'improve yourself'. Indeed, there are Queen's Awards for cigarette exports and the Department of Industry gives grants and subsidies to the manufacturer – a classic example of the state's fundamental role in maintaining the social and economic conditions of capitalism.

The lifestyle approach not only misses the role of capital and the state in the reproduction of smoking but also excludes the societal pressures that create the need to seek solace in such consumption. Jacobson (1981), in her discussion of the recent and rapid rise in the number of female smokers, argues that women have turned to smoking as a safety valve, as a means of expressing frustration and anxiety and coping with the realities of life. In Graham's (1984, p. 173) telling phrase, 'we begin to see the responsibility of irresponsible behaviour'. Moreover, Stark (1982, p. 66) has documented how smoking increases as women enter the job market and as school-leavers become unemployed, and the individual-based explanation also ignores the growing evidence for the interaction between cigarette smoking and hazards in the general environment and workplace. For example, there is evidence (Chapter 2) that bronchitis is produced by the interaction of smoking and air pollution while Sterling (1978) has asked the important question 'does smoking kill workers or working kill smokers?', finding that it is occupation and not smoking that appears to be the major cause of lung cancer. Finally, Phillips and Bruhn (1981) have found that social support in differing communities has an important role in mediating the link between smoking and coronary heart disease.

Hypertension

Our second example necessitates a somewhat longer account. Our aim is to show that while extensive empirical research has found certain regularities associated with hypertension that are difficult to explain by an individual-level explanation, a materialist approach appears to provide a more coherent explanation. Hypertension is high blood pressure and is seen by medicine as a serious organic disease which can have dire consequences in stroke, heart attack and kidney failure. The current approach to the problem, especially as promoted by the American National Heart, Blood and Lung Institute, is a particular form of behaviour modification, namely, 'keep taking the tablets'. The approach is well described by Syme (1984, p. 234):

> in treatment, we ask that people recognise the need for
> medical help . . . and they follow that advice. . . .
> Hypertension is a chronic condition requiring lifetime
> treatment. Treatment for hypertension requires that patients

be convinced to embark upon a long-term course of drug treatment even though they feel well.

From this medical viewpoint, the major problem of dealing with hypertensives is their failure to comply with this drug regime; indeed, the variations in outcome between the social classes have been ascribed to class differences in following medical authority. But a general unwillingness to comply is not surprising when studies find that nearly 70 per cent of patients receiving treatment report side-effects, with over 20 per cent experiencing mental depression and sexual impotence; many of these side-effects occur among people who were previously symptomless.

The current medical emphasis on treatment rather than prevention is due to the lack of understanding of the causes of the disease. Medical science has concentrated its efforts at the individual level, with the clinical approach seeking a specific abnormality in a body organ (none has been found) and epidemiology trying to identify individual risk factors that lead to the disease. Indeed, in a striking example of compartmentalisation, the majority of epidemiological research has cast hypertension as risk factor for heart disease, and has therefore taken it as a 'given', not to be explained but to be used as an explanation of other diseases. However, in terms of hypertension, three variables have been suggested as causal factors: salt, age, and genetic inheritance. The evidence on salt is not very strong, and indeed experimental salt doses given to human subjects have not induced higher blood pressure. It has been generally accepted in the West that increasing age means higher blood pressure but recent surveys have shown that not all Americans experience higher pressure as they grow older, while international studies have found societies where this relationship does not generally occur. Moreover, medical science has found it difficult to explain how males in the USA have higher rates than females between the ages of 25 to 45 years, but for the oldest age-groups this is reversed. The genetic explanation is based on the differences between racial groups with, for example, blacks in the USA having on average a rate that is over twice as high as whites. However, there are again rather intriguing findings in that there is no difference between blacks and whites until after the age of 24, and blacks in tribal Africa do not show elevated levels. Syme and Torfs (1978, p. 43) in their review of the literature conclude that 'research has reached a dead end . . . it is puzzling that we have failed to discern systematic and patterned relationships among these variables'. Positivist, empirical epidemiology cannot explain the regularities that it has found.

From a critical viewpoint, we must question the self-evident facts of hypertension and attempt to understand the social production of this particular form of ill-health. Schnall and Kern (1982) have followed this approach by asking why high blood pressure is defined in this way, and by trying to develop an explanation that uses the basic variables of age, sex and race not as physical, reified attributes but as social relationships.

The clinical definition of hypertension is remarkably clear: essential hypertension is chronic high blood pressure above 140/90, that is a pressure of 140 systolic, the highest pressure when blood is pumped out of the heart, and over 90 diastolic, the resting pressure between pulses. Accordingly, a very large number of people are classified as hypertensive and it is estimated that some 50 million American adults have the disease. The definition is essentially, however, an arbitrary one and epidemiological studies have not found a natural dividing line between those with and those without the disease. Moreover, blood pressure has been found to vary continuously, increasing during arguments and exercise and decreasing in sleep and relaxation. Consequently, medical practitioners have urged caution with the diagnosis of hypertension, and Page (1979, p. 1897) has warned that

> many subjects receive a misdiagnosis by careless screening. . . . It is wicked to label a person hypertensive only because one or two measurements show mild systolic hypertension. This may mean a lifetime of taking a drug in useless quantities and a lifetime of anxiety.

It may be wicked but it is also good business; in 1974 the drug company Merck, Sharp and Dohme were advising that 'no doctors should miss an opportunity to record the blood pressure of a young or middle-aged patient who is a rare visitor to the surgery'; if hypertension is found then *Aldomet* (methyldopa) should be prescribed. Such advertising and forming of opinion has undoubtedly contributed to what can be called an epidemic of hypertension and associated prescribing. Yet according to Melville and Johnston (1982), there is no scientific evidence for any benefit of such drugs in treating mild hypertension. The apparently sharp scientific fact is once again seen to be a subjective judgement that generates a considerable number of consumers for the medical profession and a great deal of business for the drug companies.

While the extent of hypertension may have been over-emphasised by defining it in this way, there is little doubt that there is a real problem; there is strong evidence that elevated blood pressure leads to premature death. From the materialist viewpoint, an explanation is required that links society with the production of this condition and this has been achieved by using the mediating concept of 'stress'. Orthodox epidemiology has also used the notion of stress, but in the reified form of the A and B personality. People with an A personality are supposed to have a disposition to hypertension caused by their ruthless, ambitious, driving personality and obsession with work and urgency. Such individuals are illness-prone because of the way they choose, or are forced by their personality, to live; the broader social system is taken as given and therefore as unproblematic. Prevention means individuals changing their behaviour and bio-feedback techniques and relaxation have been suggested as appropriate forms of treatment. In a materialist epidemiology, stress is seen not as a trait of individual personality but as induced by the conflicting roles and demands that society makes upon individuals.

Eyer (1975, 1984), on the basis of detailed epidemiological evidence, identifies hypertension as a disease of capitalist society. He compares the relationship between age and blood pressure among different types of society; the increase of blood pressure with age, taken as normal in our society, is less steep in the agricultural communities and non-existent in hunter-gatherers. This, of course, does not mean traditional tribal societies are disease-free, but that illness is expressed in different ways such as famine, infectious disease and infant mortality. For Eyer, the high blood pressures found in capitalist society can be explained by

1) the disruption of social communities and the transfer of power from a kin-based extended family to state or private capitalists;
2) the rise of hierarchically-controlled, time-pressured work that characterises the social relations of capitalist production.

Such an explanation is supported by Karasek's (1981) findings that blood pressure was highest for those men working with close supervision in time-pressured work who had little control over their working conditions. According to such an explanation, the observed increase with age is due to the cumulative nature of work-induced stress, but it must not be forgotten that old age in our society is a social position which brings its own stresses. In particular, there is a general lack of prestige and privilege attached to the elderly, and many of them live a lonely life based on low income.

As discussed earlier, the major difference in prevalence of hypertension between blacks and whites in the USA has been taken by epidemiology as strong evidence for the importance of genetic inheritance. Recent gene studies, however, have shown that there is no genetic basis for distinguishing between races and there is greater genetic variation within races than between races (Rose *et al.*, 1984). The notion that races are biologically different is important, however, to the ideology of capitalism for, in what has been called 'super-exploitation', it is deemed natural and appropriate that certain ethnic groups receive very low wages, while by the 'class effect of racism', ethnic differences are used to divide the working class, thereby limiting the power of what could be a united front committed to change (Cooper, 1983). Consequently, while colour as race may make little sense biologically, colour as ethnicity is very important, for ethnic inequalities reflect the underlying class structure of capitalism. It is well documented that blacks in general have lower pay, more dangerous jobs, higher levels of unemployment and attendant poorer diets and housing, and it is not surprising therefore that blacks are subjected to higher levels of stress and experience higher blood pressure than whites. According to this argument the finding that there is no difference between whites and blacks until their mid-twenties and a widening difference thereafter is explained by blacks of this age being forced to recognise the reality of their social situation. Harburg (1973) in a study of blacks and whites in Detroit found a greater amount of 'repressed anger' in those blacks with the highest levels of blood pressure. He argued that blacks faced with a bleak future of low pay, uncertain employment and racial discrimination 'internalised' their anger because of their socialisation and their fears of expressing their rage, the result being elevated blood pressure. Syme and Torfs (1978) in their review note that the differences between black and white blood pressure is substantially diminished when socio-economic variables are statistically controlled. But such studies miss the point for race is socio-economic and not only biological; race cannot be separated from other socio-economic variables. Race in the form of discrimination and exploitation is not, therefore, an attribute but a causal relationship. Recently, racialism has been proposed to account for the high level of heart-disease mortality experienced by British

Asians despite their low consumption of saturated fat, cholesterol and cigarettes (Russell, 1986).

In a similar fashion, it is possible to examine the differences between men and women not as biological or sex differences but as social or gender ones. It is a puzzling finding for epidemiology that men have higher rates for age-groups below 45 years but women have the higher rates thereafter. Schnall and Kern (1982) suggest that a plausible explanation is the decreasing social and cultural value with which society views the ageing woman. Indeed, they reference research that shows that women who devote their earlier lives exclusively to unpaid domestic labour are more likely to become ill after their children are grown and leave home. But they also argue that these hypotheses must remain speculative because of the dearth of research on women that characterises orthodox epidemiology.

While the individual approach of epidemiology has resulted in a dead end and puzzling findings, the materialist perspective makes intuitive sense. Moreover, this perspective suggests areas for future research. For example, Eyer's hypotheses could be examined by comparing workplaces run on autocratic lines with those with collective decision making and organisation. But if the materialist position is really to mean anything, research has to consider simultaneously conditions in the workplace and the home and the way that these interconnect with social positions based on age, gender and ethnicity. While such research has not really begun, the approach adopted here at least casts doubt on the medical reliance on drug treatment; if high blood pressure is a form of 'release' for an individual under social stress, 'attempts to keep blood pressure down through indefinite medication, without attempting to deal with its causes, is like trying to cool a room by chilling a thermometer' (Melville and Johnston, 1982, p. 80).

Teenage drinking

As our third example, we examine a particular aspect of lifestyle that has been increasingly conceived, in recent years, as a health problem: alcohol consumption. We have chosen this example because the approach taken by one researcher, Dorn (1980, 1983), is the closest illustration of the critical approach that we have been able to find. In order to appreciate the distinctiveness of Dorn's approach, it is appropriate to examine, albeit briefly, two examples of the type of approach that he roundly criticises. The first example is by McGuinness (1979) who uses a time-series analysis to relate alcohol consumption (1956–75) to a set of social and economic factors such as average income, number of licensed premises, expenditure on alcohol advertising. A regression-type model is then used to predict what would happen if the independent variables are changed. Dorn (1983, p. 262) is trenchant:

His statistical analysis assumes that the UK total population may be treated as one aggregate of individuals not clustering into specific groups (class, gender, etc.) that may react in different ways; that unwaged and dependent persons (e.g. housewives) respond to changes in social conditions in the same way as waged persons . . . that such relationships do not take different forms within the sociologically arbitrary period 1956–75 and that such relationships operate similarly in the 1980's.

The second example of the epidemiological approach is the 'distribution of consumption' model of alcohol problems which has received widespread support from, among others, the WHO. The model assumes that the adverse consequences of drinking are directly related to the amount consumed which is directly related to the cost of consumption. For Dorn (1983, p. 250) the model suffers from several inadequacies:

atomisation of the population, neglect of social class and other divisions and associated cultures and practices, neglect of the social forms of drinking practices within specific cultures, neglect of problems inherent in using medical statistics. . . . It nevertheless retains considerable ideological importance because it generates data that is used to legitimise advocacy of restrictions on alcohol production.

Dorn's (1983) study is very different and consists of two interrelated parts. The first consists of an historical investigation of the emergence of drink-related problems. He concludes that there have been 'moral panics' and efforts by the state to control drinking at certain times not necessarily because of increased consumption but as an expression of the need for social discipline during periods of general social and economic crisis. He recognises the post-war redefinition of alcohol consumption in medical terms and argues that health education as a means of inculcating self-discipline in the individual has failed to reduce youthful drinking. He attributes this failure to a concentration on the individual and the quantity of drink, and the disregard of the social context and the qualitative form that drinking takes. In the second part, he researches this form in a specific setting (a declining inner city London borough) for a particular social group (service-sector youth).

Dorn (1980) argues that it is impossible to lump all teenagers (and certainly not all adults) as a 'taxonomic collective'. Teenage drinking can only be understood in terms of teenagers' roles and positions in society. In particular, he contends that their 'recreational' behaviour must be seen in the context of the class structure, sexual divisions and traditions of the local economy. Following a study of the local labour market, talks with employers and career teachers, and interviews with teenagers, he used observational and participant research in the street and in one pub to study one qualitatively heterogeneous group intensively. This particular group consisted of upper working-class whites (both males and females) who worked in the service sector; other groups such as blacks, lower-class teenagers from poorer estates, and middle-class children who tended to be still at school, had different cultures and different drinking patterns. For the service-sector group, drinking took the form of 'standing their round' with each person, irrespective of age, gender and financial circumstances, striving to take his or her turn at buying the drinks. Dorn interprets this form of consumption as both a collective strategy for dealing with the difficulties the individuals were experiencing in the transition from school to work, and as a means of showing their independence and (especially for the girls) their equality. For him 'learning to drink is also learning to labour and vice-versa', the round being a collective and public affirmation of the particular social relations facilitated by service-sector employment: socio-economic independence, sociability, and (temporary) sex equality. Most importantly, Dorn concludes that the traditional approach of health education, that of prevention by promoting individual responsibility and self-control, was largely irrelevant for this group, for drinking was a means of asserting and celebrating such independence.

From this research, Dorn constructs a materialist model of health (Figure 9.3) in which there are two major links between a social group's socio-economic circumstances and health. Firstly, there are direct effects and this entails physical conditions (such as industrial noise and toxins) and social relations (such as working with very little control over the job). Secondly, there are indirect effects whereby social groups produce particular local cultures which, in turn, shape 'leisure' activities such as smoking and drinking. Such a model clearly places an individual within a wider social context of material conditions of work and the home,

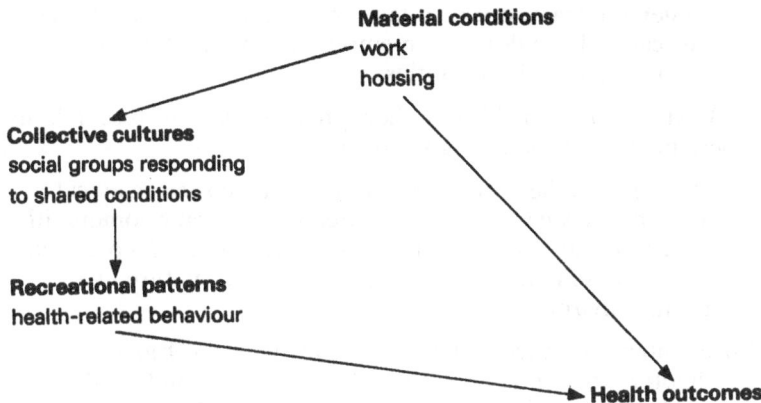

Figure 9.3 A model for a materialist epidemiology
Source: Dorn (1980)

and at the same time integrates cultural and materialist explanations which have often been seen as competing explanations (DHSS, 1980, Chapter 6).

And geography?

Explicit in Dorn's work is the importance of geography and, moreover, space in terms of 'locality' is seen as much more than just an organising framework for recognising regular associations. He writes (1983, p. 128):

> Since areas in different regions of the country differ considerably in their economic, demographic, social and cultural histories, it follows that contemporary cultures may differ from area to area . . . what are required are investigations into specific areas.

And again (1983, p. 134):

> In the first place, the basic material conditions . . . of life in a locality may influence the type and amount of wage work that comes to be performed there, and this affects the development of relations in and around the labour market . . . in the second place, social classes and national and state agencies (such as schools) respond to the conditions in which they find themselves and these may differ for historical reasons from locality to locality . . . one has to examine how local variations in material circumstances and class and state responses, fill out

the general frameworks of labour market and sexual divisions. This can only be done in an empirical manner, by looking closely at particular localities.

Blaxter (1985, p. 371) in her prescription for research into 'socio-medical' problems goes further:

attitudes and the social environment create socio-medical problems in ways which are particular to specific communities . . . any programme directed at the alleviation of a problem must be based on a knowledge of its specific nature in a specific environment.

Such statements have a clear resonance with the framework used in Chapter 7 to explain health care inequalities, and what is being called the 'new regional geography' (Massey, 1984). In the latter, the argument is that while there are general tendencies or causal processes in capitalist societies, they do not operate in isolation from each other and from the previous operation of such processes. The result is variety and uniqueness with particular local combinations and developments of the tendencies. Capitalism may create and foster such geographical variations for its own ends, and while each area is unique it also fits into the wider scheme of capitalist production and social relations. This has particular implications for understanding and action:

General laws are about causation, not empirical correlation. They are as well if not better established in causal studies of the particular, the much-maligned case study. It is indeed time that regional and local particularities were reinstated as a central focus of geographical thinking. (Massey, 1984, p. 120)

British economy and society can only be understood by recognising its fundamentally capitalist nature. But it can only be changed – changed politically – in its specific form. Both the general and the specific are essential, both to analysis and to action. (Massey, 1984, p. 7)

Such notions are explicit in the work of Dorn and implicit in the work of Smith (1981) on black lung, but so much of the work in epidemiology and medical geography has been concerned with establishing universal, regular (and usually biological) relationships that capitalist social relations and their specific geographical manifestations in terms of health remain unresearched and often unacknowledged. If a geography of health that considers the production of ill-health (and not just the consumption of health care resources) is to be developed, then such aspects of a materialist epidemiology must be central to the task.

The politics of change

Thus far, this book has been mainly concerned with problems and ways of examining problems and it is tempting to stop here and leave solutions to others. But one of the major themes of this chapter is that one must look at reality in its entirety. To examine solutions, of necessity requires us to enter politics, for not only do political standpoints colour the solutions but they also determine the nature of the problems. In order to illustrate these different standpoints we will examine three contrasting perspectives which we can label 'free-market', 'welfarist' and 'marxist'. Our aim is to portray the fundamentals of these different positions and to consider their implications for health policy in the light of our previous discussion.

The free-market approach

From this viewpoint, the 'crisis' in the health service is not seen as the major inequalities of outcome nor as the lack of long-term improvement in people's health, but rather as the debilitating cost of providing a state-subsidised medical and welfare service. Consequently, free-market theorists and politicians have adopted a two-part strategy: a reorganisation and privatisation of medical care, and a concomitant shift of the burden to individuals and voluntary organisations. This shift to the individual and away from societal responsibility coincides with the free-market delight in individual freedom, for in what has been called the 'moral reason for capitalism' it has been argued that the real world is too complicated to plan; society is best served by the innumerable decisions of individuals acting in their own best interest.

For the USA, Navarro (1984) has discussed the guiding policies of the Reagan administration, identifying three major problems and solutions. First, it is believed that the size and growth of federal spending has contributed to the current recession with, for example, General Motors spending more money on health care than on steel in 1975. Second, it is argued that previous health and safety legislation was restricting the competitiveness of American industry and that there needed to be a reduction in such controls. Third, the governmental health schemes are regarded as inefficient and requiring a large bureaucracy; the proposed solution was further privatisation of medical services, and there have been substantial cut-backs in public programmes like Medicaid.

In Britain too, the problems are perceived in a similar way and the Conservative government is especially worried about the cost of the growing proportion of elderly people in the population. Mohan and Woods (1985) have examined the policies of the Thatcher Conservative governments and the spatial implications of their implementation. The major policies since 1979 have been the restriction of NHS finance and the encouragement of private medical care and the voluntary sector. In the period 1979–83 there was a 4 per cent increase in public health expenditure while it was estimated that 6.5 per cent was needed to maintain current services given the growth in the elderly population and medical inflation. Consequently, and because of the workings of RAWP, inner city District Health Authorities have suffered major decreases in their real income. There has been an accompanying increase in private activity in the medical field with changes in tax incentives, relaxation of planning controls, allowing NHS consultants to do more private work without affecting their NHS salary, and attempts to ensure competitive tendering for NHS domestic, catering and laundry services. There has been a major growth of private hospitals and in some areas the number of private beds now exceeds the number of public. Overall there has been a major move away from a health service planned on social need to allow for the increased importance of the market. As Navarro (1984) has argued, while the welfare state is not being dismantled it is being restructured in an attempt to restore a situation more conducive to capital accumulation.

While free-market policy has been much concerned with the reorganisation of medical provision, it has also emphasised that individuals will have to do more for themselves in place of expensive medical care. This view is well expounded by John Knowles (1977), a former President of the Rockefeller Foundation. For him the causes of ill-health are clear: 'Over 99% of us are born healthy and are made sick as a result of personal misbehavior', and the onus rests with the individual: 'the primary critical choice facing the individual is to change his personal habits or stop complaining' (p. 2). Behind this invective is the major concern of cost: 'the cost of sloth, gluttony, alcoholic intemperance, reckless driving, sexual frenzy have now become a national and not an individual responsibility. . . . But one man's freedom in health is another man's shackle in taxes and insurance premiums' (p. 3). Given these viewpoints, Knowles makes recommendations for action:

> For example, in dealing with . . . heart attacks, preventive measures would include screening for high-risk factors (high

> blood pressure, . . . fat levels, cigarette smoking . . .) and
> making available emergency services for rapid transit to
> hospital-based coronary-care units; environmental measures
> would include altering food supply to reduce the intake of fat
> . . . and individual and mass-educational efforts would include
> encouraging the use of screening examinations, the cessation
> of smoking, the maintenance of optimal weight with a
> balanced low-fat diet, and obtaining regular exercise. (p. 11)

Such views have found outlets in official government documents on both sides of the Atlantic. In the USA, the Surgeon General's (1979) report urges us to stop smoking, to exercise regularly, to seek proper medical advice and to eat well and properly. Similarly, in Britain the DHSS's (1976) pamphlet *Prevention and Health: Everybody's Business* concludes that 'many of the current major problems in prevention are related less to man's outside environment than to his personal behaviour' (p. 11), and again stress is placed on individuals taking care of themselves by changing their lifestyle.

The critics of free-market policies have argued that public provision of medical services is relatively efficient and that societal changes must accompany individual changes. Navarro (1984), in a series of international comparisons, attacks the economic arguments of free-marketeers. He contends that high public expenditure has not contributed to economic recession by showing that Japan, West Germany and Sweden have a faster economic growth rate than the USA but also a larger public expenditure on health. Navarro also states that 'waste on the one hand and insufficient coverage on the other are the major trademarks' of the largely free-market US medical system, while for Conrad and Kern (1982, p. 574), 'the NHS delivers better care to most parts of the population at less cost (and with no discernible difference in health status) than is accomplished in the USA.' While Britain spends about 6 per cent of its GDP on health, America spends 10 per cent. Moreover, in the USA where market forces are the primary allocator of resources, there are large administrative structures which take 5 per cent of total medical expenditure; in Britain this figure is below 1 per cent.

Mohan and Woods (1985) in their review believe that the NHS is becoming a two-tier service in terms of quality and social and geographical provision. The growth of the private sector has been highly concentrated in the south-east and this goes directly against the RAWP aims of redistributing resources to the north and west. While voluntary and charitable organisations are being encouraged to take over services and hospitals, the charities themselves doubt their ability to maintain the quality of provision. Moreover, as the scope for fund-raising varies geographically, there are again likely to be more resources in the relatively affluent south-east. They argue that freeing of market forces will inevitably lead to worse care for those that are unable to pay.

The free-market view of health places the major burden of change on the individual and ignores the vital societal constraints that we discussed earlier. Even Knowles (1977, p. 3) recognises the difficulties of urging people to eat properly when 'twenty-six million Americans, 11 million of whom receive no federal food assistance, live below a level which does not support an adequate diet'. But so pervasive has the ideology of individual responsibility become, that when a couple complain of damp in their bedroom, the housing officer suggests that they modify their behaviour by 'reducing their lovemaking' because of the excessive condensation it generates (*Radical Community Medicine*, Summer 1983, p. 9). From a critical perspective, lifestyle changes are not just pragmatic, stop-gap solutions, but they are a means of legitimating the status quo and reproducing the major inequalities that are so much part of the present system. The onus for change is placed on the individual and the nature of society is rendered unproblematic. Moreover, it is deeply ironic that choice in relation to the type of health care is presented in such a limited way in the free-market proposals. The emphasis is firmly on the provision and support of technological, biomedical interventionist medicine; little mention is made of alternative medicine. Of course, currently there are large profits to be made in mass-produced drug treatments that are not possible in an alternative medicine that is based on individual-specific treatment and 'renewal of the internal environment' (Salmon, 1984).

A welfarist approach

We have chosen to represent this approach by focusing on the Black Report (DHSS, 1980) and in particular on the views of one of its authors: Peter Townsend, the author of the monumental *Poverty in Britain* (1979). The essence of his approach is contained in the title of his 1981 article, 'Toward equality in health through social policy', the basic argument being that if the market is allowed to operate without control and intervention, then unacceptable and debilitating health inequalities will result. Therefore, and because no individual person nor individual corporation can prevent these inequalities, the state has to intervene. Here we will outline the findings and recommendations of the Black Report and consider subsequent reaction and criticism.

After evaluating alternative explanations, the Black Report favours a materialist account of health inequalities: 'there remains much that is not explicable in any direct fashion and meanwhile must be attributed to the pervasive effects of the class structure' (Townsend and Davidson, 1982, p. 200). In particular, Townsend (1981) highlights the material deprivation of a huge number of young working-class families with an inadequate standard of living, the material impoverishment that results from becoming disabled, and the deprivations experienced at work by many manual workers.

The thirty-seven recommendations that are contained in the Black Report are based on pragmatism and concentrate on doing something for particular social groups (especially children, the elderly and the disabled) and for particular areas (especially for inner city); indeed many of the suggestions have an overtly geographical flavour. The recommendations can be classified into two types: firstly, in the health area, and secondly, a wider social and economic strategy. In terms of the health field the report argued for a move away from hospitalised care towards prevention, community and primary health care. More specifically, the report recognised the spatial inequalities of provision and suggested a modification of RAWP so that it would become based on social measures and include an element of redistribution between the different sectors to increase community provision. Access to primary-care facilities was seen as a problem and they were disturbed by the quality and geographical coverage of general practitioners; they suggested that community nurses should be deployed and higher remuneration given to those areas which were understaffed and had a high need (based on SMRs and social variables). They also suggested that the school health service be increased in size and scope and, in particular, that the surveillance of certain types of family in areas of special need be increased. For the elderly and disabled there should be an expansion in domiciliary care and home-help to allow them to live in their own homes.

In terms of prevention, they suggested that health education should be expanded and focused on schools. Screening should also be encouraged but they argued that blanket coverage was unlikely to be cost-effective. They preferred geographical targeting for certain diseases (e.g. serious hypertension) in high-risk areas. They recognised that prevention was not just an individual responsibility for government and the producers of harmful goods were also involved. In particular they advocated increasing the tax on tobacco, declaring more non-smoking areas, and industrial diversification so that cigarette manufacture could eventually be phased out.

They suggested that the ten most unhealthy areas of the country should receive additional funds (£30 million at 1981 prices) and this was not only to tackle the serious problems in these areas but also to act as an example of what could be achieved. The extra money was to be spent on such services as mobile clinics, clinics with smaller catchment areas, clinics open at weekends and offering counselling for pregnant women and young families, domiciliary services to support those discharged from hospital and to follow up missed clinic appointments, and better services for the disabled in their own homes.

The second major set of recommendations provided the wider strategy of reducing health inequalities through reducing social inequalities. They proposed a comprehensive anti-poverty strategy, concentrating on young families and disabled children. The specific policy for families included an increased child benefit, an increased grant for parents at childbirth, an extra allowance for children under five, a statutory obligation for the provision of pre-school day-care facilities, free and improved school meals, and child-accident prevention programmes aimed at planners and designers. In other areas the unions and management were urged to co-operate and set minimum standards for the work environment. For housing they recognised the need to extend the privileges of owner occupiers to the rented sector and they contended that there must be a substantial increase in spending on house improvements. They advocated the broadening of education to provide better access to knowledge and information, and indeed control, over what is happening in a specific community.

Fundamental to this strategy is the need for planning and it was recognised that there would have to be a change in governmental planning so that all ministries played their role in diminishing health inequalities. They recommended an independent body to promote and monitor such social policies. The money for this substantial increase in government intervention was to come from redistribution of income, the current distribution being judged as very unequal. According to Townsend (1983, p. 23) these measures were chosen with a 'sporting chance of mobilising the support of the political parties. After all, a capitalist society like

Japan was capable of providing material security for its workforce.' In this social welfare approach, the basic societal structure of capitalism is retained but the inequalities and debilities that are produced are diminished by the state intervening to redistribute wealth and to provide facilities for the disadvantaged.

This approach has been criticised by the free-marketeers for going too far, and by marxists for not going far enough. The Conservative government's frosty reaction to the Black Report was discussed in Chapter 6. Patrick Jenkin, the then Secretary of State for Social Services, in his Foreword to the report stated that the additional expenditure on the scale recommended was 'quite unrealistic in present or any foreseeable economic circumstances'. In a subsequent speech at Cardiff he outlined the three principal shortcomings of the report. First, the report was unable to pinpoint the mechanisms linking disease, poverty and the class structure; second, there was new evidence that the poor were not doing as badly as the report maintained; and finally, there was no evidence that spending more money would make any difference.

Townsend's reaction to these criticisms can be read in the introduction to the Black Report which was eventually published in a revised format in 1982. First, while recognising the limitations of their rudimentary materialist explanation, he does not doubt that the explanation is developed enough to identify the kind of strategy to be pursued. Second, he dismisses the 'new' evidence for reduced inequalities as methodologically poor, and cites numerous studies supporting the original empirical evidence. On the issue of cost, he argues that the implementation of the proposals would be less than 3 per cent of total public expenditure and that Britain is still a wealthy country. Moreover, he rejects that public expenditure is a burden on growth and contends that welfare spending can contribute to growth when it is realised that economic growth always involves social costs. Thus health expenditure is a 'necessary part of a wise national investment in a thriving national economy'. The money for health is there but it depends on your social, economic and political priorities.

From the marxist viewpoint, it is generally accepted that the implementation of the report could bring some improvements in people's health, and the linkage of health and material conditions is to be welcomed. But as Greaves (1982) argues, the materialist explanation of the causes of inequality has not been carried through to the recommendations and, in particular, the report represents, in the terminology of Waitzkin (1976), a 'reformist' reform when what is required is 'non-reformist' reform. Reformist reforms make small incremental improvements without challenging the current distribution of political power and economic domination. That is, they are a means of legitimation; defusing discontent while preserving the system as it is. Historically, Waitzkin argues, the reformist changes have been subverted or cutback when not supported by continued social protest. In the terms of our earlier discussion, reformist reforms are concerned with mediating the effects of mechanisms rather than tackling and changing the underlying structures. Non-reformist reforms, however, aim to achieve true changes in power and control by increasing political tension and activism, and by exposing and changing structural inequalities.

In this marxist context, the Black Report is seen as a reformist reform for while it accepts that the present economic system may promote inequality, it nevertheless assumes that changes can be made that substantially reduce inequality while leaving the framework of 'economic and socio-structural factors unchanged'. In particular, while in Chapter 6 the report does not even consider the maldistribution of resources as one cause of inequality, a large number of the recommendations are concerned with the improvement of geographical redistribution of care. From a critical position it is like applying a sticking plaster when the whole body is in trouble; it is treating the superficial geographical inequalities while ignoring the underlying, and more fundamental, class inequalities. Moreover, while the report does use a materialist position to explain current inequalities, it does not attempt to use the same approach to explain why (and despite increased public expenditure and previous reforms like the NHS) the inequalities have not been reduced (Gray, 1982). The report is at its weakest in discussing how effective reform can be achieved, for it appears to accept that rational argument and an independent watchdog of technical experts will be sufficient to overcome sectional and class interests. For example, the recommendation that capitalists and workers should collaborate in improving workers' conditions assumes a consensus politics that ignores conflicting interests. Similarly the report presumes that a shift to community care can be achieved by merely altering the redistributive formula and thereby fails to recognise the power of the medical profession and the class interests they represent in maintaining the status quo. The recommendation that cigarette production should be phased out simply ignores the power that the producers will employ to retain their profits and assumes that the state will be somehow neutral to these pressures. In the whole of the report there is very little discussion on changing this imbalance of power and on how a reformed health service could be made more democratic and responsive to the needs of people.

A marxist approach

The use of 'a' in our heading is again deliberate for we are going to illustrate the marxist approach by the writings of one researcher, Vincente Navarro, editor of the *International Journal of Health Services*. It is important at the outset to appreciate the bases of his approach. First, his analyses are informed by empirical considerations of specific historical struggles, for example the USSR (1977), Great Britain (1978), Chile and the USA (1976). Second, he argues that medicine has developed a particular form under capitalism, and to understand the nature and distribution of medical resources requires the study of the world outside medicine and, in particular, an analysis of the role of the dominant capitalist class in determining the nature of medicine. Third, for him dominance does not mean complete control and reality is seen as the product of the dialectic process of class struggle in which the capitalist class, the bourgeoisie, is currently dominant. Fourth, he recognises the need to penetrate beneath superficial appearances for capitalism wishes to mystify what it is about; many institutions and procedures (such as science, medicine, the electoral process and the state) are presented as neutral and above class interests when in fact they legitimate and preserve this dominance of the bourgeoisie. Fifth, he argues for a different kind of medicine which can only be achieved by the mass mobilisation and direct involvement of the working class. Finally, he rejects the view that sees the world as dichotomised into separate, independent things (e.g. body/mind, individual/environment, fact/value), preferring the dialectic position that sees reality as an interconnected totality and change as a continuous process which results from the internal conflict of opposing forces. Such ideas have permeated many areas of our previous discussions, particularly Chapter 7 and the earlier part of this present chapter.

For Navarro, the nature of medicine depends on the dominant mode of production so that he recognises a feudal, capitalist and communist medicine, each with a different approach as to what is health and how it can be achieved. The current dominant form is capitalist medicine which is characterised by an emphasis on positivism, biology, individuals, intervention, technology, and specialism. The biomedical model of our Chapter 1 is the capitalist medicine of Chapter 7. This particular form is not a result of progress, nor an accident, nor of the demands of technology, nor even of the professional power of the medical practitioners, but instead it is the result of a continuing class struggle in which the capitalist class is dominant.

The essence of Navarro's view of the development of capitalist medicine is contained in a key diagram (Figure 9.4). Capitalist economic organisation is seen to lead to industrialisation which in turn requires a specific type of state intervention to maintain industrialisation and to deal with the disbenefits that would be harmful to capitalism. This, in turn, requires a specific form of medicine which is not a threat to capitalism and the result is a form which has a dual, and simultaneous, function: (1) the useful liberating function of care and some cure; (2) the dominating or control function of legitimation. The dual roles exist in contradictory unity; they cannot be divided into this bit good and therapeutic and that bit bad and repressive, for capitalist medicine is controlling because it is effective, or at least it is seen to be so. Indeed, so successful has this been that many working-class demands for change have been expressed in the form of more capitalist medicine. The class struggle produces this dual form with the working class demanding medical care because they need it, and the bourgeoisie providing it in an individualised form that dominates. This pattern of dominance is determined not by conspiracy but by the internal logic of capitalism with the dominating bourgeoisie having a vision of reality that promotes some positions while excluding others.

Clearly for Navarro a different form of medicine is required but this form should not merely be statist or nationalised medicine. The state takeover is only socialist if the very nature of the state is changed with the degree of socialism being measured by the amount of direct mass participation. Thus, while the British NHS and the Soviet medical system are nationalised medicine, they are not socialist for they are still dominated by the demands of the capitalist class in one case and the party elite in the other. Moreover, socialist medicine involves not just the better geographical spread of resources, for the very nature and form of medicine has to be changed. For example, we may tackle the geographical maldistribution of doctors not by financial incentives to relocate but by the radical alternative of changing the patterns of class and academic dominance of the control of production and distribution of doctors. This does not mean merely changing the class character of medical personnel without

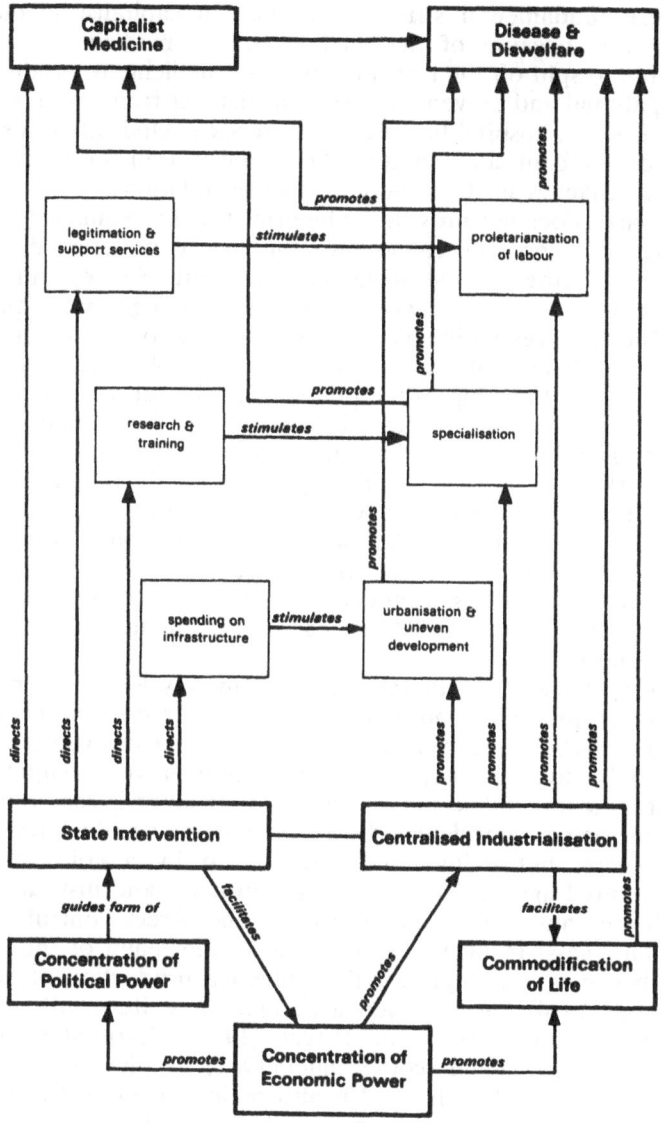

Figure 9.4 The development of capitalist medicine
Source: Navarro (1976)

a more fundamental structural change in capitalist medicine. Moreover, because of the dual nature of medicine it is not possible to split off the liberating parts of medicine to produce an occupational and preventive service as distinct from the dominating parts of a hospital-based curative service. Medicine is a social relation in contradiction and the socialist transformation will require changes in the structure of social relations.

Navarro does not provide a blueprint for the required form of medicine (indeed, he cannot, for this has to be decided in the future by further class struggle and democratic choice), but there can be little doubt as to what fundamental changes are required. For Navarro, revolutionary change is needed to democratise the political, ideological and economic aspects of society, so that there is a 'dictatorship of the proletariat', the self-government of the masses. Socialism has elements of both capitalism and communism but the key element in the transition is the need for the working class to seize state power and then to direct the economic transformation. Socialism is therefore a new form of the state with the proletariat in control, but to achieve communism the state must wither away and so too must the division of labour based on classes; the result is a social form of life for which the motto is 'from each according to ability, to each according to need'.

Socialist medicine, therefore, is not merely lowering environmental pollution nor more occupational health care but primarily a redistribution of power that could make these changes possible. Not only must priority be given to preventive, occupational, environmental and social medicine but a change is required such that medicine is not defined by a dominant class and administered by experts but defined and reproduced by a collectivity of unexploited agents. The knowledge, practice and institutions of medicine must be submerged under the direct control of the popular mass; medicine ceases to be in the realm of the expert and becomes a struggle for the collective production of health. Deacon (1984) has surveyed a number of writers with marxist visions including Navarro, Doyal (1979), Carpenter (1980), Thunhurst (1982) and Scott-Samuel (1981), and he recognises the twin emphases of a changed health policy to prevent avoidable disease and a changed caring and curing health service for unpreventable disease. Table 9.2 details the changes that are needed before a socialist and communist health policy can be achieved, but as Deacon admits such a table fails to reflect the way in which issues of *medical* care must be re-defined into issues of *health care*. Communist medicine entails a fundamental break

Table 9.2 Expectations of socialist and communist medical care policy

Aspect of social policy	Aspect of medical care policy	Socialism	Communism
Priority	1. Outcomes in terms of health	Less and more equal morbidity and infant mortality than capitalism; greater and more equal life expectancy	Less and equal morbidity and infant mortality; greater and equal life expectancy
	2. Resources in terms of money	Higher expenditure than capitalism	Need for higher expenditure may no longer exist
	3. Resources in terms of person-power/facilities	Higher level of resources than capitalism	Need for higher level of resources may no longer exist
	4. Priorities in terms of cure, care, prevention	Prevention and care prioritised	Prevention and care central
Control over welfare provision	5. Central control	Central direction with political cadre influences	Centre provides democratically resolved planning guidelines only
	6. Local control	Democratic worker and user involvement	Mass participation in policy resolution and implementation

continued

Table 9.2 continued

Aspect of social policy	Aspect of medical care policy	Socialism	Communism
	7. Control of medical technology industry	Nationalised and progress toward socialised relationships	Socialised working relationships within industry and between it and the health service
Agency	8. Agency of provision	State, workplace, family and market giving way to community provision	Community provision
Relationships between provider and users	9. Status of doctors	Lower than under capitalism	Equal status with all workers
	10. Division of labour in medicine	Reduction in vertical and horizontal divisions	Abolition of vertical divisions; movement between horizontal divisions
	11. Nature of medical technology	Progress toward new forms	New forms of medical technology reflecting communist social relations
	12. Status of patients	Higher, accompanying deprofessionalisation of doctors	Equal status with providers

Aspect of social policy	Aspect of medical care policy	Socialism	Communism
Rationing systems	13. Region and class access, usage, and outcome	Progress toward equality	Equal access, usage and outcome
	14. Rationing procedures between individual patients	Free usage with access rationed by work and need according to democratically determined formulae	Free usage with access according to self-perceived need
Sexual divisions	15. Sexual division in medical care employment	Progress toward no division	No sexual division of medical labour
	16. Sexist content of medical practice	Progress toward no sexist content	No sexist content of medical practice

Source: Deacon (1984).

with capitalist medicine in that it involves itself with all aspects of social and productive life in so far as they affect health.

Navarro is not a maximalist and he does not believe that improvement requires a complete transformation and the end of capitalism. The process can begin under capitalism and Assennato and Navarro (1980) have considered how occupational epidemiology has been transformed in Italy with the workers assuming control. Prior to 1969, if a health problem was thought to exist in a factory, the management could call in outside experts. They would try to be as objective as possible, placing the greatest faith in quantitative measurements; workers would be regarded as passive objects of research whose knowledge of their own job and health was not to be accepted until confirmed by the expert. This approach underwent a major change following the workers' revolt of 1969. The factory workers organised themselves into homogeneous groups on the basis of their work environments and each group provided a delegate for the workers' committee. These groups have been instrumental in the creation of local occupational health services and, in particular, there is an occupational health physician whose job as a public official is to provide technical help to support the workers' struggle against the management.

Each homogeneous group keeps records and discusses collectively their working environment and health hazards. They collect two types of information: objective information details the toxins, chemicals, shifts and physical efforts of the working environment, while subjective information consists of the disease, unease and illness that they are experiencing. Each worker keeps a book of their work experience and there is also a book in which collective understanding of the problem is detailed. The workers can call on outside expertise and scientific measurements can be made. Crucially, however, the scientific findings are discussed by the assembled workers and are accepted or rejected according to the workers' experience. When the problem and its causes have been identified by the workers, the assembly decides on the appropriate strategy. Thus, there is a collective generation of the hypotheses, the data, the probable cause and the solution. Moreover, scientific objective data is not automatically accepted but is evaluated in the light of subjective experience.

As important as these changes are for producing improved well-being for Italian workers, they cannot be regarded as an island of socialism within the sea of capitalism. Social reforms are possible within capitalism but socialist ones are not; for Navarro the transition to socialism cannot be achieved by reforming capitalism. He believes that the only route to achieving the socialist transformation is mass mobilisation, a seizure of state power and massive and direct participation in the political generation of policies and not merely the implementation of them. To accomplish this there must be long-term opposition and consciousness raising with the aim of a continuous cultural revolution whereby values are questioned and the reproduction of dominating relationships is avoided. Navarro (1980b) concentrates on two areas to achieve this change: work and the neighbourhood. Both areas require that there is more direct and local control, so that in the neighbourhood strategy there is community control over community organisations with collective discussion and secret vote. Thus, health institutions need direct participation in the running of these institutions by those who are affected or served by them as well as by those who work in them. In summary (Navarro, 1980a, p. 203):

there is a need for . . . a direct form of participation on a daily basis by the working class and popular masses in all economic, political, and social institutions (including the medical and scientific institutions). It is only in this way that the democratisation of our institutions will imply a massive transformation of the majority of the working population from being passive subjects to active agents in the redefinition of those societies, a transformation that takes place as part and parcel of their becoming the agents and not the objects of history.

At first sight geography appears to be downplayed if not entirely removed from Navarro's vision. For example, he argues that areal differences are secondary to class ones and he applies the same diagram (Figure 9.4) to the development of medicine in Britain as in the Soviet Union. This, however, would be a serious misinterpretation of his work, for while recognising that there are general tendencies in capitalist societies, he implicitly recognises that differing material conditions together with tendencies operating in combination can produce distinctive differences between countries. Indeed, that is why he has written a number of books which deal with unique history and geography of health and medicine. Moreover, he explicitly recognises the uniqueness of place in relation to the need for a transformed medicine. For example, he contends that the Chinese barefoot doctor cannot be simply transferred from one area to another and be expected to work automatically. Instead, the new form of medicine must grow out of the particular needs of the local area if it is to be more than just a transplanted managerial system. Finally, space is seen as more than just a framework to describe differences, for it is an 'action' space with small geographical areas (neighbourhoods) being an important part of achieving mass democracy.

The critics of the Navarro and the marxist position have used a number of arguments (see, for example, Cox, 1983). They contend that there are now no fundamental divisions in society along class lines but rather there is a plurality of interests with so many different groups that the owners of the means of production are not dominant. They fear that revolutionary societal change may bring disastrous irreversible consequences and prefer instead that reform should be pragmatic, trial-and-error or 'bureau-incrementalist' (Walker, 1984), so that any mistakes are likely to be minor. They insist that the socialist revolution will lead to a highly controlled, centralised totalitarian state in which personal freedom will be severely restricted. Moreover, they regard countries with an avowedly marxist government as crucial test grounds for theory, and they point to continued inequalities in health in such countries. If this is not a sufficient argument, they conclude that it could not occur in the West.

In response to these criticisms it is possible to construct a reply using, in particular, Navarro (1985). Navarro rejects pluralism and in his 1976 book (pp. 82–97) he empirically investigated the distribution of power in the USA, concluding that 'one can see a pattern of consistent and continuous class dominance in our corridors of power in which the few control much and the many very little'. The bureau-incrementalist position, however, is based on pluralism and therefore ignores that power is distributed unequally between different groups; in the marginal choice between one policy and another, the weak are again likely to be squeezed. Moreover, if we believe that capitalist social structures are the real problem, small-scale changes will and can only lead to their continuation.

Navarro (1977) has, in effect, devoted a whole book to the Soviet analogy, and the charge that fundamental change inevitably leads to totalitarianism. For him the specific problems facing that country after the revolution help to explain the development of a centralised and bureaucratic form of government. In particular we need to understand the previous, and necessary, clandestine activity of the Bolshevik party and the priority placed on capital accumulation (via massive industrialisation) as the best means for rapidly developing a country subject to external hostility. For the leaders of the revolution, socialism was not possible in a poor country and, instead of changing the nature of the state, they believed that massive industrialisation could only be achieved by using the existing Czarist state machinery to centralise power and direct the economic transformation. Accordingly, the analogy between Britain, USA and revolutionary Russia is a false one for the West is already well developed in material terms and there is already democracy (albeit only in a partial form). It is another success for bourgeois ideology if we accept the Soviet model as the only possible exemplar. Indeed, as Chapter 6 has shown, there have been genuine improvements in such countries even under an 'imperfect' socialism. Moreover, Navarro aims not for centralisation of power but the reverse; human liberty is placed above centralisation and the aim is democratisation and decentralisation. While he admits that it may take a long time, he also regards (1985) 'reality as a source of optimism rather than pessimism'; citing eleven successful revolutions in the 1970s, he contends that worldwide revolutionary praxis is the trademark of our era. The image of powerlessness when faced with apparently unmovable structures is at least partly transformed when we realise that although the structure and constraints of capitalism are enduring, they also have to be continually reproduced.

Conclusions

In the last fifteen years, medical geography has undoubtedly undergone a major shift in emphasis from a geography of disease to a geography of health care. In the former approach, geography was seen as a supporting discipline (if it was seen at all); the problems came from physicians, and geographers used the skills of disease ecology and spatial analysis to suggest possible causal relationships. In the latter approach, geography has helped to investigate the delivery of health care (or rather ill-health care), concentrating on such issues as efficiency, accessibility, inequality and allocation of resources. While not denying the usefulness of both these approaches, we feel, like Eyles and Woods (1983), that the focus of research should now shift to the totality of health and society. This shift requires us to examine economic, social and political processes which, in turn, necessitates the removal of academic boundaries to provide a full understanding of the social world. In the different approaches and viewpoints revealed in this chapter (and indeed in the whole book) geography and space have different meanings and uses. For the orthodox epidemiologist, space can be regarded as an ordering framework (along with other primary variables such as age, sex and time) to reveal general patterns and anomalies. From the critical position we need to understand health in relation to the specific materialist conditions existing in specific localities, that is we need to study the unique geography of particular places. From the free-market standpoint, space can again be used as an ordering framework for recognising and improving inefficiency, while from the welfare position, space, in the form of distance, can be regarded as a barrier to the utilisation of services, a framework for recognising spatial inequalities and, with areal targeting of resources, a means of ameliorating these inequalities. For the marxist these areal differences while real enough are produced by class inequalities which must be tackled directly; space in the form of locality, however, is a vital element in the understanding of unique outcomes of general social processes while space in the form of neighbourhood is at a human scale which allows direct participation in a transformed mode of care.

You will have to make your own personal evaluation of the

three political strategies that have been outlined, and of the appropriateness of the different methodological positions we have considered. We will end with two quotations:

medicine is assigned an impossible task – that of taking care of solving, and administering the disease, unease, and diswelfare created by the process of production and consumption in capitalism; or in other words, solving the unsolvable . . . the much-heralded crisis of effectiveness of the Western system of medicine . . . is a result of the inability of medicine to deal with and even change the economic and social forces that determine most of the prevalent mortality and morbidity in the first place. (Navarro, 1978, p. 91)

the nature of society in which we live profoundly affects our biology as well as our behaviour. In a healthier and more socially just society, even though pain, illness and death can never be eliminated, our individual biologies will nonetheless be different and healthier. (Rose *et al.*, 1984, p. 195)

Guided reading

A realist methodology for the social sciences is developed by Sayer (1984); a shortened account is given in Sayer (1985); Keat and Urry (1982) compare the realist approach with other methodologies. The development of a critical epidemiology is considered by Paterson (1981), Cameron and Jones (1982), Davies (1982) and Navarro (1980a).

For an individualist free-market position see Knowles (1977); Conrad and Kern (1982) is an excellent reader with many original articles from different political positions. The welfarist approach can be seen in the work of Blume (1982) and Townsend (1981); critiques of the Black Report are provided by Gray (1982) and Greaves (1982). Navarro is a prolific and clear writer and his (1980a) and (1983) articles are good starting-points. His marxism is criticised by Reidy (1984) while there is a rebuttal in Navarro (1985). Other marxist accounts include Gough (1979) on the welfare state, Doyal (1979) which has a British flavour, and Waitzkin (1983) which is American-based. For a thoughtful account of the socialist approach that uses the experiences of real people to develop the argument, the early chapters of Mitchell (1984) can be recommended.

The pages of *Radical Science Journal*, *Radical Community Medicine*, *Social Science and Medicine* and the *International*

Journal of Health Services should be scanned for continuing debate and reactions.

References

Armstrong, D. (1980), *An Outline of Sociology as Applied to Medicine*, Bristol, Wright.

Assennato, G. and Navarro, V. (1980), 'Worker's participation and control in Italy: the case of occupational medicine', *International Journal of the Health Services*, vol. 10, pp. 217–32.

Berliner, H. S. and Salmon, J. W. (1979), 'The new realities of health policy and influence of holistic medicine', *Journal of Alternative Human Services*, vol. 5, pp. 13–16.

Bhaskar, R. (1975), *A Realist Theory of Science*, Leeds, Leeds Books.

Blaxter, M. (1985), 'Research into sociomedical problems', in W. W. Holland (ed.), *Oxford Textbook of Public Health*, pp. 361–72, London, Oxford University Press.

Blume, S. (1982), 'Explanation and social policy, the problem of social inequalities in health', *Journal of Social Policy*, vol. 11, pp. 7–32.

Bridford, K. (1978), *Estimates of the Fraction of Cancer in the US Related to Occupational Factors*, New York, National Cancer Institute.

Cameron, D. and Jones, I. (1982), 'For discussion: theory in community medicine', *Community Medicine*, vol. 4, pp. 3–11.

Carpenter, M. (1980), 'Left orthodoxy and the politics of health', *Capital and Class*, vol. 11, pp. 73–98.

Chalmers, A. F. (1982), 'Epidemiology and the scientific method', *International Journal of Health Services*, vol. 12, pp. 36–40.

Chouinard, V., Fincher, R. and Webber, M. (1984), 'Empirical research in scientific human geography', *Progress in Human Geography*, vol. 8, pp. 347–80.

Conrad, P. and Kern, R. (1982), *The Sociology of Health and Illness: Critical Perspectives*, New York, St Martin's Press.

Cooper, R. (1983), 'Race and the social origins of disease', *Radical Community Medicine*, vol. 16, pp. 5–19.

Cornwell, J. (1984), *Hard-Earned Lives: Accounts of Health and Illness from East London*, London, Tavistock.

Cox, C. (1983), *Sociology: an Introduction for Nurses, Midwives and Health Visitors*, London, Butterworths.

Crombie, D. L. (1984), *Social Class and Health Status: Inequality or Difference?*, Occasional Paper no. 25, Royal College of General Practitioners.

Davies, C. (1982), 'Criticising epidemiology: some notes on the debate', *Radical Community Medicine*, vol. 12, pp. 6–15.

Davies, C. and Roche, S. (1980), 'The place of methodology: a critique of Brown and Harris', *Sociological Review*, vol. 28, pp. 641–56.

Day, S. (1980), 'Is obstetric technology depressing?', *Radical Science Journal*, vol. 4, pp. 17–45.

Deacon, B. (1984), 'Medical care and health under state socialism', *International Journal of Health Services*, vol. 14, pp. 453–80.

Department of Health and Social Security (1976), *Prevention and Health: Everybody's Business*, London, DHSS.

Department of Health and Social Security (1980), *Inequalities in Health*, Report of a Research Working Group chaired by Sir Douglas Black, London, DHSS.

Dorn, N. (1980), 'Alcohol in teenage cultures: a materialist approach to youth cultures, drinking and health education', *Health Education Journal*, vol. 39, pp. 67–73.

Dorn, N. (1983), *Alcohol, Youth and the State*, London, Croom Helm.

Doyal, L. (1979), *The Political Economy of Health*, London, Pluto Press.

Draper, P., Best, G. and Dennis, J. (1977), 'Health and wealth', *Royal Society of Health Journal*, vol. 97, pp. 121–6.

Elson, D. (1979), *Value: the Representation of Labour in Capitalism*, London, CSE Books.

Eyer, J. (1975), 'Hypertension as a disease of modern society', *International Journal of Health Services*, vol. 5, pp. 539–58.

Eyer, J. (1984), 'Capitalism, health and illness', in J. B. McKinlay (ed.), *Issues in the Political Economy of Health*, pp. 23–59, New York, Tavistock.

Eyles, J. and Woods, K. J. (1983), *The Social Geography of Medicine and Health*, London, Croom Helm.

Gardner, M. J. (1973), 'Explaining and predicting mortality', *Journal of the Royal Statistical Society*, vol. 136, pp. 421–40.

Gough, I. (1979), *The Political Economy of the Welfare State*, London, Macmillan.

Graham, H. (1984), *Women, Health and the Family*, Brighton, Harvester.

Gray, A. M. (1982), 'Inequalities in health: the Black Report, a summary and comment', *International Journal of Health Services*, vol. 12, pp. 349–80.

Greaves, D. (1982), 'The Black Report and materialism', *Radical Community Medicine*, vol. 11, pp. 2–6.

Habermas, J. (1970), *Towards a Rationalist Society*, Boston, Beacon.

Habermas, J. (1974), *Theory and Practice*, London, Heinemann.

Harburg, E. W. (1973), 'Socio-ecological stress, suppressed hostility, skin colour and black-white male blood pressure', *Psychomatic Medicine*, vol. 35, pp. 4–11.

Harré, R. (1979), *Social Being*, Oxford, Blackwell.

Jacobson, B. (1981), *The Lady Killers: Why Smoking is a Feminist Issue?*, London, Pluto Press.

Karasek, R. (1981), 'Job decision, latitude, job demands and CVD: a prospective study of Swedish men', *American Journal of Public Health*, vol. 71, pp. 694–705.

Keat, R. and Urry, J. (1982), *Social Theory as Social Science*, London, Routledge & Kegan Paul.

Kleinbaum, D., Kupper, L. and Morgenstern, H. (1982), *Epidemiologic*

Research: Principles and Quantitative Methods, California, Wadsworth.

Knowles, J. (1977), *Doing Better and Feeling Worse: Health in the United States*, New York, Norton.

McGuinness, T. (1979), *An Econometric Analysis of Total Demand for Alcohol Beverage in the UK, 1956–1975*, Edinburgh, Scottish Health Education Unit.

Mandel, E. (1979), *Introduction to Marxism*, London, Pluto Press.

Massey, D. (1984), *Spatial Divisions of Labour: Social Structures and the Geography of Production*, London, Macmillan.

Melville, A. and Johnston, C. (1982), *Cured to Death*, New York, Stein & Day.

Mitchell, J. (1984), *What is to be Done About Illness and Health?*, Harmondsworth, Penguin.

Mohan, J. and Woods, K. J. (1985), 'Restructuring health care?: the social geography of public and private health care', *International Journal of Health Services*, vol. 15, pp. 197–215.

Moss, R. W. (1982), *The Cancer Syndrome*, New York, Grove Press.

Navarro, V. (1976), *Medicine Under Capitalism*, New York, Prodist.

Navarro, V. (1977), *Social Security and Medicine in the USSR*, Lexington, D. C. Heath.

Navarro, V. (1978), *Class Struggle, The State and Medicine*, London, Martin Robertson.

Navarro, V. (1980a), 'Work, ideology and science: the case of medicine', *Social Science and Medicine*, vol. 14c, pp. 191–205.

Navarro, V. (1980b), 'The nature of democracy in the core capitalist countries: meanings and implications for class struggle', *The Insurgent Sociologist*, vol. 10, pp. 3–15.

Navarro, V. (1983), 'Radicalism, marxism, and medicine', *International Journal of Health Services*, vol. 13, pp. 179–202.

Navarro, V. (1984), 'Selected myths guiding the Reagan administration's health policies', *International Journal of Health Services*, vol. 14, pp. 321–7.

Navarro, V. (1985), 'Double standards in the analysis of marxist scholarship: a reply to Reidy's critique of my work', *Social Science and Medicine*, vol. 20, pp. 441–51.

Oakley, A. (1981), 'Interviewing women: a contradiction in terms', in H. Roberts (ed.), *Doing Feminist Research*, pp. 30–61, London, Routledge & Kegan Paul.

Open University (1979), *Data Collection Procedures*, Milton Keynes, Open University Press.

Page, I. M. (1979), 'The continuing failure to understand and treat hypertension', *Journal of the American Medical Association*, vol. 241, pp. 1897–8.

Paterson, K. (1981), 'Theoretical perspectives in epidemiology', *Radical Community Medicine*, vol. 8, pp. 21–9.

Pawson, R. (1978), 'Empiricist explanatory strategies: the case of causal modelling', *Sociological Review*, vol. 26, pp. 615–40.

Phillips, B. U. and Bruhn, J. G. (1981), 'Smoking habits and reported illness in two communities with different systems of social support', *Social Science and Medicine*, vol. 15a, pp. 625–31.

Reid, B. L., Hagan, B. E. and Coppleson, M. (1979), 'Homogeneous Hetero Sapiens', *Medical Journal of Australia*, vol. 86, pp. 377–80.

Reidy, A. (1984), 'Marxist functionalism in medicine: a critique of the work of Vincente Navarro on health and medicine', *Social Science and Medicine*, vol. 19, pp. 897–910.

Renaud, M. (1978), 'On structural constraints to state intervention in health', in J. Ehrenreich (ed.), *The Cultural Crisis in Modern Medicine*, pp. 101–23, New York, Monthly Review Press.

Rose, S., Kamin, L. J. and Lewontin, R. C. (1984), *Not in Our Genes*, Harmondsworth, Penguin.

Russell, J. (1986), *Coronary Heart Disease and Asians in Britain*, London, Coronary Prevention Group.

Salmon, J. W. (1984), *Alternative Medicines*, New York, Tavistock.

Saunders, P. (1981), *Social Theory and The Urban Question*, London, Hutchinson.

Sayer, A. (1979), 'Explanation in economic geography', *Progress in Human Geography*, vol. 6, pp. 66–68.

Sayer, A. (1984), *Method in Social Science: A Realist Approach*, London, Hutchinson.

Sayer, A. (1985), 'Realism in geography', in R. J. Johnston (ed.), *The Future of Geography*, London, Methuen.

Schnall, P. L. and Kern, R. (1982), 'Hypertension in American society: an introduction to historical materialist epidemiology', in P. Conrad and R. Kern (eds), *The Sociology of Health and Illness*, pp. 97–122, New York, St Martin's Press.

Scott-Samuel, A. (1981), 'Towards a socialist epidemiology', *Radical Community Medicine*, vol. 7, pp. 13–18.

Smith, B. E. (1981), 'Black lung: the social production of disease', *International Journal of Health Services*, vol. 11, pp. 343–59.

Stark, E. (1982), 'Doctors in spite of themselves: the limits of radical health criticism', *International Journal of Health Services*, vol. 12, pp. 419–59.

Sterling, T. D. (1978), 'Does smoking kill workers or working kill smokers?', *International Journal of Health Services*, vol. 8, pp. 437–52.

Surgeon General (1979), *Healthy People*, Washington DC, Government Printing Office.

Susser, M. (1973), *Causal Thinking in the Health Sciences*, New York, Oxford University Press.

Syme, S. L. (1984), 'Education and modification of behaviour', in W. W. Holland (ed.), *The Oxford Textbook of Public Health*, London, Oxford University Press.

Syme, S. L. and Torfs, C. P. (1978), 'Epidemiologic research in hypertension: a critical appraisal', *Journal of Human Stress*, vol. 4, pp. 43–8.

Thunhurst, C. (1982), *It Makes You Sick: the Politics of the NHS*, London, Pluto Press.

Tobler, W. (1970), 'A computer movie simulating urban growth in the Detroit region', *Economic Geography*, vol. 46, pp. 234–40.

Townsend, P. (1979), *Poverty in Britain*, Harmondsworth, Penguin.

Townsend, P. (1981), 'Toward equality in health through social policy', *International Journal of Health Services*, vol. 11, pp. 63–75.

Townsend, P. (1983), A lecture on the Black report reported in *Radical Community Medicine*, vol. 13, pp. 22–3.

Townsend, P. and Davidson, N. (1982), *Inequalities in Health: The Black Report*, Harmondsworth, Penguin.

Waitzkin, H. (1976), 'A marxist view of medical care', *Annals of Internal Medicine*, vol. 89, pp. 264–78.

Waitzkin, H. (1983), *The Second Sickness: Contradictions of Capitalist Health Care*, New York, Free Press.

Walker, A. (1984), *Social Planning: A Strategy for Socialist Welfare*, Oxford, Blackwell.

Willis, P. (1980), 'Notes on method', in S. Hall (ed.), *Culture, Media and Language*, pp. 88–97, London, Hutchinson.

Author Index

Subject Index